Nonresponse in
Household Interview Surveys

Nonresponse in Household Interview Surveys

ROBERT M. GROVES
MICK P. COUPER

University of Michigan
Ann Arbor, Michigan

and

Joint Program in Survey Methodology
College Park, Maryland

A Wiley-Interscience Publication
JOHN WILEY & SONS, INC.
New York • Chichester • Weinheim • Brisbane • Singapore • Toronto

This text is printed on acid-free paper. ⊗

Copyright © 1998 by John Wiley & Sons, Inc.

All rights reserved. Published simultaneously in Canada.

Library of Congress Cataloging in Publication Data:

Groves, Robert M.
    Nonresponse in household interview surveys / Robert M. Groves,
Mick P. Couper.
        p.   cm. — (Wiley series in probability and statistics.
Survey methodology section)
    "Wiley-Interscience publication."
    Includes bibliographical references and index.
    ISBN 0-471-18245-1 (cloth : alk. paper)
    1. Household surveys.   I. Couper, Mick.   II. Title.   III. Series.
HB849.49.G757   1998
                                                    97-39223
                                                    CIP

Printed in the United States of America

10 9 8 7 6 5 4 3 2

Contents

Preface

This book was written out of frustration. Its genesis came in 1986–1988 when a review of the then extant research literature of survey nonrepsonse yielded few answers to the question, "How important is nonresponse to surveys?"

In teaching courses in survey methodology, it was common for us to emphasize that once a probability sample had been drawn, full measurement of the sample was crucial for proper inference to apply. Bright students would sometimes question, "How do we know when nonresponse implies error and when it doesn't? Is it cheaper and more effective to reduce nonresponse error by decreasing nonresponse rates or by adjusting for it *post hoc*? Is it more important to reduce nonresponse due to noncontact or nonresponse due to refusals? Why, after all, do people choose not to cooperate with survey requests?" We felt unprepared for such questions and, indeed, grew to believe that the lack of answers was a pervasive weakness in the field, not just a result of our ignorance.

Gathering information *post hoc* about nonrespondents from diverse surveys, which formed one of the central databases of this book, was an attempt to address a critical weakness in the area—the lack of common information about nonresponse across several surveys. (This was an idea stolen from Kemsley, who in 1971, mounted such a study in Great Britain.) Around 1988, the major U.S. federal household surveys were beginning to consider redesign efforts related to incorporating new population distribution data from the 1990 decennial census. We approached Maria Gonzalez, of the Statistical Policy Office of the Office of Management and Budget, who was leading an interagency group developing those redesign research plans. Our idea was to draw samples of nonrespondents and respondents from household surveys conducted about the time of the decennial census, and match their records to the decennial census data. We would thus have at our disposal all variables on the census form to describe the nonrespondents.

This was an idea whose time had clearly not come to the interagency group. We received tough criticism on what practical lessons would be learned; how would those surveys be improved because of the work, and so on. Maria should be credited with quietly listening to the criticism, but forcefully arguing the merits of our case to the survey sponsors. We dedicate this book to her memory.

What appear to the reader as 11 chapters of theory and analysis are based on many person-years of effort, during which the perspectives on the conceptual foundations of survey participation evolved. Some history of the project may provide a sense of that process.

We sought to develop a diverse set of surveys to match to the decennial records. Ideally, we wanted to represent all major survey design variations among the matched surveys. However, the match tool was to be a unit's address, so we were limited to area frame surveys, most often conducted in face-to-face mode. We failed to get cooperation from the commercial surveys we approached. We failed to get extra funds to add some academic surveys to the set.

In the end we established a consortium of funders including the Bureau of the Census, Bureau of Justice Statistics (BJS), Bureau of Labor Statistics (BLS), National Center for Health Statistics (NCHS) and the National Institute on Drug Abuse [NIDA, later called the Substance Abuse and Mental Health Services Administration (SAMSHA)]. Research Triangle Institute and the National Opinion Research Center also provided documentation on surveys sponsored by NIDA and Census, respectively, to facilitate the match and administered questionnaires to their interviewers. At each agency there were key contact people, who facilitated our work. These were William Nicholls, Robert Tortora, and Jay Waite (Census Bureau), Cathryn Dippo and Clyde Tucker (BLS), Michael Rand (BJS), Steve Botman (NCHS), and Joseph Gfroerer (SAMHSA).

Completely independent of this research program, in early 1990, Groves took on a temporary post as an Associate Director at the Bureau of the Census, as the project was nearing its implementation. Couper simultaneously took a post as visiting researcher at Census. This permitted Couper to focus full time on the project between 1990 and 1994.

In 1989, samples were drawn from the Census Bureau surveys, mostly by staff in the Statistical Methods Division of the Bureau, under the direction of Jay Waite. John Paletta coordinated the selection of match cases. After the Census, in 1991–1992, Couper began a commuting life between Washington, DC, and Jeffersonville, Indiana, the vast processing complex for the U.S. Census Bureau. There he worked with a team headed by Judith Petty. The leader of the match team, Maria Darr, collaborated in defining the match methods, training and supervising staff, and implementing quality control procedures. Couper directed the match effort, living out of a suitcase, eating too many meals at the Waffle House in Jeffersonville (whose broken sign read "affle House"). Matching survey and census records was a tedious, slow process, but the care and professionalism of the Jeffersonville staff produced a match data set that we believe is as complete and accurate as possible. Acquiring the completed survey data, cleaning data, merging files, determining weighting schemes, variance estimators, and appropriate modeling techniques took some time after the completion of the match in 1993.

We are both indebted to the executive staff of the Census Bureau, which provided a research cocoon at Suitland, permitting Couper to focus entirely on the research activities at crucial times during the match process, and Groves to join him after ending his stint as associate director in 1992.

However, the work of the decennial match project was not our only focus during the years 1988–1992. Even while the match project was being discussed, two other lines of research were developing. The first was a refinement of conceptual thinking on the process of survey participation. This was partially funded by the Census Bureau and was a collaborative effort with Robert Cialdini, a social psychologist who has made important contributions to understanding helping behavior and compliance. We collaborated in a series of focus groups with interviewers from different organizations, seeking insights from their expertise in gaining the cooperation of persons in surveys. This led to a basic framework of influences on survey participation that forms the structure of this book. Cialdini provided important insights about how survey participation decisions might resemble to other decisions about requests and, more broadly, to attitude change. We are in his debt, especially for the insight one Saturday morning that most decision making in the survey context must be heuristically based, ill-informed by the central features of the respondent's job in a survey.

When our interest grew concerning the effect of the social environment of survey participation, we joined with Lars Lyberg, our friend and colleague, to organize a set of international workshops on household survey nonresponse, starting in 1990. These gave researchers in different countries a chance to compare notes on survey participation across societies. The workshops have stimulated the replication of nonresponse research across countries. Our own research has benefitted from such replication. We have also learned much from the interactions and enjoyed the camaraderie. We thank the regulars at the meetings, including Lars, Bab Barnes, Sandy Braver, Pam Campanelli, Cathy Dippo, Wim de Heer, Lilli Japec, Seppo Laaksonen, Clyde Tucker, and many others.

The other line of research that arose in 1990 involved chances to test empirically our ideas with new data collection efforts. Through the good graces of our colleague Ron Kessler, we smuggled into the National Comorbidity Survey a set of interviewer observations that permitted key initial tests of our notions of the influence of contact-level interactions. This survey was supported by the National Institutes on Mental Health (Grants MH46376 and MH 49098). Later we recieved support from the National Institute on Aging (Grant RO1 AG31059) to add similar measures to the AHEAD survey, which permitted tests of the ideas on a survey of the elderly. Bill Rodgers and Tom Juster were very supportive of including these in AHEAD. Both of these grants were important to Chapters 8, 9, and Chapter 11 of this text.

After the match project data were available, Trivellore Raghunathan became a collaborator when he joined the Survey Methodology Program at Michigan. He collaborated in translating our ideas and findings into a statisitcal modeling strategy for postsurvey adjustment. Raghu deserves full credit for the two-stage adjustment procedures in Chapter 11.

Audience. We've written the book for students of survey methodology: those in school, practicing in the field, and teaching the subject. We assume basic knowledge of survey design, at a level comparable to most initial undergraduate survey methods courses. The statistical models are kept simple deliberately

Those readers with limited time should read Chapters 1 and 2 in order to understand the conceptual framework. Then they should read the summary sections of each chapter, as well as Chapter 11.

Those readers most interested in the practical implications of the work should read the last sections of Chapters 4–9, labeled "Practical Implications for Survey Implementation" as well as Chapters 10 and 11.

In using the book as a text in a course on survey nonresponse we have used Chapters 2 and 4–10.

Collaborators.   In addition to those mentioned above, other stimulating colleagues helped shape the research. These include Toni Tremblay and Larry Altmayer at the U.S. Census Bureau, and Joe Parsons, Ashley Bowers, Nancy Clusen, Jeremy Morton, and Steve Hanway at the Joint Program in Survey Methodology. Lorraine Mc-Call was responsible for the interviewer surveys at the Census Bureau. Teresa Parsley Edwards and Rachel Caspar at Research Triangle Institute worked with us on parts of the analysis of the National Household Survey on Drug Abuse. Brian Harris-Kojetin, John Eltinge, Dan Rope, and Clyde Tucker examined various features of nonreponse in the Current Population Survey. Judith Clemens, Darby Miller-Steiger, Stacey Erth, and Sue Ellen Hansen provided assistance at various points during the work, especially on the Michigan Survey Research Center surveys. We appreciate the criticisms of a set of students in a summer course on survey nonresponse in 1994 offered through the SRC Summer Institute in Survey Research Techniques. Finally, the administrative staff of the Joint Program in Survey Methodology, including Jane Rice, Pam Ainsworth, Nichole Ra'uf, Christie Nader, and Heather Campbell, provided help at many crucial points.

We are members of the Survey Methodology Program (SMP) at the University of Michigan's Institute for Social Research, a research environment that stimulates theoretical questions stemming from applied problems. We thank our SMP colleagues for helping us think through much of the material we present in this book. Jim House, as director of the Survey Research Center, has been a consistent supporter of bringing science to survey methodology and we thank him for being there.

We have profited from critical reviews by Paul Biemer, John Eltinge, Robert Fay, Brian Harris-Kojetin, Lars Lyberg, Nancy Mathiowetz, Beth-Ellen Pennell, Stanley Presser, Eleanor Singer, Seymour Sudman, Roger Tourangeau, and Clyde Tucker. Errors remaining are our responsibility.

We are especially indebted to Northwest Airlines, whose many delayed and cancelled flights between Detroit Metro and Washington National airports permitted long and uninterrupted discussions of the research.

Finally, we thank our editor at Wiley, Steve Quigley, for making the publication process as trouble-free as possible.

Robert M. Groves
Mick P. Couper

Ann Arbor, Michigan
College Park, Maryland

Acknowledgments

We are grateful to various copyright holders for permission to reprint or present adaptations of material previously published. These include the University of Chicago Press, on behalf of the American Association for Public Opinion Research, for adaptation of material from Groves, Cialdini, and Couper (1992) and Couper (1992), appearing in Chapters 2 and 10, and for reprinting a table from Dillman, Gallegos, and Frey (1976), as Table 10.1; the Minister of Industry of Canada, through Statistics Canada for adaptation of Couper and Groves (1992) in Chapter 7; Statistics Sweden, for adaptation of Groves, R.M., and Couper, M.P. (1995) "Theoretical Motivation for Post-Survey Nonresponse Adjustment in Household Surveys," 11, 1, 93–106, in Chapter 9; and "Contact-Level Influences on Cooperation in Face-to-Face Surveys," 12, 1, 63–83, in Chapter 8; Kluwer Academic Publishers, for adaptations of Couper, M.P., and Groves, R.M. (1996) "Social Environmental Impacts on Survey Cooperation," 30, 173–188, in Chapter 6; and Jossey-Bass Publishers for adaptation of Couper and Groves (1996) in Chapter 5.

Nonresponse in
Household Interview Surveys

CHAPTER ONE

An Introduction to Survey Participation

1.1 INTRODUCTION

This is a book about error properties of statistics computed from sample surveys. It is also a book about why people behave the way they do.

When people are asked to participate in sample surveys, they are generally free to accept or reject that request. In this book we try to understand the several influences on their decision. What influence is exerted by the attributes of survey design, the interviewer's behavior, the prior experiences of the person faced with the request, the interaction between interviewer and householder, and the social environment in which the request is made? In the sense that all the social sciences attempt to understand human thought and behavior, this is a social science question. The interest in this rather narrowly restricted human behavior, however, has its roots in the effect these behaviors have on the precision and accuracy of statistics calculated on the respondent pool resulting in the survey. It is largely because these behaviors affect the quality of sample survey statistics that we study the phenomenon.

This first chapter sets the stage for this study of survey participation and survey nonresponse. It reviews the statistical properties of survey estimates subject to nonresponse, in order to describe the motivation for our study, then introduces key concepts and perspectives on the human behavior that underlies the participation phenomenon. In addition, it introduces the argument that will be made throughout the book—that attempts to increase the rate of participation and attempts to construct statistical adjustment techniques to reduce nonresponse error in survey estimates achieve their best effects when based on sound theories of human behavior.

1.2 STATISTICAL IMPACTS OF NONRESPONSE ON SURVEY ESTIMATES

Sample surveys are often designed to draw inferences about finite populations, by measuring a subset of the population. The classical inferential capabilities of the survey rest on probability sampling from a frame covering all members of the population. A probability sample assigns known, nonzero chances of selection to every member of the population. Typically, large amounts of data from each member of the population are collected in the survey. From these variables, hundreds or thousands of different statistics might be computed, each of which is of interest to the researcher only if it describes well the corresponding population attribute. Some of these statistics describe the population from which the sample was drawn; others stem from using the data to test causal hypotheses about processes measured by the survey variables (e.g., how education and work experience in earlier years affect salary levels).

One example statistic is the sample mean, an estimator of the population mean. This is best described by using some statistical notation, in order to be exact in our meaning. Let one question in the survey be called "Y," and the answer to that question for a sample member, say the ith member of the population, be designated by Y_i. Then we can describe the population mean by

$$\overline{Y} = \sum_{i=1}^{N} Y_i/N$$

where N is the number of units in the target population. The estimator of the population mean is often

$$\overline{y} = \left(\sum_{i=1}^{r} w_i y_i\right)\bigg/\left(\sum_{i=1}^{r} w_i\right)$$

where r is the number of respondents in the sample and w_i is the reciprocal of the probability of selection of the ith respondent. (For readers accustomed to equal probability samples, as in a simple random sample, the w_i is the same for all cases in the sample and the computation above is equivalent to $\sum y_i/n$.)

One problem with the sample mean as calculated above is that is does not contain any information from the nonrespondents in the sample. However, all the desirable inferential properties of probability sample statistics apply to the statistics computed on the *entire* sample. Let's assume that in addition to the r respondents to the survey, there are m (for "missing") nonrespondents. Then the total sample size is $n = r + m$. In the computation above we miss information on the m missing cases.

How does this affect our estimation of the population mean, $\overline{Y}$? Let's first make a simplifying assumption. Assume that everyone in the target population is either, permanently and forevermore, a respondent or a nonrespondent. Let the entire target population, thereby, be defined as $N = R + M$, where the capital letters denote numbers in the total population.

Assume that we are unaware at the time of the sample selection about which stratum each person belongs to. Then, in drawing our sample of size n, we will likely select some respondents and some nonrespondents. They total n in all cases but the actual number of respondents and nonrespondents in any one sample will vary. We know that, in expectation, the fraction of *sample* cases that are respondent should be equal to the fraction of population cases that lie in the respondent stratum, but there will be sampling variability about that number. That is, $E(r) = fR$, where f is the sampling fraction used to draw the sample from the population. Similarly $E(m) = fM$.

For each possible sample we could draw, given the sample design, we could express a difference between the full sample mean, $\bar{y}_n$, and the respondent mean, in the following way:

$$\bar{y}_n = \left(\frac{r}{n}\right)\bar{y}_r + \left(\frac{m}{n}\right)\bar{y}_m$$

which, with a little manipulation becomes

$$\bar{y}_r = \bar{y}_n + \left(\frac{m}{n}\right)[\bar{y}_r - \bar{y}_m]$$

that is,

Respondent Mean = Total Sample Mean + (Nonresponse Rate)
× (Difference between Respondent and
Nonrespondent Means)

This shows that the deviation of the respondent mean from the full sample mean is a function of the nonresponse rate (m/n) and the difference between the respondent and nonrespondent means.

Under this simple expression, what is the expected value of the respondent mean, over all samples that could be drawn given the same sample design? The answer to this question determines the nature of the *bias* in the respondent mean, where "bias" is taken to mean the difference between the expected value (over all possible samples given a specific design) of a statistic and the statistic computed on the target population. That is, in cases of equal probability samples of fixed size the bias of the respondent mean is approximately

$$B(\bar{y}_r) = \left(\frac{M}{N}\right)(\bar{Y}_r - \bar{Y}_m)$$

or

Bias(Respondent Mean) = (Nonresponse Rate in Population)
× (Difference in Respondent and
Nonrespondent Population Means)

where the capital letters denote the population equivalents to the sample values. This shows that the larger the stratum of nonrespondents, the higher the bias of the respondent mean, other things being equal. Similarly, the more distinctive the nonrespondents are from the respondents, the larger the bias of the respondent mean.

These two quantities, the nonresponse rate and the differences between respondents and nonrespondents on the variables of interest, are key to the studies reported in this book. Because the literature on survey nonresponse does not directly reflect this fact (an important exception is the work of Lessler and Kalsbeek, 1992), it is important for the reader to understand how this affects nonresponse errors.

Figure 1.1 shows four alternative frequency distributions for respondents and nonrespondents on a hypothetical variable, y, measured on all cases in some target

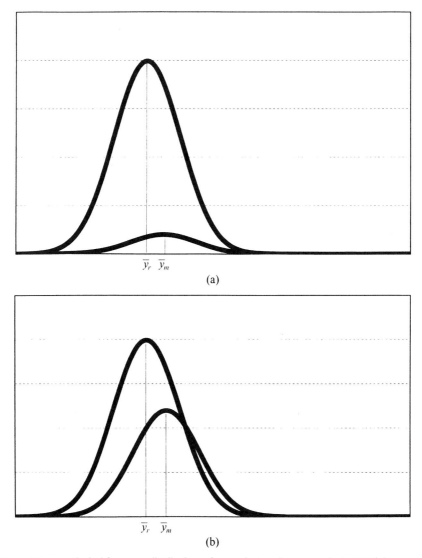

Figure 1.1. Hypothetical frequency distributions of respondents and nonrespondents. (a) High response rate, nonrespondents similar to respondents. (b) Low response rate, nonrespondents similar to respondents.

population. The area under the curves is proportional to the size of the two groups, respondents and nonrespondents.

Case (a) in the figure reflects a high response rate survey and one in which the nonrespondents have a distribution of y values quite similar to that of the respondents. This is the lowest-bias case—both factors in the nonresponse bias are small. For example, assume the response rate is 95%, the respondent mean for reported ex-

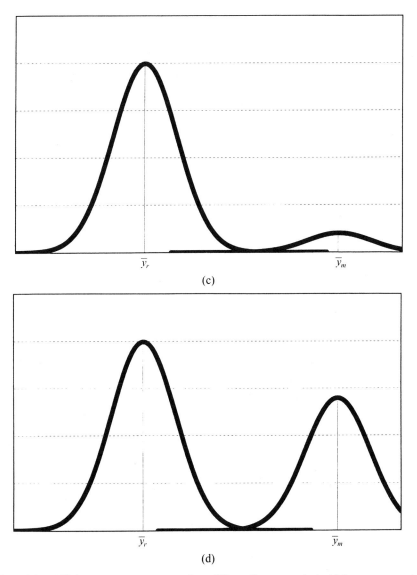

Figure 1.1. (c) High response rate, nonrespondents different from respondents. (d) Low response rate, nonrespondents different from respondents

penditures on clothing for a quarter was $201.00, and the mean for nonrespondents was $228.00. Then the nonresponse error is $0.05(\$201.00 - \$228.00) = -\$1.35$.

Case (b) shows a very high nonresponse rate (the area under the respondent distribution is about 50% greater than that under the nonrespondent—a nonresponse rate of 40%). However, as in (a), the values on y of the nonrespondents are similar to those of the respondents. Hence, the respondent mean again has low bias due to nonresponse. With the same example as in (a), the bias is $0.40(\$201.00 - \$228.00) = -\$10.80$.

Case (c), like (a), is a low nonresponse survey, but now the nonrespondents tend to have much higher values than the respondents. This means that the difference term, $[\bar{y}_r - \bar{y}_m]$, is a large negative number—the respondent mean underestimates the full population mean. However, the size of the bias is small because of the low nonresponse rate, about 5% or so. Using the same example as in (a), with a nonrespondent mean now of $501.00, the bias is $0.05(\$201.00 - \$501.00) = -\$15.00$.

Case (d) is the most perverse, exhibiting a large group of nonrespondents, who have much higher values in general on y than the respondents. In this case, m/n is large (judging by the area under the nonrespondent curve) and $[\bar{y}_r - \bar{y}_m]$ is large in absolute terms. This is the case of large nonresponse bias. Using the example above, the bias is $0.40(\$201.00 - \$501.00) = -\$120.00$, a relative bias of 60% of the respondent-based estimate!

To provide another concrete illustration of these situations, assume that the statistic of interest is a proportion, say, the number of adults who intend to save some of their income in the coming month. Figure 1.2 illustrates the level of nonresponse bias possible under various circumstances. In all cases, the survey results in a respondent mean of 0.50; that is, we are led to believe that half of the adults plan to

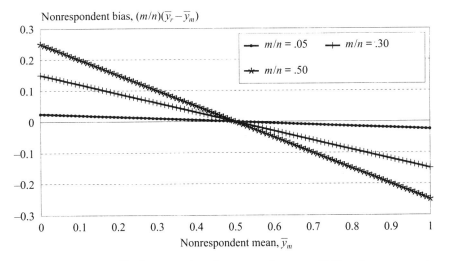

Figure 1.2. Nonresponse bias for a proportion, given a respondent mean of 0.50, various response rates, and various nonresponse means.

save in the coming month. The *x*-axis of the figure displays the proportion of nonrespondents who plan to save in the coming month. (This attribute of the sample is not observed.) The figure is designed to illustrate cases in which the nonrespondent proportion is less or equal to the respondent proportion. Thus, the nonrespondent proportions range from 0.50 (the no bias case) to 0.0 (the largest bias case). There are three lines in the figure, corresponding to different nonresponse rates: 5%, 30%, and 50%.

The figure gives a sense of how large a nonresponse bias can be for different nonresponse rates. For example, in a survey with a low nonresponse rate, 5%, the highest bias possible is 0.025. That is, if the survey respondent mean is 0.50, then one is assured that the full sample mean lies between 0.475 and 0.525.

In the worst case appearing in Figure 1.2, a survey with a nonresponse rate of 50%, the nonresponse bias can be as large as 0.25. That is, if the respondent mean is 0.50, then the full sample mean lies between 0.25 and 0.75. This is such a large range that it offers very little information about the statistic of interest.

The most important feature of Figure 1.2 is its illustration of the dependence of the nonresponse bias on both response rates and the difference term. The much larger slope of the line describing the nonresponse bias for the survey with a high nonresponse rate shows that high nonresponse rates increase the likelihood of bias even with relatively small differences between respondents and nonrespondents on the survey statistic.

1.2.1   Nonresponse Error on Different Types of Statistics

The discussion above focused on the effect of nonresponse on estimates of the population mean, using the sample mean. This section briefly reviews effects of nonresponse on other popular statistics. We examine the case of an estimate of a population total, the difference of two subclass means, and a regression coefficient.

The Population Total. Estimating the total number of some entity is common in government surveys. For example, most countries use surveys to estimate the total number of unemployed persons, the total number of new jobs created in a month, the total retail sales, the total number of criminal victimizations, etc. Using notation similar to that in Section 1.2, the population total is ΣY_i, which is estimated by a simple expansion estimator, $\Sigma w_i y_i$, or by a ratio-expansion estimator, $X(\Sigma w_i y_i / \Sigma w_i x_i)$, where X is some auxiliary variable, correlated with Y, for which target population totals are known. For example, if y were a measure of the number of criminal victimizations experienced by a sample household, and x were a count of households, X would be a count of the total number of households in the country.

For variables that have nonnegative values (such as count variables), simple expansion estimators of totals based only on respondents always underestimate the total. This is because the full sample estimator is

$$\sum_{i=1}^{n} w_i y_i = \sum_{i=1}^{r} w_i y_i + \sum_{i=r+1}^{n} w_i y_i$$

that is,

Full Sample Estimate of Population Total = Respondent-Based Estimate
+ Nonrespondent-Based Estimate

Hence, the bias in the respondent-based estimator is

$$-\sum_{i=r+1}^{n} w_i y_i$$

It is easy to see, thereby, that the respondent-based total (for variables that have non-negative values) will always underestimate the full sample total, and thus, in expectation, the full population total.

The Difference of Two Subclass Means. Many statistics of interest from sample surveys estimate the difference between the means of two subpopulations. For example, the Current Population Survey often estimates the difference in the unemployment rate for Black and nonBlack men. The National Health Interview Survey estimates the difference in the mean number of doctor visits in the last 12 months between males and females.

Using the expressions above, and using subscripts 1 and 2 for the two subclasses, we can describe the two respondent means as

$$\bar{y}_{1r} = \bar{y}_{1n} + \left(\frac{m_1}{n_1}\right)[\bar{y}_{1r} - \bar{y}_{1m}]$$

$$\bar{y}_{2r} = \bar{y}_{2n} + \left(\frac{m_2}{n_2}\right)[\bar{y}_{2r} - \bar{y}_{2m}]$$

These expressions show that each respondent subclass mean is subject to an error that is a function of a nonresponse rate for the subclass and a deviation between respondents and nonrespondents in the subclass. The reader should note that the nonresponse rates for individual subclasses could be higher or lower than the nonresponse rates for the total sample. For example, it is common that nonresponse rates in large urban areas are higher than nonresponse rates in rural areas. If these were the two subclasses, the two nonresponse rates would be quite different.

If we were interested in $\bar{y}_1 - \bar{y}_2$ as a statistic of interest, the bias in the difference of the two means would be approximately

$$B(\bar{y}_1 - \bar{y}_2) = \left(\frac{M_1}{N_1}\right)[\bar{Y}_{1r} - \bar{Y}_{1m}] - \left(\frac{M_2}{N_2}\right)[\bar{Y}_{2r} - \bar{Y}_{2m}]$$

Many survey analysts are hopeful that the two terms in the bias expression above cancel. That is, the bias in the two subclass means is equal. If one were dealing with two subclasses with equal nonresponse rates that hope is equivalent to a hope that the difference terms are equal to one another. This hope is based on an assumption that nonrespondents will differ from respondents in the same way for both subclasses. That is, if nonrespondents tend to be unemployed versus respondents, on average, this will be true for all subclasses in the sample.

If the nonresponse rates were not equal for the two subclasses, then the assumptions of canceling biases is even more complex. But to simplify, let's continue to assume that the difference between respondent and nonrespondent means is the same for the two subclasses. That is, assume $[\bar{y}_{r1} - \bar{y}_{m1}] = [\bar{y}_{r2} - \bar{y}_{m2}]$. Under this restrictive assumption, there can still be large nonresponse biases.

For example, Figure 1.3 examines differences of two subclass means where the statistics are proportions (e.g., the proportion planning to save some of their income next month). The figure treats the case in which the proportion planning to save among respondents in the first subclass (say, high-income households) is $\bar{y}_{r1} = 0.5$ and the proportion planning to save among respondents in the second subclass (say, low-income households) is $\bar{y}_{r2} = 0.3$. This is fixed for all cases in the figure. We examine the nonresponse bias for the entire set of differences between respondents and nonrespondents. That is, we examine situations where the differences between respondents and nonrespondents lie between −0.5 and 0.3. (This difference applies to both subclasses.) The first case of a difference of 0.3 would correspond to

$$[\bar{y}_{r1} - \bar{y}_{m1}] = 0.5 - 0.2 = 0.3$$

$$[\bar{y}_{r2} - \bar{y}_{m2}] = 0.3 - 0.0 = 0.3$$

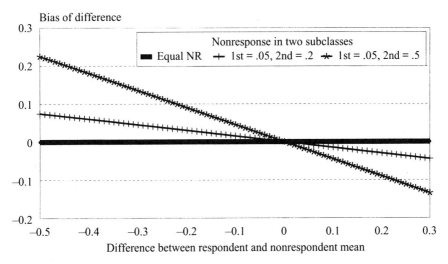

Figure 1.3. Nonresponse bias for a difference of subclass means, for the case of two respondent subclass means (0.5, 0.3) by various response rate combinations, by differences between respondent and nonrespondent means.

The figure shows that when the two nonresponse rates are equal to one another, there is no bias in the difference of the two subclass means. However, when the response rates of the two subclasses are different, large biases can result. Larger biases in the difference of subclass means arise with larger differences in nonresponse rates in the two subclasses (note the higher absolute value of the bias for any given $[\bar{y}_r - \bar{y}_m]$ value for the case with a 0.05 nonresponse rate in subclass 1 and a 0.5 in subclass 2 than for the other cases).

A Regression Coefficient. Many survey data sets are used by analysts to estimate a wide variety of statistics measuring the relationship between two variables. Linear models testing causal assertions are often estimated on survey data. Imagine, for example, that the analysts were interested in the model

$$y_i = \beta_0 + \beta_1 x_i + \epsilon_i$$

which, using the respondent cases to the survey, would be estimated by

$$\hat{y}_{ri} = \hat{\beta}_{r0} + \hat{\beta}_{r1} x_{ri}$$

The ordinary least squares estimator of β_{r1} is

$$\hat{\beta}_{r1} = \frac{\sum_{i=1}^{r}(x_i - \bar{x}_r)(y_i - \bar{y}_r)}{\sqrt{\sum_{i=1}^{r}(x_i - \bar{x}_r)^2}}$$

Both the numerator and denominator of this expression are subject to potential nonresponse bias. For example, the bias in the covariance term in the numerator is approximately

$$B(s_{rxy}) = \frac{M}{N}(S_{rxy} - S_{mxy}) - \left(\frac{M}{N}\right)\left(1 - \frac{M}{N}\right)(X_r - X_m)(Y_r - Y_m)$$

This bias expression can be either positive or negative in value. The first term in the expression has a form similar to that of the bias of the respondent mean. It reflects a difference in covariances for the respondents (S_{rxy}) and nonrespondents (S_{mxy}). It is large in absolute value when the nonresponse rate is large. If the two variables are more strongly related in the respondent set than in the nonrespondent, the term has a positive value (that is the regression coefficient tends to be overestimated). The second term has no analogue in the case of the sample mean; it is a function of crossproducts of difference terms. It can be either positive or negative depending on these deviations.

As Figure 1.4 illustrates, if the nonrespondent units have distinctive combinations of values on the x and y variables in the estimated equation, then the slope of the regression line can be misestimated. The figure illustrates the case when the pat-

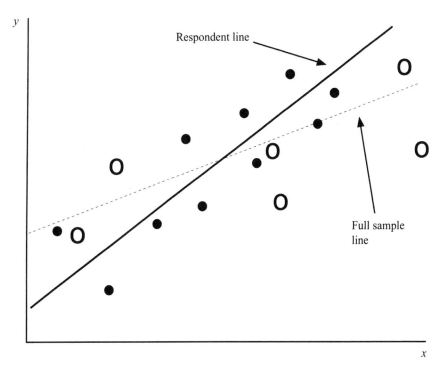

Figure 1.4. Illustration of the effect of unit nonresponse on estimated slope of regression line.

tern of nonrespondent cases (designated by "O") differ from that of respondent cases (designated by "●"). The result is that the fitted line on the respondents only has a larger slope than that for the full sample. In this case, the analyst would normally find more support for an hypothesized relationship than would be true for the full sample.

1.2.2   Considering Survey Participation a Stochastic Phenomenon

The discussion above made the assumption that each person (or household) in a target population either is a respondent or a nonrespondent for all possible surveys. That is, it assumes a fixed property for each sample unit regarding the survey request. They will always be a nonrespondent or they will always be a respondent, in all realizations of the survey design.

An alternative view of nonresponse asserts that every sample unit has a probability of being a respondent and a probability of being a nonrespondent. It takes the perspective that each sample survey is but one realization of a survey design. In this case, the survey design contains all the specifications of the research data collection. The design includes the definition of the sampling frame, the sample design, the questionnaire design, choice of mode, hiring, selection, and training regimen for interviewers, data collection period, protocol for contacting sample units, callback rules, refusal conversion rules, and so on. Conditional on all these fixed properties of the sample survey, sample units can make different decisions regarding their participation.

In this view, the notion of a nonresponse rate must be altered. Instead of the nonresponse rate merely being a manifestation of how many nonrespondents were sampled from the sampling frame, we must acknowledge that in each realization of a survey different individuals will be respondents and nonrespondents. In this perspective the nonresponse rate above (m/n) is the result of a set of Bernoulli trials; each sample unit is subject to a "coin flip" to determine whether it is a respondent or nonrespondent on a particular trial. The coins of various sample units may be weighted differently; some will have higher probabilities of participation than others. However, all are involved in a stochastic process of determining their participation in a particular sample survey.

The implications of this perspective on the biases of respondent means, respondent totals, respondent differences of means, and respondent regression coefficients is minor. The more important implication is on the variance properties of unadjusted and adjusted estimates based on respondents.

1.2.3   The Effects of Different Types of Nonresponse

The discussion above considered all sources of nonresponse to be equivalent to one another. However, this book attempts to dissect the process of survey participation into different components. In household surveys it is common to classify outcomes of interview attempts into the following categories: interviews (including complete and partial), refusals, noncontacts, and other noninterviews. The other noninterview category consists of those sample units in which whoever was designated as the respondent is unable to respond, for physical and mental health reasons, for language reasons, or for other reasons that are not a function of reluctance to be interviewed. Various survey design features affect the distribution of nonresponse over these categories. Surveys with very short data collection periods tend to have proportionally more noncontacted sample cases. Surveys with long data collection periods or intensive contact efforts tend to have relatively more refusal cases. Surveys with weak efforts at accommodation of nonEnglish speakers tend to have somewhat more "other noninterviews." So, too, may surveys of special populations, such as the elderly or immigrants.

If we consider separately the different types of nonresponse, many of the expressions above generalize. For example, the respondent mean can be described as a function of various nonresponse sources, as in

$$\bar{y}_r = \bar{y}_n + \frac{m_{rf}}{n}(\bar{y}_r - \bar{y}_{rf}) + \frac{m_{nc}}{n}(\bar{y}_r - \bar{y}_{nc}) + \frac{m_{nio}}{n}(\bar{y}_r - \bar{y}_{nio})$$

where the subscripts *rf, nc,* and *nio* refer to refusals, noncontacts, and other noninterviews, respectively.

This focuses attention on whether when survey designs vary on the composition of their nonresponse (i.e., different proportions of refusals, noncontacts, and other non-interviews), they produce different levels of nonresponse error. Do persons difficult to contact have distinctive values on the survey variables from those easy to contact? Do persons with language, mental, or physical disabilities have distinctive values from others? Are the tendencies for contacted sample cases to sort themselves into either interviews or refusals related to their characteristics on the survey variables?

Consider a practical example of these issues. Imagine conducting a survey of criminal victimization, where respondents are asked to report on their prior experiences as a victim of a personal or household crime. As will be seen in later chapters, some of the physical impediments to contacting a sample household are locked gates, no-trespassing signs, and intercoms. These are also common features that households who have experienced a crime install in their unit. They are preventative measures against criminal victimization. This is a situation in which early contacts in a survey would be likely to have lower victimization rates than late contacts. At any point, the noncontacts will tend to have higher victimization rates than contacted cases.

Now consider the causes of cooperation or refusal with the survey request. Imagine that the survey is described as an effort to gain information about victimization in order to improve policing strategies in the local community. Those for whom such a purpose is highly salient will tend to cooperate. Those for whom such a goal is less relevant will tend to refuse. Thus, refusals might tend to have lower victimization rates than cooperators, among those contacted.

This situation implies that the difference terms move in different directions:

$$\frac{m_{rf}}{n}(\bar{y}_r - \bar{y}_{rf}) > 0 \quad \text{and} \quad \frac{m_{nc}}{n}(\bar{y}_r - \bar{y}_{nc}) < 0$$

Now let's add to the situation the typical process of field administration. Initial effort by interviewers is concentrated on contacting each sample unit. This initially reaches those with low victimization rates, who disproportionately then refuse to be interviewed. Initial refusal rates are quite high. As contact rates increase, victims, who are interested in responding, are disproportionately contacted. They disproportionately move into the interviewed pool, increasing the victimization rate among respondents. Alternatively, if efforts at higher response rates are concentrated on the initial refusal cases, through refusal conversion, the interviewed pool will increasingly contain nonvictims, lowering the respondent victimization rate.

This is a case where the final nonresponse error is a function of the balance between the noncontact and the refusal rate. For any given overall response rate, the higher the refusal rate, the more likely the survey will overestimate the population's victimization rate. For any given overall response rate, the higher the noncontact rate, the more likely the survey will underestimate the rate.

This example illustrates the need to dissect the causes of nonresponse into constituent parts that share relationships with the key survey variables. Considering only the overall response rate ignores the possible counteracting biases of different types of nonresponse. This process of dissection is one of the purposes of this book.

1.2.4   Reducing Nonresponse Rates

There are two traditional reactions to survey nonresponse among practitioners: reducing nonresponse rates and using estimators that include adjustments for nonresponse. As we discuss in more detail in Chapter 10, various survey design features act to reduce specific sources of nonresponse.

There is a well-documented set of techniques to increase the likelihood of contacting sample cases. These include advance contacts by mail or telephone in face-to-face surveys in order to schedule convenient times to visit. They include setting the number of days or weeks in the data collection period so that those households that are rarely at home will nonetheless be contacted. In addition, interviewers are trained to call repeatedly on sample units, seeking contact with the household. As the field period progresses, calls on cases tend to be at different times of day or evening; interviewers may be trained to attempt telephone contact, etc.

There are many design features chosen to reduce refusals as a source of nonresponse. These include the use of advance letters, attempting to communicate that the survey is conducted by an organization with legitimate need for the information. The advance communication sometimes contains a cash or in-kind incentive. The interviewer attempts to make appointments with the sample person at times convenient for them to provide the interview. Repeated attempts to persuade reluctant respondents may involve switches to a different interviewer, persuasion letters, or visits by supervisors—all intended to communicate the importance of cooperating with the survey request.

Finally, the design features to reduce the rate of "other noninterviews" include the use of nonEnglish speaking interviewers, translation of the instruments into various languages, and the use of proxy respondents.

Most of these efforts to reduce nonresponse rates are aimed at different potential causes of nonresponse, not directly different characteristics of nonrespondents. They attack the rate term (m/n) in the expression, not the difference terms, $[y_r - y_m]$. This means they exert no direct control over the nonresponse error itself, but only on one term of the error expression.

Since design decisions are made under cost constraints, designs often tend to use the cheapest means possible to reduce the nonresponse rate. Usually, noncontact rates can be reduced most cheaply, merely by making more calls on cases not yet contacted. If at any one point in a field period, the current noncontacts are quite different (on the survey measures) from the current refusals, then it is possible that this strategy would not reduce nonresponse error. That is, if $[\bar{y}_r - \bar{y}_{nc}]$ is small, but $[\bar{y}_r - \bar{y}_{rf}]$ is large, then moving cases from a noncontact status to an interview status may do little to reduce overall nonresponse error. This observation underscores how blindly the researcher must often make decisions on efforts to reduce nonresponse components.

Our work described in this book attempts to uncover differences in the mechanisms producing noncontacts and refusals, so that investigators might build survey designs that employ more intelligence about differences among nonrespondents. This intelligence can then be used either to reduce nonresponse during the data collection efforts or to mount more effective postsurvey adjustments for nonresponse.

1.2.5   Using Postsurvey Adjustment for Nonresponse Error Reduction

The other traditional approach to nonresponse is a statistical one, using estimation procedures that attempt to reduce the effects of missing observations. In practice the procedures used in postsurvey adjustment for missing data depend on how much information is available about the nonrespondent cases. At one extreme, if every survey variable of interest, except one, is known about the nonrespondents, then using those variables to form an imputation model is common. The imputation model predicts a value for the missing variable for the case, conditional on values of all the known variables. If the predictive model reflects strong relationships among the variables, then the imputed values tend to be close to the value that would have been obtained in the interview. Imputation is common in unit nonresponse in longitudinal surveys, when full data records from a prior wave are available for a nonrespondent to a current wave. In one-time surveys, it is rare to impute for unit nonresponse because little information is typically known about the nonrespondent cases.

For unit nonresponse (versus item missing data) imputation is less often used than is case weighting. In weighting adjustments, some respondent cases (those resembling the nonrespondents) are given larger weights in the sample estimators than are other respondent cases. "Weighting classes" (a group assigned the same weight) are formed among the respondent cases. When cases in a weighting class share similar likelihoods of participation *and* similar values on the survey variables, then nonresponse error in the weighted estimator is lower than in the unweighted estimator. Sharing similar values on the survey variable must occur within a weighting class both for respondents and nonrespondents. It is common that the reduced bias of the weighted estimator is accompanied by somewhat higher sampling variances. Thus, adjustment decisions are often tradeoffs between bias and variance properties of unadjusted and adjusted estimators.

For purposes of this book, we focus on the common features of these adjustment schemes, the specification of observed attributes of a sample that can inform the researcher about the unobserved attributes. Specifically, we seek to identify influences on survey participation that can be observed on all sample cases and used as predictors in postsurvey adjustment models. Identifying the variables to observed requires more understanding of the decision-making process of survey participation than we had prior to mounting this research.

1.3   HOW HOUSEHOLDERS THINK ABOUT SURVEY REQUESTS

Over the years of studying survey participation we have learned the importance of viewing the phenomenon from the sample householder's perspective. Survey designers and methodologists sometimes find it difficult to take this vantage point. However, repeated contacts with householders, monitoring of survey introductions, and focus groups with interviewers have convinced us that taking a survey researcher's viewpoint risks misunderstanding. This section presents one plausible perspective that sample householders may take. (We use the term "householder" throughout this book to include both those sample persons who become respondents and those who remain nonrespondents.)

The contrast between this perspective and that of survey researchers is, first, that none of the statistical requirements for complete enumeration of the sample are either understood or valued by the householders. Second, the importance to the sponsor or to society of obtaining the survey information is generally not shared by the householders.

Householders may see survey requests as a specific type of request from a stranger. There are several categories of those, which tend to sort themselves by the medium of communication, the physical location of the request, and the nature of the relationship between the requestor and person.

1.3.1 Requests from Others in Day-to-Day Life

It is useful to compare various characteristics of survey requests to householders with requests by other types of organizations. Table 1.1 presents some characteristics of requests of unsolicited sales agents, business contacts, charities, and surveys.

By "sales" we mean all contacts with a household by a person attempting to sell some good or service to the household. This would include approaches for telephone service, credit card services, home improvement products, encyclopedias, vacuum cleaners, lawn services, and investment services. By "service calls" we mean contacts with an unknown functionary of an organization that is already providing services or products to the household. This would include public utilities, newspaper delivery services, cable television services, insurance agencies, or medical care services. The distinction between "sales" and "service calls" is thus whether the household already has some relationship with the organization, even though it has no relationship with the given person who makes contact with the household. By "religions, charities" we mean any contact by an agent of an organization seeking funds from or actions by the household for its cause. This would include proselytizers for specific churches, collectors for contributions to volunteer fire departments, medical research societies, public radio or television stations, school fundraisers, environmental action groups, or societies aiding the poor. Finally, by "surveys" we mean any request for information for statistical purposes. This would include government, academic, or commercial studies of the household population.

Table 1.1 compares these requests on several dimensions, including the likely frequency of a household experiencing such a request, the level of public knowledge of the organization generating the request, the likely media of communication of the request, the likelihood of prior contact with the requesting organization, the use of incentives to the households associated with the request, the persistence at contact of the requestor when dealing with those rarely at home or those reluctant to grant the request, and the likelihood of ongoing contact.

At the current time in the United States, sales and service calls on households probably are more common than charitable and survey requests. Name recognition by large segments of the household population would be high for business contacts (because the household is involved in an economic exchange with them) and for

Table 1.1. Selected characteristics of householder encounters with sales, business, charities, and surveys requests

Characteristic	Sales	Service calls	Religions, charities	Surveys
Frequency of experience	High	High	Medium	Low
Level of public knowledge of sponsor	Usually low	High	Usually high	Usually low
Usual medium of communication	Phone, mail	Mail, phone	Face-to-face, phone, mail	Mail, phone, Face-to-face
Prior contact with sponsor	No	Yes	Sometimes	No
Use of incentives	Rare	Never	Sometimes	Sometimes
Persistence at contact	Absent	High	Low	High
Nature of Request	Money for goods or services	Money, information	Money, time	Time, for information
Likelihood of ongoing contact	Low	Depends	Depends	Low

some national, long-standing charitable organizations (e.g., American Cancer Society). When survey sponsors are universities or government agencies, sometimes the population may have prior knowledge of the requestor. Surveys and charitable requests use all three media of communication, but sales and service calls usually rely on telephone and mail communication. Even with surveys and charities, the telephone and mail modes predominate over face-to-face contact.

In contrast to service calls and some charities, it is common that the sales and surveys' approach is the first time for contact with the householder. Service calls can refer to the past transactions with the household as a way to provide context for the purpose of the request. A few charities and surveys use incentives as a way to provide some token of appreciation to the householder for granting the request. Charities send address labels, calendars, kitchen magnets, and offers of listing donors' names publicly. Surveys sometimes offer money or in-kind gifts. Sales and business requests rarely offer such inducements.

Sales and charity requests rarely utilize multiple attempts. If reluctance is expressed on first contact with the household in a sales call on the telephone, for example, the caller tends to dial the next number to solicit. Profit is generally not maximized by effort to convince the reluctant to purchase the product or service. Surveys and service calls are quite different. Probability sample surveys often make repeated callbacks to sample households attempting to obtain participation of the household. Service-call communication will generate repeated calls until the issue is resolved.

Finally, service calls and charitable requests are often made by persons who have had or will have ongoing relationships with the householder. When the requestor is known by the householder, that prior knowledge can influence initial householder behavior. Sales and survey calls are most often made by persons unknown to the householder. In the early moments of interaction with the requestor, the householder may be attempting to determine whether the requestor is or is not known by them.

1.3.2   Participation in Surveys and in Other Social Activities

The fact that there are different reasons for requests, different purposes of requests, and different institutions that are making requests that householders receive routinely may lead to standardized reactions to requests. These might be "default" reactions that are shaped by experiences over the years with such requests.

Service calls for clarification of orders, billing issues, and other reasons are shaped by the fact that the requestor provides products or services valued by the household. Charities and religious requests may be filtered through opinions and attitudes of householders about the group. Sales requests may generate default rejections, especially among householders repeatedly exposed to undesired sales approaches.

Survey requests, because they are rare relative to the other requests, might easily be confused by householders and misclassified as sales calls, for example. When this occurs, the householder may react for reasons other than those pertinent to the survey request. The fact that surveys often use repeated callbacks is probably an effective tool to distinguish them from sales calls. When surveys are conducted by well-known institutions that have no sales mission, interviewers can emphasize the sponsorship as a means to distinguishing themselves from salespersons.

Government and academic surveys are *de facto* conducted by agents of major institutions in the society. Once the householder discerns such sponsorship of the survey request, it is likely that past contacts with the institution, knowledge about the institution, or attitudes about its value to the householder or important reference groups of the householder become relevant. That is, the householder uses knowledge of the sponsor to guide behavior. Once it is clear that the request concerns a survey interview, then prior experiences with social research, interviews, polls, and scientific studies may become salient to the decision of the householder. Finally, reactions to the interviewer provide input to the decision to cooperate.

1.4   HOW INTERVIEWERS THINK ABOUT SURVEY PARTICIPATION

Interviewers are "request professionals." They are the agents of the survey designer who deliver the request for the survey interview. All of the design features that can affect interest of householders in responding and willingness to provide information and time to the interviewer are implemented by interviewers. We would thus suspect that interviewers can have large effects on householders' reactions to survey requests.

Interviewers, however, have many other duties in most surveys. They must identify and document sample units. They must determine housing units' eligibility for the sample. In many surveys they must select respondents within the household. After the householder grants the survey request, the interviewer must administer the questionnaire, with care to communicate correctly the intent and meaning of each question, to encourage candid and thorough responses from the householder, and to record accurately the responses of the householder. Thus, contacting and gaining participation of sample households is but one job in an interviewer's portfolio.

1.4.1 How Interviewers are Trained and Evaluated Regarding Response Rates

It is common for interviewers to receive two sorts of training prior to a survey. The first type of training is generic to all survey work for their employing organization. The second is specific to the survey they will soon begin.

General interviewer training tends to have several components. First, the administrative aspects of the job must be communicated. These include recording work time, receiving and returning sample materials that identify sample households, and communicating with supervisors. Second, the process of identifying sampling housing units assigned to the interviewer, correcting any errors in the identity, and documenting the outcome of calls on sample cases, must be described.

Next, the training often turns to issues of contacting sample units. It is common to instruct interviewers to call on sample units at different times of the day and different days of the week. Sometimes, rigid guidelines for call patterns are given (e.g., first call on a weekday day, then an evening, then a weekend, until first contact is made). Interviewers are sometimes instructed to ask neighbors when members of a noncontacted household would be at home. Some organizations forbid interviewers to seek such information, for fear of violating the confidentiality of the sample household. In centralized telephone surveys, call scheduling is often handled by software embedded in the computer assisted interviewing systems. However, little useful information can be observed about sample households by telephone interviewers prior to the first contact.

Finally, the interviewers are instructed in the administration of a structured questionnaire. Usually this entails guidelines to read the questions exactly as written. They are taught how to discern whether the response provided to a question is adequate for the purposes of the research. They are taught how to probe nondirectively when given an inadequate answer. They are taught how to record responses to open questions.

The training then moves to study-specific training. Study-specific training often has some material on seeking cooperation of the sample household. It is common for interviewers to be instructed in the larger purposes of the survey, to be supplied with answers to commonly asked questions about the survey, and to be trained in issues about the confidentiality of provided data. Some organizations have interviewers role-play situations with different types of reluctant respondents. The purpose of this is to prepare interviewers with quick responses to objections to survey participation.

The bulk of the study-specific training, however, focuses on the administration of the questionnaire. Interviewers are instructed in key definitions of terms, in the intent of each question, in what constitutes adequate answers. They are instructed in how to handle unusual circumstances that arise for some respondents.

In short, training interviewers to contact sample households and to obtain their cooperation in the survey is but one part of their training. In many organizations, the time devoted to the survey participation step in interviewer training is only a small fraction of the time.

Once the interviewers begin work on a survey, most organizations will use the individual response rates they achieve as important performance indicators. Many organizations will produce daily or weekly statistical summaries of sample cases assigned, cases not yet contacted, cases interviewed, cases with initial refusals, cases with other types of noninterviews, cases assigned for refusal conversion, etc. When individual interviewers achieve lower than expected response rates, they are often given remedial training to ameliorate the situation.

1.5   HOW SURVEY DESIGN FEATURES AFFECT NONRESPONSE

Over the years of the development of the survey method, various antidotes to nonresponse have been developed and proven valuable in diverse settings. These are features chosen by the survey designer prior to mounting the data collection step. Most have cost implications for the survey. Some may increase the cost per completed interview; some may move costs from interviewer salaries to other components of the budget (i.e., they reduce interviewer effort to obtain participation of sample units but increase other staff or material costs).

A simple way to classify the design features is by what portion of nonresponse they address: noncontact, refusals, or other noninterviews. The other relevant classification of techniques concerns when in the temporal order of a survey the design feature is present.

1.5.1 Features to Enhance the Rate of Contact

There are essentially four methods of improving the rate of contact of sample households. The first is to increase the number of calls on previously uncontacted units. In area-based household surveys, this is at times implemented through rules on number of visits to a sample neighborhood. In telephone surveys, this might be implemented through software that controls the maximum number of calls. The second approach is to control the timing of repeated calls to sample units. Often, these direct the interviewer to mix the time of calls between daytime, weekend, and evening calls, in order to contact households with different at-home patterns. The third approach is to increase the length of the data collection period. In a way, this permits more variation of time of calls and more calls, but also addresses households who are temporarily absent because of out-of-town travel. The fourth design feature is to permit interviewers to seek supplemental information about the noncontacted unit. This includes obtaining telephone numbers so that calls can be made in another mode and seeking information about at-home patterns from neighbors, doormen, or building managers. The latter technique is generally limited to face-to-face surveys.

1.5.2 Features to Enhance the Rate of Cooperation

There are many more techniques that are used in practice to reduce refusals than to reduce noncontacts. They are usefully divided into precontact methods, methods during contact, and refusal conversion methods.

Agencies of data collection convey different levels of authority and legitimacy to different populations. For example, it is common that government agencies obtain higher cooperation rates than other survey organizations. Thus, one design decision is the sponsorship of the survey and the affiliation and stature of the data collection organization relative to the sample population. Related to this is describing the survey's purposes in ways that heighten the attention to uses that might benefit the sample household or groups to which the household is affiliated.

Prior to an interviewer seeking an interview in a face-to-face survey, advance letters are sent to sample households in order to convey the sponsorhip and purpose of the survey and to alert the household to an upcoming visit by the interviewer. In telephone surveys, letters can be sent only to listed household numbers. These generally act to increase the willingness of the household to consider the request and to heighten the confidence of interviewers in seeking participation. Sometimes the advance contact with the sample household contains some incentive, to increase the benefits of participation. These might be prepaid monetary incentives or small gifts, which attempt to increase willingness to grant the interview.

Upon contact with the sample household, all interviewer behavior affecting participation becomes relevant. In face-to-face surveys, this involves giving any descriptive brochures or material about the survey. The interviewers provide description of the purposes of the survey, the nature of the interview, the confidentiality provisions affecting the data, etc. They react to any questions or signs of reluctance from the householder according to training guidelines in an attempt to gain cooperation. When the contact is complete, the interviewers document the results of the call and make notes about the nature of the interaction.

If the initial contacts result in a refusal to participate in the survey, various design features are used to urge the household to reconsider its decision. These might include mailing a letter to the household reiterating the importance of their participation to the success of the survey and the importance of the survey to the community. These letters may contain incentives to respond that attempt to increase the benefits of cooperation by the household.

A different interviewer may make a "refusal conversion" call on the reluctant household. Alternatively, a supervisor may call the household, in an attempt to convey the importance of the household's participation. Sometimes the mode of contact will be changed to one offering greater likelihood of success (e.g., from telephone to face-to-face contact).

When all else fails, the survey designers may radically reduce the burden of the request, seeking a very short interview from the household in order to collect the basic information useful to postsurvey adjustments.

1.6   THE FOCUS OF THIS BOOK

This book attempts to document pervasive influences on persons' participation in surveys. In doing so, it is concerned both with the statistical impact of nonresponse on survey statistics, with survey design features that might decrease or increase participation, with the nature of interviewer–householder interaction producing the participation decision, and with postsurvey adjustments to correct for nonresponse.

The book was written with the judgement that at this point in the development of the field, basic research on the participation decision was needed. To be helpful, the research would inform survey designers about what features of surveys tended to increase participation and why they did so. To be useful, it would identify design features that would improve the power of postsurvey adjustments for nonresponse. Finally, to be valuable to survey managers, it would identify principles underlying differential abilities of interviewers to obtain participation from householders.

In short, the book attempts to link several cultures within the survey field. It seeks to build and test theoretical constructs influencing survey participation, but it wishes to draw implications for practitioners. It seeks to find insights into ways to reduce nonresponse in field administrations, but it attempts also to discover ways to improve statistical adjustments by analysts of survey data.

1.7   LIMITATIONS OF THIS BOOK

Although our goal is a widely useful set of theories and findings about survey participation, the book has clear limitations. It focuses on surveys only of the U.S. household population. It contains no direct investigation of participation in establishment, business, or organization surveys. Further, although we believe some of the underlying principles apply to many survey designs of persons, it does not directly investigate the process of participation in surveys of memberships, social networks, or employee groups, where issues of group identity might be stronger.

All of the empirical data used in the book come from face-to-face surveys. There are two reasons for this: a) we take advantage of a unique matching of survey respondent and nonrespondent records to the 1990 U.S. decennial census sponsored by agencies conducting several face-to-face surveys; and b) the ability to observe characteristics of nonrespondent households is greatly enhanced in face-to-face surveys versus telephone or self-administered surveys. The fact that face-to-face surveys are studied implies that we have limited ability to draw inferences about survey modes not using an interviewer as the agent of the survey request. Because telephone surveys are increasingly common (especially in the United States), we comment throughout the book on the applicability of the results to the telephone survey mode.

This is a book about "unit nonresponse," not "item nonresponse." We are interested in what induces people to grant a request for a survey interview from a stranger who appears on their doorstep. We do not study the process by which respondents who begin an interview fail to supply answers to some questions. We believe that the influences toward this behavior are quite different from those of the initial acceptance of the interview request.

We examine the process of decisions to participate in one-time surveys or the first wave of a longitudinal survey. We do not examine the influences on dropping out of a panel survey after initial response or the factors that influence long-term panel retention. We suspect that the length of the initial-wave's interview, the sensitivity of questions in the first wave interview, the cognitive demands of the respondent task in the first wave interview, and the rapport built with the first wave interviewer make the process of continued cooperation in a longitudinal survey quite distinctive from that of granting first-time survey requests.

Finally, the surveys we study are either collected directly by U.S. government agencies or sponsored by U.S. government agencies. Government surveys throughout the world tend to have higher response rates than surveys conducted in the academic or commercial sectors. Some of the surveys we study intensively have unusually high contact and cooperation rates with sample households. At key points in the book we discuss any limitations this poses on the inference from our findings.

We hope that the book provides a blend of conceptual frameworks that have widespread utility and empirical tests that provide persuasive evidence. The next chapter maps out the conceptual framework that guides the thinking about survey participation throughout the book.

1.8   SUMMARY

This book seeks to illuminate the behavioral foundations of a statistical error property of sample statistics—that arising from unit nonresponse. We address the topic by first imposing a conceptual structure on our search for understanding, which is described in full in Chapter 2. This chapter described the major building blocks in models of the process of deciding to participate in a survey.

In studying the phenomenon, we are attracted to the viewpoint that the process of granting a survey request is a stochastic one; that is, although it is subject to consistent and powerful influences, there are, in general, few deterministic features to the process. There are, therefore, negligible proportions of truly "hard-core" nonrespondents.

We have found that dissecting the nonresponse phenomenon into one of noncontacts, refusals, and other causes, sensitizes us to considering alternative causes of each outcome. Since these processes are mixes of ones under the control of the researchers (e.g., number of callbacks), and ones out of their control (e.g., the urbanicity levels of the target populations), studying each separately is important both for practical implications of field administration and specification of postsurvey adjustment models.

We will consistently search for causes of different types of survey nonresponse, seek observable proxy indicators of those causes, and suggest that they be used both to guide targets for nonresponse reduction during data collection and be used for postsurvey adjustment models.

The next two chapters lay out the theoretical orientation that guides the analysis (Chapter 2) and review the data resources we bring to bear to address survey participation (Chapter 3). Then we begin presenting the results of empirical analysis. In Chapter 4 we examine the process of contacting sample households. In Chapter 5 we look at household-level influences on survey cooperation among contacted households. In Chapter 6 we turn to the social environmental influences on survey participation. Chapter 7 studies how interviewers act to influence cooperation when they contact householders. Chapters 8 and 9 turn to the interaction level, studying what interviewers and householders say and do during contacts that portend later cooperation with or refusal of the survey request. In Chapter 10 we review all the survey design features that researchers control to affect levels of response rate. Finally, in Chapter 11 we review step-by-step the process of survey design, implementation, and analysis, and apply the knowledge we learned to each of the steps, with the goal of producing overall survey statistics with minimal nonresponse error.

CHAPTER TWO

A Conceptual Framework for Survey Participation

2.1 INTRODUCTION

This chapter examines the components of survey participation that must be explained by a theory of survey participation, in order to develop effective methods to reduce nonresponse or to construct useful postsurvey compensation schemes. It is a multilevel conceptual framework that includes influences from the levels of the social environment, the household, the survey design, and the interviewer. It describes the role of the interaction between interviewer and householder in affecting the decision regarding the survey. The chapter ends with a discussion of implications of the theory for survey practices.

For surveys to be useful information gathering devices, sampling frames of households and individuals must be possible, strangers must be able to visit or telephone sample housing units and gain access to their households, persons must be willing to participate in an interview with the stranger and to trust pledges of confidentiality made regarding personal data provided to the interviewer. When many decades have passed since this writing, it may be the case that the social ingredients necessary for surveys to be useful tools of information assembly were present in a fairly limited period of historical time. The increasing difficulty of gaining participation may be the beginning of the disintegration of the necessary ingredients permitting surveys to function. Even without such a dire future, in order to understand the statistical implications of nonresponse, we must understand its behavioral bases.

2.2 PRACTICAL FEATURES OF SURVEY NONRESPONSE NEEDING THEORETICAL EXPLANATION

Before examining the influences on survey participation from a theoretical viewpoint, it is important to be quite specific about the details of the phenomenon of survey participation itself. What do we mean by participation in household surveys?

Perhaps this is best answered by dissecting the process temporally. In the next section we review those steps that yield interview data on sample units. We focus on the case of a household survey, either based on an area frame or a telephone frame. As shown in Figure 2.1, we move from locating and contacting the sample household, to identifying persons in the household, to choosing a respondent/informant, to seeking their participation in the survey. Survey nonresponse can arise at all of these points.

2.2.1 Contacting the Sample Household

Theoretically, the process of contacting a sample household is rather straightforward. As Figure 2.2 shows, the success at contacting a household should be a simple function of the times at which at least one member of the household is at home, the times at which interviewers call, and any impediments the interviewers encounter in gaining access to the housing unit. In face-to-face surveys, the latter can include locked apartment buildings, gated housing complexes, no-trespassing enforcement, as well as intercoms or any devices that limit contact with the household. In telephone surveys, the impediments include "caller ID," "call blocking," or answering machines that filter or restrict direct contact with the household.

In most surveys the interviewer has no prior knowledge about the at-home behavior of a given sample household. In face-to-face surveys interviewers report that they often make an initial visit to a sample segment (i.e., a cluster of neighboring housing units sampled in the survey) during the day, in order to gain initial intelligence about likely at-home behaviors. During this visit the interviewer looks for bicycles left outside (as evidence of children), signs of difficulty of accessing the unit (e.g., locked apartment buildings), small apartments in multiunit structures (likely to be single-person units), absence of automobiles, etc. Sometimes, when neighbors of the sample household are available, interviewers seek their advice on a good time

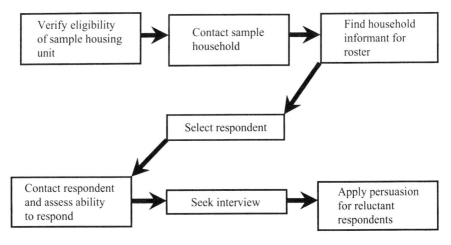

Figure 2.1. The process of survey participation.

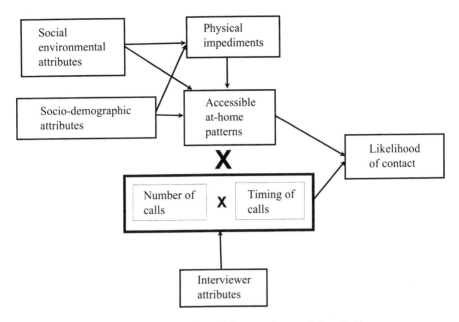

Figure 2.2. A conceptual model for contacting sample households.

to call on the sample unit. This process is the practical method of gaining proxy information about what call times might successfully encounter the household at home. In telephone surveys, no such intelligence gathering is possible. The only information about at-home practices of a sample household is obtained by calling the number. (This imbalance leads to the larger number of calls required to make first contact with a household in telephone surveys.)

In face-to-face surveys physical impediments to access are sometimes so strong that they literally prevent all contact with a sample unit. For example, some higher-priced multiunit structures have doormen that are ordered to prevent entrance of all persons not previously screened by a resident. Such buildings may be fully nonrespondent to face-to-face surveys. Similarly, although there is evidence that the majority of owners of telephone answering machines use them to monitor calls to their unit when they are absent (see Tuckel and Feinberg, 1991; Tuckel and O'Neill, 1995), some apparently use them to screen out calls when they are at home, thus preventing telephone survey interviewers from contacting the household.

Other impediments to contacting households may offer merely temporary barriers, forcing the interviewer to make more than the usual number of calls before first contacting them. For example, an apartment building whose entrance is controlled by a resident manager may require negotiations with the manager before access to sample households is given.

For units without any physical impediments to contact, the challenge to the interviewer is finding a time when the household is at home. Information from time use surveys, which ask persons to report on their activities hour by hour, have shown common patterns of at-home behavior by weekday mornings and afternoons, weekday evenings, and weekends. Those in the employed labor force are commonly out of the house, with the lowest rates of occupancy during the day. Interviewers make repeated calls to households they do not contact on the first call. Their choice of time for those callbacks can be viewed as repeated samples from a day-of-week, time-of-day frame. They base their timing of successive calls on information they obtain on prior unsuccessful visits and on some sense of consistency. For example, interviewers are often trained to make a callback on a unit not contacted at the last visit on a weekday afternoon by visiting on an evening or weekend.

2.2.2 Determining Eligibility, Obtaining a Household Roster

The next step in the process of survey participation can vary greatly by the target population of the study and by what person or persons in the households are used as respondents to the survey interview. Once contact is made with the household, if there is no named householder designated as respondent (for example using sampling frames of persons), the interviewer must identify the appropriate respondent(s) for the survey.

There are three common types of respondent rules: a) a household informant rule, sometimes seeking the most knowledgeable on the topics of interest; b) a randomly selected adult respondent; and c) a rule specifying that all householders are to be interviewed. The first and third rules more often have provisions for using a proxy respondent when the preferred respondent cannot or will not provide the interview. In some studies, only householders of a particular age, gender, or occupational status are eligible. For such surveys, the three rules above are used within the eligible set.

In all respondent rules, one of the first tasks of an interviewer is to obtain knowledge of what persons are eligible to be a respondent within the household (e.g., in a survey of those between the ages of 55–70, the interviewer would ask questions about the ages of household members). In some designs, interviewers are free to use any adult who appears to be competent to answer screening questions about the household.

In other designs, interviewers provide a brief introduction of themselves and the survey and then seek to obtain a listing of all household members. This listing often consists of a name or a relationship to the informant, gender, and age of the householder. The listing is then used to identify eligible persons and to provide a small sampling frame of the eligibles for selection of respondents. After the listing and selection of respondents, the interviewer seeks to interview the chosen respondent(s).

Reluctance from householders often first arises during the attempt at listing the household. This step commonly takes place after only a short introduction to the purposes of the interviewer's visit. To some householders, these questions may seem quite intrusive. The perceived violation of privacy may be exacerbated by the apparent lack of rationale for such questions. For example, when the interviewer has stated a purpose of learning about the expenses and consumer purchases of the household, it may not be clear why the interviewer begins with questions about each member of the household.

2.2.3   Repeated Callbacks

Sometimes the attempt to obtain a household listing or otherwise identify or speak to the respondent is unsuccessful in the first contact. In these cases, the interviewer ends the initial conversation by asking for a time to call again in order to contact the chosen householder.

This step is quite different from the attempts to find a call time prior to the first contact. Interviewers rely greatly on the information provided by the informant on the call. Interviewers will seek to make as firm an appointment as possible, to reduce the need for further calls.

Nonresponse may be threatened when a different respondent is chosen than the person providing the household listing. At the very least, it is common that another call must be made, based on an appointment time that is uncertain.

2.2.4   Refusal Conversion

Either at the initial contact (in which the household listing may have been obtained) or a later contact with a chosen respondent, householders' reluctance to participate in the survey may be so strong that they explicitly refuse to participate. They may provide no reason for this at all (e.g., "I don't want to do this"); they may provide superficial answers (e.g., "I'm too busy for this"); or they may tell the interviewer more detailed reasons why they are refusing (e.g., "I don't want to do anything to help the Federal government").

It is common for survey organizations to set such cases aside for a period of time and then attempt another contact. In face-to-face surveys, the survey organization might send a letter urging the householder to reconsider. A different interviewer or perhaps the supervisor might be assigned the case for recontact.

2.2.5   Nonresponse Because of Incapacitation

At any of the contacts with the sample household, the interviewer may learn that the chosen respondent is unable to provide the interview, regardless of how willing he or she might be. This can arise from the inability to speak the languages the survey organization is prepared to use to administer the interview or failure to find a suitable translator. Alternatively, the chosen respondent may suffer from physical health problems that rob them of the energy necessary to answer the survey questions. Finally, the chosen respondents may suffer from depression, mental retardation, or a variety of other mental health disorders that prevent them from comprehending the questions or otherwise attending to the respondent task.

2.3   A CONCEPTUAL STRUCTURE FOR SURVEY PARTICIPATION

Once the interviewer contacts a sample household, we believe that the influences on the householder's decision to participate arise from both relatively stable features of the environment and their backgrounds, fixed features of the survey design, as well as quite transient, unstable features of the interaction between the interviewer and the householder. This conceptual scheme is portrayed in Figure 2.3, listing influences of the social environment, householder, survey design features, interviewer attributes and behavior, and the contact-level interaction of interviewers and householders.

The influences on the left of the figure (social environment and sample household) are features of the population under study, out of control of the researcher. The

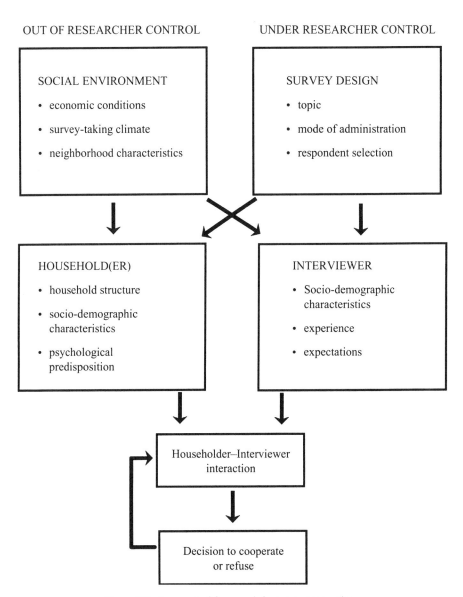

Figure 2.3. A conceptual framework for survey cooperation.

influences on the right are the result of design choices by the researcher, affecting the nature of the survey requests and the attributes of the actors (the interviewers) who deliver them. The bottom of the figure, describing the interaction between the interviewer and the householder, is the occasion when these influences come to bear. Which of the various influences are made most salient during that interaction determines the decision outcome of the householder.

2.3.1   Social Environmental Influences on Survey Participation

Since surveys are inherently social events, we would expect that societal and group-level influences might affect their participation rates. There are a set of global characteristics in any society that affect survey participation. These factors serve to determine the context within which the request for participation takes place, and constrain the actions of both householder and interviewer. For example, the degree of social responsibility felt by a sample person may be affected by such factors as the legitimacy of societal institutions, the degree of social cohesion, and so on. Such factors influence not only the expectations that both interviewer and respondent bring to the interaction, but also determine the particular persuasion strategies (on the part of the interviewer) and decision-making strategies (on the part of the respondent) that are used. More specific to the survey-taking climate are such factors as the number of surveys conducted in a society (the "oversurveying" effect) and the perceived legitimacy of surveys.

We would expect, therefore, to the extent that societies differ on these attributes, to observe different levels of cooperation for similar surveys conducted in different countries. There is evidence for this (see de Heer and Israëls, 1992), but the evidence is clouded by different design features used across countries, especially intensity of effort to reduce nonresponse. These include different protocols for advance contact with sample households, for repeated callbacks on noncontacted cases, and for dealing with initial refusals.

There are also environmental influences on survey cooperation below the societal level. For example, urbanicity is one of the most universal correlates of cooperation across the world. Urban dwellers tend to have lower response rates than rural dwellers. This contrast has been commonly observed in part because the urbanicity variable is often available from the sampling frame. The nature of urbanicity effects on response rates has been found to be related to crime rates (House and Wolf, 1978) but also may be related to population density, the type of housing structures, and household composition in urban areas. The effect may also be a function of inherent features of urban life—the faster pace, the frequency of fleeting single-purpose contacts with strangers, and the looser ties of community in such areas. We explore the issue of environmental influences (both at the societal and subnational levels) in greater depth in Chapter 6.

2.3.2   Characteristics of the Sample Householder

The factors affecting nonresponse most widely discussed in the survey literature are socio-demographic characteristics of the householder or sample person. These include age, gender, marital status, education, and income. Response rates have been shown to vary with each of these, as well as other, characteristics (see Chapter 5).

There are other factors associated with these which also have been studied for their relationship to response rates. These include household structure and characteristics, such as the number and ages of the household members and the quality and upkeep of housing; and the past experience of the respondent, such as exposure to situations similar to the interview interaction or a background that provided information or training relevant to the survey topic.

We do not believe these factors are directly *causal* to the participation decision. Instead, they tend to produce a set of psychological predispositions that affect the decision. Some of them are indicators of the likely salience of the topic to the respondent (e.g., socioeconomic indicators on income-related surveys); others are indicators of reactions to strangers (e.g., single-person households).

The socio-demographic factors and household characteristics all may influence the householder's psychological predispositions. Feelings of efficacy, embarrassment, or helpfulness, and moods of depression, elation, or anger will all be affected by these factors. All of these characteristics will then influence the cognitive process that will occur during the interaction with the interviewer.

As we note in Chapter 1, we believe that few householders have strongly preformed decisions about survey requests. Rather, these decisions are made largely at the time of the request for participation. Much social and cognitive psychological research on decision making (e.g., Eagly and Chaiken, 1984; Petty and Cacioppo, 1986) has contrasted two types of processes. The first is deep, thorough consideration of the pertinent arguments and counterarguments, of the costs and benefits of options. The second is shallower, quicker, more heuristic decision making based on peripheral aspects of the options.

It is our belief that the survey-request situation most often favors a heuristic approach because the potential respondent typically does not have a large personal interest in survey participation and, consequently, is not inclined to devote large amounts of time or cognitive energy to the decision of whether or not to participate. Further, little of the information typically provided to the householder pertains to the details of the requested task. Instead, interviewers describe the purpose of the survey, the nature of the incentive, or the legitimacy of the sponsoring organization. All of these in some sense are peripheral to the respondent's task of listening to the interviewer's questions, seriously considering alternative answers, and reporting honestly one's judgement.

Cialdini (1984) has identified several compliance principles that guide some heuristic decision making on requests and appear to be activated in surveys. These include reciprocation, authority, consistency, scarcity, social validation, and liking. We briefly review these here (see also Groves, Cialdini, and Couper, 1992) and link them to other concepts used in the literature.

Reciprocation. This heuristic suggests that a householder should be more willing to comply with a request to the extent that compliance constitutes the repayment of a perceived gift, favor, or concession. Thus, one may choose to participate in a survey based on a perceived sense of obligation to the organization making the request, or to the broader society it represents. On a narrower level, more peripheral features of the request (e.g., incentives, interviewer compliments) may be sufficient to invoke the reciprocity heuristic.

Reciprocation, as a concept, is closely related to sociological notions of *social exchange.* Social exchange theories tend to focus on long-run relationships between individuals and groups, but contain the same influence of past favors given by another influencing similar actions by a focal person or group.

Authority. People are more likely to comply with a request if it comes from a properly constituted authority, someone who is sanctioned by the society to make such requests and to expect compliance. In the survey interview context, the immediate requester is typically not the authority figure but is seen as representing some sponsoring organization that can be judged to have varying degrees of authority status. Survey organizations with greater legitimacy (e.g., those representing Federal government agencies) are more likely to trigger the authority heuristic in influencing the householders' decision to participate.

Notions of *social isolation,* the perception by people that they are not part of the larger society or bound by its norms, may be useful here. Socially isolated groups include both those believing they have suffered historical inequities at the hands of major institutions or groups and those identifying quite strongly with a distinct subculture. These types of groups may be guided by the same norms of reciprocation or influences of authority during interactions involving institutions of the majority culture, but in such cases their effects on cooperation may be negative.

We have found concepts of reciprocation and authority very important to understanding the behavior of sample persons and in Chapter 5 we describe empirical tests of these concepts. In addition, however, there are four other compliance heuristics described by Cialdini that are relevant to surveys: consistency, scarcity, social validation, and liking.

Consistency. The consistency heuristic suggests that, after committing oneself to a position, one should be more willing to comply with requests for behaviors that are consistent with that position. This is the likely explanation for the foot in the door effect in surveys (e.g., Freedman and Fraser, 1966), where compliance with a small initial request leads to greater willingness to accede to a larger request.

Scarcity. This heuristic notes that one should be more willing to comply with requests to secure opportunities that are scarce. To the extent that the survey request is perceived as a rare opportunity to participate in an interesting and/or important activity, the scarcity principle may lead to greater likelihood of acceptance of the request.

Social Validation. Using this heuristic, one would be more willing to comply with a request to the degree that one believes similar others are likely to do so. If householders believes that most people like themselves agree to participate in surveys, they may be more inclined to do so themselves.

Liking. Put simply, one should be more willing to comply with the requests of liked others. A variety of factors (e.g., similarity of attitude, background, dress, praise) have been shown to increase liking of strangers, and these cues may be used to guide the householder's decision in evaluating the interviewer's request.

While we believe these heuristics often come to the fore when a householder is confronted with a request to participate in a survey, there are other factors, more closely associated with a rational choice perspective, that may also influence his or her decision.

For example, a common finding in research on attitude change (see, for example, Petty and Cacioppo, 1986) is that when the topic of discussion is highly salient to laboratory subjects, they tend to give careful consideration to the arguments pro and con concerning the topic. Similarly, we think that saliency, relevance and interest in the survey topic are relevant to the householder's decision process. That is, when the survey topic is highly relevant to the well-being or for other reasons of interest to the householders, they might perform a more thorough analysis of the merits of co-operating with the survey request.

However, in contrast to the laboratory experiments in the attitude change litera-ture, largely based on willing and motivated subjects, the survey setting probably lim-its cost–benefit examination of a survey request. Calls by interviewers to sample households are generally unscheduled events. The amount of discretionary time per-ceived to be possessed by the householders at the time of the contact will also affect their tendency to engage in deliberate, careful consideration of the arguments to par-ticipate in the survey. Householders who see themselves as burdened by other oblig-ations may overwhelmingly choose heuristic shortcuts to evaluate the survey request.

In Chapter 5 we examine characteristics of the household and householder in more detail, focusing primarily on the socio-demographic characteristics of the householder and how these may relate to cooperation with survey requests.

2.3.3   Attributes of the Survey Design

Much survey research practice is focused on reducing nonresponse by choosing fea-tures of the survey design that generate higher participation rates. These by and large are fixed attributes of the request for an interview that are applied to all cases. This section discusses those features in an indirect manner, by identifying and elab-orating the concepts that underlie their effectiveness.

Many of the survey design features aimed at gaining cooperation make use of one or more of the compliance heuristics reviewed above. For example, the recipro-cation heuristic probably underlies the large literature on the effects of incentives on survey participation rates. Consistent with the concept of reciprocation, there ap-pear to be larger effects of incentives provided prior to the request for the survey, compared to those promised contingent on the completion of the interview (Berk *et al.,* 1987; Singer *et al.,* 1996).

The concept also underlies the common training guideline in some surveys for interviewers to emphasize the potential benefits of the survey to the individual respondent. For example, in the Consumer Expenditure Survey, used as part of the Consumer Price Index of the United States, interviewers often tell elderly householders that their government social security payments are affected by the survey.

One implication of the consistency principle for survey design is that an interviewer who can draw a connection between the merits of particular (or general) survey participation and the respondent's committed beliefs, attitudes, and values (e.g., efficiency in government, advancement of knowledge) is likely to be more successful in gaining compliance.

Evoking authority is a common tool in advance mailings in household surveys and in the introductory script of interviewers. Advance letters are often crafted to use stationery that evokes legitimate authority for the information collection; the letters are signed, whenever possible, by persons with titles conveying power and prestige. Some social surveys (e.g., studies of community issues) seek the endorsement of associations or organizations that would aid the communication of legitimate authority to collect the data. Further, interviewers are often trained to emphasize the sponsor of their survey when the sponsor is generally seen as having legitimate authority to collect the information (e.g., government or educational institutions), but rarely to do so when that is less likely (e.g., certain commercial organizations).

The scarcity principle may underlie the interviewer tactics of emphasizing the value to a respondent of "making your voice heard" or "having your opinion count" while noting that such an opportunity is rare (e.g., "We only contact one person in every 30,000"). This principle may also help explain the decline of survey participation in Western society that has coincided with the proliferation of surveys. People may no longer consider the chance to have their opinions counted as an especially rare, and therefore valuable, event. Consequently, at the end of the interviewing period, some interviewers are known to say that "There are only a few days left. I'm not sure I'll be able to interview you if we don't do it now"—a clear attempt to make the scarcity principle apply.

Similarly, survey organizations and interviewers may attempt to invoke social validation by suggesting that "Most people enjoy being interviewed," or "Most people choose to participate," or by evincing surprise at the expression of reluctance by a householder.

The use of race or gender matching by survey organizations may be an attempt to invoke liking through similarity, as well as reducing the potential threat to the householder.

There are other survey design features that do not fit nicely into the compliance heuristics conceptualized by Cialdini. Indeed, these are much more closely aligned with rational choice, cost versus benefit tradeoff decisions. For example, there is some evidence (see Chapter 10) that longer questionnaires require the interviewer to work harder to gain cooperation. In interviewer-assisted surveys, some of the disadvantages can be overcome by interviewer action, but more work is required. Thus, other things being equal, choosing a short survey interview may yield easier attainment of high participation.

Related to burden as measured by time is burden produced by psychological threat or low saliency. Survey topics that ask respondents to reveal embarrassing facts about themselves or cover topics that are avoided in day-to-day conversations between strangers may be perceived as quite burdensome. For example, surveys about sexual behaviors or income and assets tend to achieve lower cooperation rates, other things being equal, than surveys of health or employment. On the other hand, when the topic is salient to the householders, when they have prior interests in the topic, then the perceived burden of answering questions on the topic is lower. This probably underlies the finding of Couper (1997) that householders who express more interest in politics are more easily interviewed in political surveys than those with no such interests.

In Chapter 10 we explore the issue of survey design and its effect on the cooperation decision in greater depth. We review much of the methodological literature on survey design to reduce nonresponse, in the light of the theoretical perspectives outlined above.

2.3.4 Attributes of the Interviewer

Observable attributes of the interviewer affect participation because they are used as cues by the householder to judge the intent of the visit. For example, consider the socio-demographic characteristics of race, age, gender, and socioeconomic status. At the first contact with the interviewer, the householder is making judgements about the purposes of the visit. Is this a sales call? Is there any risk of physical danger in this encounter? Can I trust that this person is sincere? Assessments of alternative intentions of the caller are made by matching the pattern of visual and audio cues with evoked alternatives. All attributes of the interviewer that help the householder discriminate the different scripts will be used to make the decision about the intent of the call. Once the householders choose an interpretation of the intent of the call—a "cognitive script" in Abelson's (1981) terms—then the householders can use the script to guide their reactions to the interviewer.

The second set of influences from the interviewer are functions of their experience. To select an approach to use, the interviewer must judge the fit of the respondent to other respondent types experienced in the past (either through descriptions in training or actual interaction with them). We believe that experienced interviewers tend to achieve higher levels of cooperation because they carry with them a larger number of combinations of behaviors proven to be effective for one or more types of householders. A corollary of this is that interviewers experiencing diverse subpopulations are even more resourceful and are valuable for refusal conversion work. We can also deduce that the initial months and years of interviewing provide the largest gains to interviewers in providing them with new persuasion tools.

The third set of attributes might be viewed as causally derivative of the first two—interviewer expectations regarding the likelihood of gaining cooperation of the householder. Research shows that interviewers who believe survey questions are sensitive tend to achieve higher missing data rates on them (Singer and Kohnke-Aguirre, 1979). Interviewers report that their emotional state at the time of contact is crucial to their success: "I do not have much trouble talking people into cooperating. I love this work and I believe this helps 'sell' the survey. When I knock on a door, I feel I'm gonna get that interview!" We believe these expectations are a function of interviewer socio-demographic attributes (and their match to those of the householder), their personal reactions to the survey topic, and of their experience as an interviewer.

In Chapter 7 we explore the role of the survey interviewer and the influence of interviewers on survey participation. We examine interviewer experience, expectations and attitudes, and behavior in affecting survey cooperation.

2.3.5   Householder–Interviewer Interaction

It is when interviewers encounter householders that the factors discussed above come to bear on the decision to participate. The strategies the interviewer employs to persuade the sample person are determined not only by the interviewer's own ability, expectations, etc., but also by features of the survey design and by characteristics of the immediate environment and broader society. Similarly, the responses that the sample person makes to the request are affected by a variety of factors, both internal and external to the respondent, and both intrinsic and extrinsic to the survey request.

We have posited that most decisions to participate in a survey are heuristically based. The evidence for this lies in the tendency for refusals to come quickly in the interaction, for interviewers to use short, generally nonoffensive descriptors in initial phases of the contact, for respondents to only rarely seek more information about the survey. This occurs most clearly when participation (or lack thereof) has little personal consequence. With Brehm (1993) we believe that the verbal "reasons" for refusals—"I'm too busy," "I'm not interested"—partially reflect these heuristics, mirroring current states of the householder but, in contrast to Brehm, we believe they are not stable under alternative cues presented to the householder. We believe that there are two constructs regarding interviewer behavior during the interaction with householders that underlie which heuristics will dominate in the householder's decision to participate. These are labeled "tailoring" and "maintaining interaction."

Tailoring. Experienced interviewers often report that they adapt their approach to the sample unit. Interviewers engage in a continuous search for cues about the attributes of the sample household or the person who answers the door, focusing on those attributes that may be related to one of the basic psychological principles reviewed above. For example, in poor areas, some interviewers choose to drive the family's older car and to dress in a manner more consistent with the neighborhood, thereby attempting to engage the liking principle. In rich neighborhoods, interviewers may dress up. In both cases, the same compliance principle—similarity leads to liking—is engaged, but in different ways.

In some sense, expert interviewers have access to a large repertoire of cues, phrases, or descriptors corresponding to the survey request. Which statement they use to begin the conversation is the result of observations about the housing unit, the neighborhood, and immediate reactions upon first contact with the person who answers the door. The reaction of the householder to the first statement dictates the choice of the second statement to use. With this perspective, all features of the communication are relevant—not only the words used by the interviewer, but the inflection, volume, pacing (see Oksenberg, Coleman, and Cannell, 1986), as well as physical movements of the interviewer.

From focus groups with interviewers, we found that some interviewers are aware of their "tailoring" behavior: "I give the introduction and listen to what they say. I then respond to them on an individual basis, according to their response. Almost all responses are a little different, and you need an ability to intuitively understand what they are saying"; or "I use different techniques depending on the age of the respondent, my initial impression of him or her, the neighborhood, etc."; or "From all past interviewing experience, I have found that sizing up a respondent immediately and being able to adjust just as quickly to the situation never fails to get their cooperation. In short, being able to put yourself at their level, be it intellectual or street wise, is a must in this business. . . ."

Tailoring need not necessarily occur only within a single contact. Many times contacts are very brief and give the interviewer little opportunity to respond to cues obtained from the potential respondent. Tailoring may take place over a number of contacts with that household, with the interviewer using the knowledge he or she has gained in each successive visit to that household. Tailoring may also occur across sample households. The more an interviewer learns about what is effective and what is not with various types of potential respondents encountered, the more effectively requests for participation can be directed at similar others. This implies that interviewer tailoring evolves with experience. Not only have experienced interviewers acquired a wider repertoire of persuasion techniques, but they are also better able to select the most appropriate approach for each situation.

Maintaining Interaction. The introductory contact of the interviewer and householder is a small conversation. It begins with the self-identification of the interviewer, contains some descriptive matter about the survey request, and ends with the initiation of the questioning, a delay decision, or the denial of permission to continue. There are two radically different optimization targets in developing an introductory strategy—maximizing the number of acceptances per time unit (assuming an ongoing supply of contacts), and maximizing the probability of each sample unit accepting.

The first goal is common to some quota sample interviewing (and to sales approaches). There, the supply of sample cases is far beyond that needed for the desired number of interviews. Interviewer behavior should be focused on gaining speedy resolution of each case. An acceptance of the survey request is preferred to a denial, but a lengthy, multicontact preliminary to an acceptance can be as damaging to productivity as a denial. The system is driven by number of interviews per time unit.

The second goal, maximizing the probability of obtaining an interview from each sample unit, is the implicit aim of probability sample interviewing. The amount of time required to obtain cooperation on each case is of secondary concern. Given this, interviewers are free to apply the "tailoring" over several turns in the contact conversation. How to tailor the appeal to the householder is increasingly revealed as the conversation continues. Hence, the odds of success are increased with the continuation of the conversation. Thus, the interviewer does *not* maximize the likelihood of obtaining a "yes" answer in any given contact, but minimizes the likelihood of a "no" answer over repeated turntaking in the contact.

We believe the techniques of tailoring and maintaining interaction are used in combination. Maintaining interaction is the means to achieve maximum benefits from tailoring, for the longer the conversation is in progress, the more cues the interviewer will be able to obtain from the householder. However, maintaining interaction is also a compliance-promoting technique in itself, invoking the commitment principle as well as more general norms of social interaction. That is, as the length of the interaction grows, it becomes more difficult for one actor to summarily dismiss the other.

Figure 2.4 is an illustration of these two interviewer strategies at work. We distinguish between the use of a general compliance-gaining strategy (e.g., utilizing the principle of authority) and a number of different (verbal and nonverbal) arguments or tactics within each strategy (e.g., displaying the ID badge prominently, emphasizing the sponsor of the survey, etc.). The successful application of tailoring depends on the ability of the interviewer to evaluate the reaction of the householder to his or her presence, and the effectiveness of the arguments presented. Note that the interviewer's initial goal is to maintain interaction (avoiding pushing for the interview) as long as the potential respondent's reaction remains neutral or noncommittal. An interviewer will continue to present different arguments until such time as the householder is clearly receptive to an interview request, or there are no more arguments to present. For inexperienced interviewers, the latter may occur before the former, forcing the interviewer to (prematurely in some cases) initiate the interview request.

There is some support from training procedures that the "maintaining-interaction" model operates as theorized. First, interviewers are typically warned against unintentionally leading the householder into a quick refusal. If the person appears rushed, preoccupied by some activity in the household (e.g., fighting among children), the interviewer should seek another time to contact the unit. A common complaint concerning inexperienced interviewers is that they create many "soft-refusals" (i.e., cases easily converted by an experienced interviewer) by pressing the householder into a decision prematurely. Unfortunately, only rarely do interviewer recruits receive training in the multiturn repartee inherent in maximizing the odds of a "yes" over all contacts. Instead, they are trained in stock descriptors of the survey leading to the first question of the interview.

We note how similar the goals of a quota sample interviewer are to those of any salesperson, but how different are those of the probability sample interviewer. Given this, it is not surprising that many attempts to use sales techniques in probability-sample surveys have not led to large gains in cooperation. The focus of the salesper-

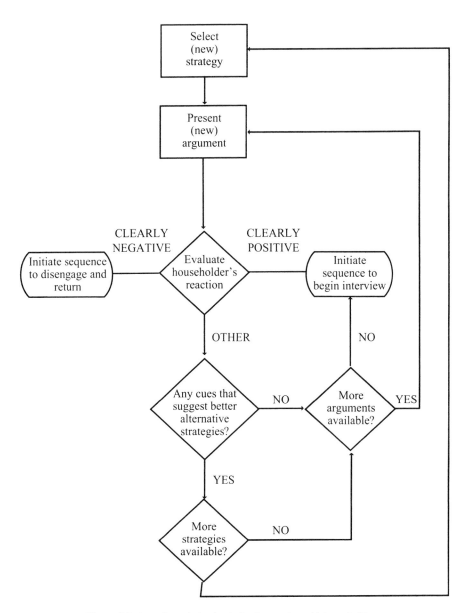

Figure 2.4. Interviewer behavior during interaction with householders.

son is identifying and serving buyers. The "browser" must be ignored when a known buyer approaches. In contrast, the probability sample interviewer must seek cooperation from both the interested and uninterested.

At the same time that the interviewer is exercising skills regarding tailoring and maintaining interaction, the householder is engaged in a very active process of determining whether there has been prior contact with the interviewer, what is the intent of the interviewer's call, whether a quick negative decision is warranted, or whether continued attention to the interviewer's speech is the right decision. Figure 2.5 describes this process.

The process has various decision points at which the householder can make positive or negative decisions regarding participation in the survey. These arise because

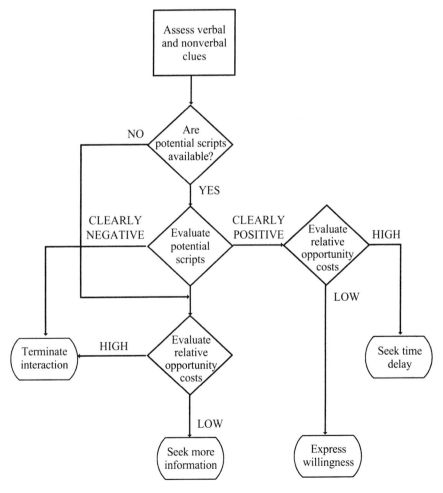

Figure 2.5. Householder behavior during interactions with the interviewer.

the householder misinterprets the visit as involving some unpleasant nonsurvey request; that is, the householder chooses the wrong script. They arise if there are very high opportunity costs for the householder to continue the interaction with the interviewer. They arise if any of the heuristics point to the wisdom of a negative or positive decision.

2.4   IMPLICATIONS FOR RESEARCH

There are several implications of this theoretical framework as an approach to survey participation. The first concerns the relationship between nonresponse rates and nonresponse error. The expression

$$\bar{y}_r = \bar{y}_n + (m/n)[\bar{y}_r - \bar{y}_m]$$

notes that the error term, $(m/n) [\bar{y}_r - \bar{y}_m]$, is a function of the nonresponse rate and a difference. If, however, the likelihood of participation is itself a function of the y values, then nonresponse error can be relatively high, even with low nonresponse rates. Such would be an example of *nonignorable* nonresponse (in Rubin's (1987) terminology).

What does the theoretical framework suggest about the likelihood of ignorable and nonignorable nonresponse mechanisms? Nonignorable nonresponse can arise if the cause of the nonresponse is the value of the survey variable itself or when causes of nonresponse are also causes of survey variables. We would speculate that the former case is most likely with self-administered questionnaires, which are presented prior to the householder committing to the survey. With this survey design, the householder can inspect each of the questions prior to making the participation decision. If, for example, they are asked to reveal embarrassing facts about themselves, they might choose not to respond. In many interviewer-administered surveys, the topic of the survey is only vaguely described, and causes of nonresponse are not the survey variables themselves. We would speculate, however, that the likelihood of nonignorable nonresponse rises, other things being equal, as a function of how completely the survey questions are described to the householder.

There are, however, specific topics in interviewer-administered questionnaires that make them specifically subject to nonignorable nonresponse. For example, time-use surveys, measuring how people spend their time, would be highly susceptible to noncontact nonresponse error. Those households who are away from home most of the time will have very different values on their statistics for "away from home" activities than others. This is an example when nonresponse (noncontact in this case) is caused by the survey variable itself.

Following the theory, any survey that attempts to measure attributes of the sample person that cause noncooperation would be subject to refusal errors. For example, surveys that measure participation in community activities and other fulfillment of civic duties would suffer from nonresponse error from refusals when the surveys are portrayed by interviewers as conducted for the good of the community or society. (We would also deduce from the theory that refusal conversion efforts could reduce this effect if other arguments for participation were made more salient in followup visits.)

It is probably more common that the survey variables measured do not cause nonresponse, but, instead, nonresponse and the survey variables share a common cause. Given the theory above, an example of this might be that fear of crime may lead persons to refuse to let the interview proceed, causing the nonresponse. If the survey variable were measuring the use of home burglar alarms, deadbolt locks, etc., there would be a tendency to underestimate their use in the target population. However, if one had measures of fear of crime, one might construct adjustment procedures reflecting these causal relationships. Conditional on one's fear of crime, there would be smaller nonresponse error from refusals.

The theory suggests that statistical errors arising from noncontact nonresponse are likely to be different from those arising from refusal nonresponse. This is deduced from the fact that the influences on noncontact are often distinctive from those for refusals.

We note that the theory above is heavily cognitive and social psychological in its orientation. However, the literature on nonrespondent characteristics is heavily socio-demographic. If the theory is correct, many of those socio-demographic differences between respondents and nonrespondents are spurious. If and when one controls for differences in the psychological influences on survey participation, those differences should disappear. This will be relevant in our later attempts to explain the pervasive lower cooperation rates in urban versus rural areas, for example (see Chapter 6).

The theory places greater importance than previous approaches on the interaction between the household and the respondent. The multiple relationships in the model suggest that interviewers, especially those interviewing diverse subgroups, would, with experience, turn to tailoring methods. From the model we might deduce steeper learning curves for interviewers serving heterogeneous subpopulations than for those in more homogeneous areas. We might deduce that initial tactics of experienced interviewers upon contacting a sample household would concentrate on "maintaining contact" or rapport building than on seeking compliance. That is, there would be identifiable ordering of tactics used by interviewers within a contact interaction. Given the relative richness of cues in the face-to-face mode, we anticipate more tailoring there than in the telephone mode. Finally, we suspect that much of the manifestation of tailoring lies in the contrast evident *among* respondents, not contrasts within interactions with individual respondents. We note that all of these aspects of tailoring are testable hypotheses and should shape the research agenda in the future.

Another deduction from the theoretical framework is that the importance of various attributes of the survey design and the enrollment process for respondents will themselves depend on the respondent and the interaction. We deduce that what features of the request are made salient to the respondent will affect their decision calculus. We also deduce that what features increase the propensity to respond for one householder may decrease the propensity for another. For example, emphasizing that the Consumer Price Index is affected by the survey may heighten the saliency of the topic for the retired, but lead to an inference that the survey is irrelevant to the self-interests of youth. This implies that we expect in models predicting the likelihood of response to find statistical interactions among subgroups. What is important for some will not be important for others.

The theory asserts that there are identifiable parts of the interaction between interviewers and sample persons, that these are predictive of the participation decision. A key research issue is whether these parts are observable at all *and* whether they are observable by the interviewer involved in the interaction. Can the interviewer document these features of the interaction in order to study their role in the participation process? Are tape recordings of the interaction (see Morton-Williams, 1993) the only method to measuring the interaction? We devote Chapters 8 and 9 to a discussion of this issue.

The model suggests a need for studies of traits of the interviewer that might be compatible with the concept of tailoring. Relatively stable attributes on "need for cognition" and "flexibility" refer to the taste for diagnosing alternative courses of action and changing strategies frequently. One theoretical question is whether these behaviors are trait-based or learned by experience in the field. If the former, it would be prudent to attempt recruiting and screening to identify interviewer candidates. If the latter, training programs could be developed to equip interviewers with these skills.

One important research item is the exploration of the role of efforts to seek cooperation and levels of measurement error. How do different compliance techniques affect later behavior of the respondent during questioning? Is there some desirability to changing tactics when the interview begins? For example, will appeals to authority at the point of gaining cooperation increase response errors on sensitive topics later in the interview?

Finally, we believe that the development and testing of theoretical models of survey participation will also lead to greater understanding of compliance behavior in humans. We believe that survey methodologists have much to learn from the literatures of compliance, persuasion, and helping behaviors. However, the benefit may be mutual. Attaining more general understanding of the influences on survey participation can help the social psychology of compliance break from the current paradigm that addresses a limited number of requests and acts of persuasion, while generally ignoring the impact of characteristics of the requestor. The simple acknowledgement of heterogeneity in the population's reaction to any particular stimulus relevant to compliance is key to the theoretical approach taken above and to its fit with the survey setting.

These and other research questions have guided our study of the nonresponse problem described in this book. We address many of these issues in the remaining chapters of the book. Some, however, remain unanswered and must await further inquiry.

2.5 PRACTICAL IMPLICATIONS FOR SURVEY IMPLEMENTATION

In addition to guiding a research agenda, if the theory is true, it has broad implications for survey design. One obvious value of this perspective is that it acknowledges that nonrespondents can have well-founded rationales for not cooperating with a survey request. In contrast to the view that nonrespondent actions are in some sense based on ignorance and lack of sophistication, this approach attempts to identify costs and benefits of responding from the sample person's perspective.

What can survey researchers do to reduce nonresponse error with this knowledge? The first possibility to pursue would be an attempt to tailor survey design to different groups that differ on relevant values. For example, those judged more "peripheral" to the goals of the survey and to societal norms relevant to survey participation might be given special benefits for their participation or some cost-reducing feature (e.g., shorter interviews). Implicit in this idea is the balancing of nonresponse error with costs of data collection.

Second, interviewer training needs to instruct the interviewer in how to read cues about the best tactics to use in approaching particular sample persons. We expect that cues observable prior to contact (e.g., nature of the neighborhood, characteristics of the housing unit) can be taught more easily than those during the interaction with the sample person. In addition to reading the cues, the interviewer must be armed with techniques found effective in each of the situations identified.

Third, the perspective heightens the value of rich frames that may have correlates of response propensity. For example, frames that identify single-person households would permit interviewers to begin their field efforts on those units because they are likely to require more effort at contact and more effort at gaining cooperation once contacted.

A fourth implication of the theory is the wisdom of tailoring survey materials to specific attributes of the sample person. Promotional materials for the survey could be customized to different lifestyles. Circumstances surrounding initial refusals could be used to effectively tailor persuasion strategies for subsequent contacts, or to inform decisions about the allocation of resources to refusal conversion.

Fifth, given initial observations on noncontact and reluctant respondents, additional field efforts might be stratified to reduce nonresponse error for a given targeted nonresponse rate. For example, consider a consumer expenditure survey for which at a specific point in the field administration 60% of the nonrespondents were noncontacts, 30% were reluctant for reasons of time constraints, and 10% reluctant because of reluctance to talk about their finances. The theory would suggest that nonresponse error of a survey that would increase the response rate merely by bringing in the noncontacts may exceed that which would bring in those reluctant because of sensitivity to the topic.

Sixth, the ability to document key features of the interaction between interviewer and respondent may permit the estimation of response propensity functions in ongoing surveys. If these are successful prediction devices, they could be employed for postsurvey adjustment procedures to reduce the bias of nonresponse.

Finally, the theory implies that using response propensity models for nonresponse adjustment should reflect the different processes of nonresponse. That is, separate models might be constructed for noncontact and refusal nonresponse, because the phenomena are so distinctive in their causes (and likely to be distinctive in their effects on nonresponse error).

These implications are in direct contradiction to the application of standardized procedures at the stage of the survey request. While standardization at the measurement phase is a basic tenet of the scientific method, it has no proven value at the cooperation phase. We suspect that much of the prepackaged survey introductions interviewers are asked to deliver evolved from the misguided applications of the tenets of standardization to that part of the survey. Successful interviewers seem to learn that fitting the approach to the sample person is wiser.

2.6 SUMMARY

Survey participation is influenced by many factors related to the survey request, the survey design, the interviewer, and the householder. Similarly, nonresponse to surveys is multifaceted. Understanding survey participation requires attention to very practical features of the stages of a survey. Nonresponse arising from noncontact appears to be heavily influenced by design features regarding number of calls and timing of calls, but will also vary greatly, conditional on those design choices, across subpopulations that differ in their at-home patterns. Refusal nonresponse is manifested in a few seconds or minutes of interaction between the interviewer and the householder.

During these interactions, the interviewer may be attempting to make salient to the householder all those features of the sponsorship, topic, and nature of the task that will enhance the perceived benefits of participation from the perspective of the individual householder. At the same time, the householder is attempting to make judgements about the real purpose of the visit and the implied burden, threat, and benefit of continuing the interaction. Both interviewers and householders can make errors of judgement during these brief encounters. Interviewers can choose to emphasize features of the design that make salient undesirable features of the request from the vantage of the householder. Householders can misjudge the visit as one involving sales, financial scams, or other undesirable requests. They can also correctly understand the nature of the request and judge that the time or effort involved in responding to the survey exceeds the benefits of responding.

The factors that influence the survey participation decision thus are a function of a very dynamic social communicative process. This process is informed by prior experiences of the householder and the interviewer. However, the circumstances of the visit, the order of presentation of material, and the sequence of conversation turns can make salient different aspects of the request over different visits. Thus, full understanding of the process of survey participation requires some insight into all levels of influences simultaneously.

CHAPTER THREE

Data Resources for Testing Theories of Survey Participation

3.1 INTRODUCTION

This chapter gives an introductory description of the data used for empirical tests of the behavioral models of survey participation. It discusses key weaknesses in the existing literature about nonresponse and describes efforts to overcome these limitations. It describes the several data sets that are used in this book, pointing out strengths and weaknesses of each and how they relate to one another.

The past several decades have produced a large literature on the correlates of nonresponse. To date, however, this literature has not been well integrated in a way that advances our understanding of the processes that produce nonresponse, and the consequences of nonresponse for survey estimates and analyses. This book, in part, represents an effort to unify the existing literature in the light of a theoretical framework, and to encourage the use of such a framework to guide future research endeavors in the field. To do so will require, we believe, a new perspective on the kinds of data that are collected to inform an understanding of nonresponse, and new analytic tools to evaluate its impact.

While voluminous, the existing body of research on survey nonresponse suffers from a number of flaws, seriously limiting the inferential impact of the work. We briefly review a number of these limitations here (see also Goyder, 1987, p. 80).

First, most studies of nonresponse are case studies, focusing on a single survey, rather than cross survey comparisons. Given the wide variety of different survey designs and implementations, this makes it difficult to separate characteristics of households that may be associated with nonresponse from survey design features that may impact on the likelihood of making contact or gaining cooperation.

Second, many of the past studies either do not differentiate among different sources of nonresponse (simply examining respondents versus nonrespondents) or present results for only one source of nonresponse (e.g., refusals). As is evident in this book, we have learned that different processes operate to produce different types of nonresponse, and it is important to distinguish between noncontacts and refusals in studying survey participation.

Third, many of the past studies use only bivariate analyses. Single correlates of nonresponse are examined one at a time, rather than exploring the joint effects of sets of variables in a multivariate context. As we note in Chapter 5, this may be one reason for the confusing findings on nonresponse among the elderly, and the effect of single-person households. Building datasets that allow the exploration of joint and interaction effects of many variables will help increase our understanding of the survey-participation process.

Fourth, most studies use predictors only if they are readily available on the sample frame or from auxiliary sources, rather than attempting to enrich the data with information from other sources or collect supplemental data during data collection. These predictors tend to be demographic or geographic information, rather than the rich social psychological variables we believe to be key to understanding survey nonresponse. For example, one reason that urbanicity is the most commonly studied correlate of nonresponse in the literature is simply because it is available in some form for almost all area probability samples.

Fifth, aside from studies exploring the effects of design changes on nonresponse, much of the literature on correlates of nonresponse are based on secondary analyses of existing datasets, rather than data collection efforts designed at the outset to collect data relevant to the study of nonresponse. This means, again, a predominance of socio-demographic variables, bivariate rather than multivariate analysis, and a tendency to look at overall nonresponse rather than components of the problem.

Finally, advancing our knowledge of survey nonresponse has been hampered, we believe, by the lack of a theoretical framework to provide context for the interpretation of these diverse findings. In a way, this is similar to the questionnaire design literature, which produced many findings of individual effects with little systematic integration until the development of theories to help explain observed phenomena (e.g., Hippler, Schwarz, and Sudman, 1987). We offer in this book a framework for exploring the process of survey participation.

This is not to say that advances have not been made in the study of nonresponse. For example, the literature on mail survey nonresponse has benefited both from meta-analyses to integrate diverse findings, and from theoretical developments, mostly within a social exchange framework (e.g., Dillman, 1978; Goyder, 1987). Other efforts to synthesize and integrate the literature have focussed on specific topics, such as incentives (e.g., Singer *et al.,* 1996). Similarly, on the statistical side, the notion of ignorability (Little and Rubin, 1987) has advanced our understanding of the statistical impacts of nonresponse. The greater variation among designs in interviewer-administered (and particularly face-to-face) surveys has made such integration harder to do.

For the reasons outlined above, the impact of the cumulative body of research on survey nonresponse has not been as large as the volume of research output may suggest. Clearly, there are exceptions to these characterizations, but all studies suffer from one or more of the above limitations. Our own data resources remedy only some of the weaknesses of past work, and the reader needs to understand their weaknesses. However, our work has been guided from the outset by an appreciation of the data requirements to fully explore nonresponse within the domain of large-scale face-to-face surveys in the United States.

3.2 APPROACHES TO STUDYING NONRESPONSE

Probably the biggest drawback in attempting to study nonresponse is that the people we are most interested in are precisely that—nonrespondents. In order to explore the extent of nonresponse error we must perforce rely on efforts to gain information from nonrespondents *that they would not normally proffer in the context of the survey*. This has been attempted in several different ways. We outline below some of the major approaches used to overcome these limitations in order to examine nonresponse.

A common approach to studying nonresponse has been the use of frame data available for both respondents and nonrespondents. In the United States, where most national personal-visit surveys are based on area or address frames, little or no information is available other than at higher levels of geography (urbanicity, region, other aggregate demographic measures). In many countries in Europe and elsewhere, the existence of a population register may provide additional information on sampled households or persons (e.g., Lindström, 1983; Heiskanen and Laaksonen, 1995). Even so, this is still limited to a few demographic items available on the frame.

A second approach is to study reluctant respondents. These are sample persons who required effort to interview (additional contact attempts, extra persuasion, or refusal conversion efforts) but who eventually provided an interview. These reluctant respondents are treated as proxies for final nonrespondents (e.g., Smith, 1984; Dunkelberg and Day, 1973). This approach is based on the key premise of a continuum of nonresponse ranging from the cooperative respondent through the reluctant or difficult-to-contact respondent, to the nonrespondent. There are some strong suggestions in the literature (e.g., Lin and Schaeffer, 1995; Guadagnoli and Cunningham, 1989) that this argument may be flawed, and that those who do not participate in a survey despite our best efforts to include them, are different from those who are eventually persuaded to cooperate, or who are found after many attempts at contact.

In recent years, we, along with other researchers (e.g., Brehm, 1993; Campanelli, Sturgis, and Purdon, 1997), have begun to collect observational data on the household or the interaction to supplement information on the sampling frame. This started with interviewers recording details of the reason for nonresponse, providing a summary of calls made to a sample household, and making notes of salient features of sample segments. These data were often collected to facilitate the data collection process, but were not available in a systematic form to explore correlates of nonresponse. More recently, these measurements have included observational measures of neighborhood characteristics, details of each call, and features of the conversations between interviewers and householders. This is an approach we have increasingly found useful in our work, and discuss at length in Chapter 8.

A fourth way to explore nonresponse is to study panel nonrespondents, using characteristics of those who responded at the first wave, but dropped out subsequently. However, we believe the process of panel attrition is qualitatively different from that of a cross-sectional study or the first wave of a panel survey. In the latter, the householder has little or no prior knowledge of the content of the survey or forewarning of the survey request. In the former, in later waves respondents are likely to base their decision to participate on their experiences in the first wave (Couper, Groves, and Raghunathan, 1996; Kalton *et al.,* 1980; Kalton and Citro, 1993; Lepkowski, 1988).

A fifth approach involves special studies of nonrespondents, usually conducted after the main study has been completed. Examples include double-sampling strategies (Rao, 1983; Singh, 1983; see also Lengacher *et al.,* 1995) and special surveys of nonrespondents (e.g., Bergman, Hanve, and Rapp, 1978). These approaches are often hampered by low response rates to the follow-up efforts.

Sixth, some have attempted surveys about survey participation (e.g., Goyder, 1986; Walker Research, 1992; Council for Marketing and Opinion Research, 1996). This approach seeks to ask respondents about their past experiences with survey requests in order to estimate frequencies of requests, acceptances, and refusals. This approach faces the problem that those who are unlikely to participate in surveys in general are less likely to be included in the survey asking about other surveys than are those who frequently accept. The Survey of Census Participation (see Section 3.4.1) is an example of this type of survey, asking about participation in the decennial census (see, e.g., Singer, Mathiowetz, and Couper, 1993).

Finally, a variety of experimental strategies have been used to study survey participation. Two broad classes of experiments can be distinguished. The first type generally involves alternative design features (e.g., incentives, advance letters, etc.) to measure main effects on response rates. While these may tell us which strategies may work better than others, they often don't identify why certain factors may work in some households but not others. For that, one again needs information on who cooperates and who does not. Much of the research discussed in Chapter 10 is based on these types of studies.

Another set of innovative experimental designs involves collecting information on social psychological dispositions *prior* to the survey request. This is exemplified in the work of Cialdini, Braver, and Wolf (1991, 1992; see also Braver and Cialdini, 1994; Hox, de Leeuw, and Vorst, 1995) in which a variety of attitudinal measures were obtained from a group of college students, following which they were given an apparently independent survey request. This approach provides some insight into the social psychological correlates of survey participation, but is limited in its generalizability by the type of population measurable in that fashion.

As far as our own work is concerned, we chose to focus our efforts on strategies that could provide us with additional information on nonresponding households in ongoing surveys. This serves the dual purpose of increasing our understanding of the nonresponse mechanism and its impact on survey statistics, and of providing data to evaluate the effectiveness of alternative adjustment strategies.

We rely on two major sources of empirical data in this book. The first is the decennial census match nonresponse project, in which 1990 U.S. decennial census data for nonrespondent (and respondent) cases were used to examine correlates of nonresponse. The second source of data is surveys in which we supplemented the data collection activities with a variety of observational measures on nonrespondent and respondent households. We focus primarily on two such surveys here: the National Survey of Health and Stress (NSHS) and the Survey on Asset and Health Dynamics of the Oldest Old (AHEAD). These are supplemented with information from focus groups, interviewer questionnaires, and ecological variables.

In the following sections of this chapter, we describe the various sources of data we use in this book. We begin with the qualitative work that initiated the series of studies on nonresponse. Then we describe the decennial nonresponse match project, the source of much of the data used here. We also describe two other surveys we use to study survey participation, focusing primarily on householder–interviewer interactions. We end with a discussion of other data sources to explore interviewer and ecological correlates of survey nonresponse.

3.3   QUALITATIVE DATA FROM INTERVIEWER GROUP DISCUSSIONS

We began our work with a series of focus groups with interviewers working for various survey organizations in the United States. These focus groups were iterative in that the initial group discussions were largely unstructured and aimed at understanding the nature of an interviewer's work (from the viewpoint of the interviewer) in gaining cooperation from sample persons. As our work progressed and we began to develop measures of some of the concepts identified in these groups, we conducted several follow-up group sessions where we replayed our theoretical developments to elicit interviewer reactions. Our early empirical findings were also shared with interviewer groups to get feedback.

In addition, we accompanied interviewers into the field on various occasions, and observed them practicing their craft. We also observed, and participated in, various interviewer training sessions over the course of several years to gain insight into how interviewers learn the task of gaining cooperation and achieving high response rates.

The insights gained from these discussions permeate much of this book. Chapter 7 focusses primarily on the role of the interviewer in survey participation, and leans heavily on the qualitative material gleaned from interviewers, as does the discussion of the householder–interviewer interaction in Chapter 8.

3.4   DECENNIAL CENSUS MATCH OF SURVEY RECORDS TO CENSUS RECORDS

The major source of data for this book was the match of survey respondent and nonrespondent cases from six surveys to the 1990 U.S. decennial census records. This was a massive multiyear undertaking, completed on-site at the U.S. Census Bureau to protect the confidentiality of individual census and survey records, with all investigators sworn to uphold the census and survey confidentiality pledges to respondents. The data from this project form the core of several chapters in this book, and we describe the key features of the decennial match project in this section.

3.4.1 Selection of Surveys for the Decennial Census Match Project

The 1990 census provided a rare opportunity to learn something about the demographic characteristics of those who choose not to participate in large-scale face-to-face surveys. We attempted to include a variety of surveys in the match project. Given the cost and effort involved, funds were sought from the individual surveys for their inclusion. This restricted the range and number of surveys that could be included.

A total of six surveys were included in the survey–census match. Four of these are ongoing surveys conducted by the Census Bureau on behalf of various other federal agencies. These are:

- Consumer Expenditure Survey—Quarterly (CEQ), sponsored by the Bureau of Labor Statistics (BLS)
- Current Population Survey (CPS), sponsored by the Bureau of Labor Statistics (BLS)
- National Health Interview Survey (NHIS), sponsored by the National Center for Health Statistics (NCHS)
- National Crime Survey (NCS), sponsored by the Bureau of Justice Statistics (BJS)

Two surveys conducted by nonprofit survey organizations, under contract to the Federal government, were also included in the match project. These are:

- National Household Survey on Drug Abuse (NHSDA), conducted by the Research Triangle Institute (RTI), sponsored by the National Institute on Drug Abuse (now called Substance Abuse and Mental Health Services Administration (SAMHSA))
- Survey of Census Participation (SCP), conducted by the National Opinion Research Center (NORC), sponsored by the Census Bureau

The selection of these surveys for inclusion in the match study was somewhat arbitrary. They omit academic and commercial surveys not sponsored by the government. (We failed to obtain permission from any commercial firm to include their data.) The six were included because they represented large national surveys conducted at the time of the 1990 decennial census, and because funds were forthcoming from each of the agencies represented by these surveys. A summary of the characteristics of the six surveys is presented in Table 3.1.

In the following sections we describe the steps taken to match the survey data for these six surveys to 1990 decennial census records.

Table 3.1. Design attributes of surveys used in the decennial match project

Survey	Topic	Sample size	Oversampling	Respondent rule	Panel component	Interview period	Secondary mode of collection
Consumer Expenditure Quarterly	Infrequent purchases	9,000 per year	None	Knowledgeable adult in consumer unit	5 interviews, once every 3 months	1 month	None
Current Population Survey	Labor force participation	60,000 per month	None	Knowledgeable adult in household	4 monthly interviews, 8-months interval, 4 monthly interviews	1 week	Telephone in later waves
National Health Interview Survey	Health conditions, behaviors	4,000 per month	Tracts with high percentage of Blacks	Adult 19 yrs or older (and others)	None	10 days	None
National Crime Survey	Victimization	60,000 per 6 months	None	Self-response for 14 years and older	7 interviews, once every 6 months	2 weeks	Telephone for late interviews in household
1990 National Household Survey on Drug Abuse	Drug usage	31,093 screened for 9,259 interviews	Washington, DC; youth; minorities	1–2 randomly selected, 12 years and older	None	14 weeks	None
Survey of Census Participation	Handling of census form	2,760 households	Small blocks, high minority blocks	Adult who had most to do with census form	None	7 weeks	Telephone as refusal conversion

3.4.2 Selection of Sample Cases for the Match to Census Records

The selection of cases for matching differed between the Census Bureau and non-Census Bureau surveys, as did procedures for collecting address and other information to facilitate the match.

The four Census Bureau surveys all have ongoing data collections throughout the year. The decennial census data collection focused on the occupants of housing units on April 1, 1990. Ideally, the reference time period for the surveys would be identical to that of the census. To obtain sufficient nonresponse cases, it was decided to include cases from a 3-month period—March, April, and May, 1990.

Most of these surveys are based on multiwave panel designs. A decision was made to include only first-wave cases, as we were most interested in studying the behavior of interviewers and respondents in the case in which there has been no prior contact between the two, when the sample person has minimal knowledge about the nature of the survey request.

Finally, cases were selected for the decennial census match project to meet multiple study needs, including an evaluation of survey noncoverage and interviewer misclassification. Thus, a number of interviewed cases were also selected, as were a subset of ineligible sample units (vacant housing units, demolished structures, etc.). To obtain sufficient nonresponse cases, all nonresponding sample units in the 3-month period were selected for inclusion in the match, along with all other cases (interviews and ineligibles) in a subset of sample segments.

Procedures differed somewhat for the two non-Census Bureau surveys. In each case, selection of cases for the census match was done after completion of the primary data collection. The NHSDA involves the random selection of one or more persons within a household as respondents. Thus, nonresponse could occur at two stages: the household screener and the selected householder(s). A random subset of both types of nonresponse cases (screener and interview nonresponse) were selected. Interviewed cases were selected to provide a comparative group of cooperative households. In addition, a small number of vacant housing units were selected to detect interviewer cheating and/or misclassification.

The SCP was a one-time survey conducted a few months after census day for the purpose of understanding the dynamics of decennial census mail-return behavior (see Fay, Bates, and Moore, 1991, and Kulka et al., 1991, for descriptions of the study). For this survey, interest in matching cases to the decennial census included not only nonresponse research but also an examination of self-reported versus actual behavior. Given these needs, and the relatively small sample size for this survey, it was decided to include all sample cases in the match.

Given the above criteria, the number of sample cases included from each of the six surveys are presented in Table 3.2. A total of 20,711 survey cases were eligible to be matched to decennial census records, of which about 50% were interviewed cases and 35% nonresponse cases. This provides us with a total of 17,047 eligible sample units from the six surveys included in the analyses. The balance were classified as ineligible for the surveys (e.g., vacant, demolished dwellings).

Table 3.2. Sample sizes for decennial match project by survey

	CEQ	CPS	NHIS	NCS	NHSDA	SCP	Total
Interviews	844	1,445	1,920	1,407	1,938	2,478	10,032
Nonresponse							
Refusals	836	451	546	520	1,273	144	3,770
Noncontacts	89	713	290	399	1,004	99	2,594
Other	33	75	49	51	404	39	651
Total	958	1,239	885	970	2,681	282	7,015
Total	1,802	2,684	2,805	2,377	4,619	2,760	17,047

3.4.3   Matching Operation

We briefly describe the match procedures here. For more detail on the match operation, see Couper and Groves (1992b). To facilitate the clerical matching operation to decennial census records, interviewers on each of the Census Bureau surveys were provided with special forms to complete on each case selected for the match project. In addition to identifying information, interviewers were instructed to provide the following:

- Complete address of sample unit, including postal address if known
- Sketch map of the immediate area, including cross-streets and other identifying features of the segment
- Names of householders and/or household information where available

The match operation itself was performed by experienced staff at the Census Bureau's Processing Office in Jeffersonville, Indiana.

In the case of the NHSDA and SCP, the selection of sample units for inclusion in the match was done after the data collection period for the surveys. Interviewers thus received no special instructions for dealing with these cases, and all the information needed to match these cases had to be obtained from existing sources such as interviewer notes, segment listing sheets, and segment maps.

It should be noted that the match operation involved matching sample *housing units*, not sample *persons*. For many sample cases, particularly nonresponding households, no person-level information was available. Given the proximity of the sample selection to census day, we make the assumption that residential mobility would have little impact on the results of the match.

The match protocol involved matching a survey address (including apartment unit designation) with the corresponding housing unit in the decennial census records, thereby permitting access to the census form for that housing unit. The primary source for locating an address in the census file was the Address Control File (ACF). This was a computerized database of approximately 100 million addresses in the United States. The ACF was the primary tool in the implementation of the

1990 census, and was used to control the mailout and return of census forms. For operational purposes, the ACF is organized in a hierarchy of geographical units. An address' position in the file is defined by the district office, address register area, and block to which it belongs. (For more detailed information on decennial census geographical concepts, see U.S. Bureau of the Census, 1990.) If the exact address of a housing unit could not be matched directly, all units in the block and adjoining blocks could be examined to find the unit.

A number of auxiliary resources were available to the match clerks to facilitate the location of a specific unit in the ACF. These included:

- Zip code directories
- Block header records, which designate the census geography for a range of house numbers on a particular street
- Address registers, which include not only descriptions of houses in rural areas, but also maps on which each housing unit is numbered and its physical location in the block designated
- A variety of other maps, including the computer-generated decennial census maps (TIGER maps) and commercially available city and county street maps.

Information on household members (e.g., surname or household roster information), where available, was also used to confirm that the correct housing unit had been found.

Note that this was a clerical match operation, although computer assistance (in the form of ACF) was used. A small group of experienced match clerks were trained to perform the matching tasks after they had obtained considerable matching experience as part of the Post Enumeration Survey (PES) effort of the census. Great emphasis was placed on the quality of the work performed rather than the speed of completion. All cases that could not be matched using the basic set of tools were set aside for supervisor review and resolution.

All matched forms were subject to a 100% verification or quality control step. Once the census-identifying information had been recorded and the match and address codes assigned, the case was sent to a different clerk for verification. This was done by reversing the procedure followed for matching a case, starting with the census identifier and working back to the address on the survey form.

The forms were keyed by a group of professional keyers, using 100% verification (independent re-keying of all forms). All keying discrepancies were resolved by a supervisor or referred back to the match unit for clarification. Check-digits in the census identifying information were also used to ensure a high degree of accuracy in keying. No decennial census information was lost due to incorrectly transcribed or keyed census identification variables.

A final quality control step was performed on a subset of cases after keying. A systematic sample of 10% of all match cases was selected. A total of 2,057 cases were selected in this manner and subjected to additional scrutiny. In 3.1% of cases, a change was made to one or more pieces of the census geocode information (permitting extraction of census data). Note that this should *not* be used as an estimate of the number of incorrectly matched forms; many of these changes were simply omissions of noncritical information or typographical errors. Check-digits were used during the keying process to detect and correct such errors.

3.4.4 Results of the Match Operation

A summary of the results of the decennial census match operation is presented in Table 3.3. Overall, 96.3% of cases were successfully matched at the housing-unit level. Matches at the census-block level were achieved for 99.4% of all sample cases, so we have at least some information for almost all cases. The match rates for nonrespondent cases are slightly lower than those for interviewed cases, probably due to the additional information available for the latter to help confirm a successful match. The difference between refusals and interviewed cases (96.6% and 97.2% respectively) is negligible. Most of the analyses we report in this book are based on these two groups, and we largely ignore the "other nonresponse" category, which includes those too incapacitated to do the interview, language barriers that preclude a successful interview, and a variety of other circumstances.

The success of the match depended in large part on the quality of the address information provided by the survey interviewer, or available on the sample frame. Thus, in cases where a street name and number (e.g., 123 Main Street) were provided, over 98% were successfully matched. At the other extreme, those cases where only a description was provided (e.g., two-story wood siding home at end of dirt road behind scrapyard), matching was less successful, with only 90.4% of such cases being matched to decennial records. However, these cases made up only 2.6% of the entire sample. An unknown proportion of the match failures can, of course, be attributed to errors on the decennial census side, with households not being enumerated (no census form available) or enumerated in a different area.

We believe that the match operation was a great success, due in no small part to the persistence and dedication of the clerical staff in Jeffersonville. We are confident that the findings we report in this book are unlikely to be affected by the small number of cases for which we could not obtain decennial census records.

Table 3.3. Percentage of survey cases successfully matched to decennial census records at household level

	CEQ	CPS	NHIS	NCS	NHSDA	SCP	Total
Interviews	98.0%	95.8%	97.3%	96.9%	97.8%	97.8%	97.2%
Nonresponse							
Refusals	95.8%	94.9%	97.1%	95.8%	97.6%	96.5%	96.5%
Noncontacts	92.1%	89.6%	96.6%	93.2%	95.7%	97.0%	93.7%
Other	81.8%	73.3%	89.8%	92.2%	96.5%	87.2%	91.5%
Total	95.0%	90.6%	96.5%	94.5%	96.7%	95.4%	95.0%
Total	96.4%	93.4%	97.0%	96.0%	97.2%	97.5%	96.3%

3.4.5 Structure of Decennial Match Data Sets

The keyed forms containing both census and survey identifiers were used to extract decennial census data and merge these data to the survey data for each household. Census short-form data (basic household and person demographics answered by all persons) were extracted for all matched cases. For those households that received the census long form (about one-sixth of all households), the responses to these additional items were also included. The decennial data we used were in a raw (preimputed) state.

While group quarters were not eligible for inclusion in the match project, during the match operation a small number of group quarters were found (less than 1% of eligible sample cases). These were dropped in all subsequent analyses. In addition, those cases that we could not successfully match were excluded from the analyses.

Given the fact that a housing-unit match was done, almost all analyses were performed at the household level. For all person-level variables (e.g., race, age), this meant creating household-level measures. In some instances this meant using data for the reference person (i.e., the first who completed the census form, often a household head), while in other cases it meant aggregating data over all household members (e.g., number of unemployed adults in household, presence of young children).

3.4.6 Inferential and Statistical Estimation Issues

The theories reviewed in Chapter 2 describe influences on decisions to participate in a survey that occur at different levels of aggregation. That is, we specify that effects may lie at the societal, city, block, interviewer, and contact levels. Key hypotheses concern whether the effects at a higher level of aggregation (e.g., urbanicity effects) can be explained by differences across areas in lower-level characteristics (e.g., household composition effects). Most of these hypotheses will be tested using variants of multivariate linear models. Most often, we will specify a logistic regression equation, with the propensity of contact or cooperation as the dependent measure. We will test whether various coefficients in the model change their values in the presence of other predictors. In the vast majority of analyses, the unit of analysis is the sample household.

Because the sample household is the unit of analysis, the different observations in the data set fail normal assumptions of independence. From the theoretical point of view, households within the same neighborhoods share similar local ecological influences; households in the same counties share higher-order ecological influences; households approached by the same interviewer share the influences of that interviewer. From the sample design point of view, households in the same primary sampling unit (i.e., counties or county groups) are subject to among-unit covariances, as are households in the same secondary sampling units. Both of these viewpoints lead to estimation concerns in both simple means and proportions, as well as coefficients in multivariate models. Assuming independence across observations when estimating descriptive and analytic statistics generally leads to underestimation of standard errors of the statistics.

Two analytic approaches are candidates for this type of data. The first are random-effect models or multilevel models (Bryk and Raudenbush, 1992; Goldstein, 1987; Hox, 1995). The predictive model is specified at the lowest level of aggregation on which data exist. However, the model provides explicit measurement of a) the contribution of higher-level units to the variability in lowest-level coefficients, and b) the contribution of the higher-level unit characteristics to explaining levels of the dependent variable. These models normally use some form of maximum-likelihood estimation, permitting specification of fixed or random effects at various levels of aggregation.

Multilevel models are most easily estimated when the different levels of aggregation are perfectly clustered. For example, they have had major impacts in explaining relative effects of school environments versus student characteristics on academic performance. In that case, students are members of one and only one student body.

In the case of the decennial match data structure, the relevant levels theoretically are the social environment, the interviewer, the householder, and the interactional level. Most of these are nicely nested. That is, interactions on contacts can be nested within households uniquely, households can be nested within environmental groupings uniquely. The interviewer level, however, presents some complications. Some interviewers will work in more than one of the environmental groupings (e.g., sample areas). Some households may be assigned to multiple interviewers during the course of the enrollment process (e.g., for dealing with reluctant respondents).

Further complicating the structure of the data is the fact that the selection of individual housing units was not independent. Instead, all the decennial match data sets are stratified cluster samples, based on selection of units in a set of primary sampling units (PSUs) (i.e., counties or county groups). Within primary sampling areas, secondary units are selected, inducing clustering within the primary selections. The mapping of primary sampling units onto the different levels is a complicated one. In multicounty PSUs some of the social environmental data are on a county level, some on a block level. In large PSUs, multiple interviewers would be assigned to sample units. In small PSUs, it is possible that one interviewer would handle the whole workload; in some cases, one interviewer might be assigned to multiple PSUs. In short, the clustering of the sample design does not perfectly coincide with the different levels of theoretical aggregation.

In multilevel models with imperfect nesting, the estimation of effects at various levels is plagued by lack of degrees of freedom for the estimation of effects. For example, if one interviewer worked both in a multi-interviewer PSU and completely worked another PSU, how best can we estimate the component of variance due to interviewers?

Another approach to the estimation of linear models on survey data of this sort attempts to reflect the major sample design factors, with their effects on the variance of coefficients in the models. There are three important sources of impact on the variance of the coefficients: a) variation in the probabilities of selection (e.g., the match project's oversample of nonrespondent cases), b) stratification prior to selection (i.e., grouping PSUs into sets relatively homogeneous on the survey variables), and c) clustering of sample housing units into primary and secondary selection areas. When applied to the data sets in the decennial match survey, this approach will estimate fixed-effects models but reflect the component of variance due to clusters within strata of the sample design. Fixed-effect models will be fit at the household level. Attributes of the social environment and of the interviewers assigned will sometimes be included in the model, as fixed effects (e.g., measuring the effect of urbanicity levels of the environment of the household).

This approach, sometimes labeled "design-based" estimation, is similar to that used to produce descriptive statistics from the surveys by the government agencies that conduct them. It reflects the lack of independence across units of analysis in the estimators by reflecting the clustering and stratification of the sample design. This approach, like the multilevel approach, probably misestimates the standard errors of interviewer effects, because of the problem of interviewers handling cases in multiple PSUs. However, it has the benefit of directly reflecting the impacts on standard errors of coefficients of the clustering and stratification in the sample design.

This latter approach (fixed-effects models using design-based estimators) is used throughout the book. However, at several points in the analyses conducted, we fit variants of multilevel models for key equations. When we did so, no changes in substantive conclusions resulted. That is, we expect that had multilevel modeling been presented throughout the book, substantially the same conclusions would have been made.

3.4.7   Case Weighting and Variance Estimation

In early analyses (see Groves and Couper, 1993a) we examined nonresponse correlates in each of the six surveys separately. However, given our primary interest in variables that proved to be stable correlates of cooperation or contact across surveys, and based on these early results which showed similar effects across surveys, we decided to combine the surveys into a single data file for analysis. Hence, the reader needs to note the strategy we used for case-level weighting in the pooled data set.

The case weights we use reflect the probabilities of selection in the original surveys, the selection probability of the case into the decennial census match project, and a factor to reflect the combined contribution of the six surveys. Given that the six surveys we examine have minor variations in definitions of the eligible population, these last sets of factors are only approximate. The sum of the weights for each of the six surveys ranges from 82.6 million for NHSDA to 94 million for the NHIS. Thus, each survey in the pooled data set is weighted to roughly the same sum. For example, even though we have a larger sample of respondents and nonrespondents from NHSDA than NHIS, the surveys will contribute roughly equally to our overall findings. We used this weighting scheme so that our analyses reflect a replication of the nonresponse study over six different surveys.

Variance estimation treated each survey as a domain with distinct stratification, clustering, and sample size features. We describe the CPS case to illustrate the procedures followed. The cases selected into the match study consisted of an equal probability subsample of segments using a systematic selection technique and a "take-all" sample of nonrespondent cases among first wave sample cases for the period in question. Thus, the overall sample is an unequal probability, stratified, clustered sample.

Many of the original surveys' primary sampling units (PSUs) in the realized match sample have only one (often nonrespondent) case in them. In addition, all surveys consist of self-representing (SR) and non-self-representing (NSR) areas. In the decennial match sample, segments in the SR area are roughly equivalent in size to PSUs in NSR areas. Hence, we use individual segments in the SR areas as sampling error computing units (SECUs), creating a single stratum for the between SECU component of variance. In the NSR areas we use PSUs and treat states as the primary stratifier.

To tackle the problem of single nonrespondent cases in PSUs or segments, we created zones for computing variances. Following the sort order used for the selection of the original CPS sample, we grouped into a single zone the selected full segment and any individual cases between that segment and the next-selected full segment. These zones were then used as sampling error computing units. This collapsing was necessary to obtain sufficient units for variance estimation, but had the expected effect of overestimating to some unknown degree the standard errors of estimates.

A variety of simulations were performed using alternative collapsing schemes (e.g., two-zone and eight-zone collapsing), and alternative strata and SECU formulations were tested. We decided upon the one-zone collapsing scheme, as it produced variance estimates similar to the two-zone scheme, but more closely resembles the actual clustering of the design. Similar procedures were followed for the remaining surveys, with slight differences reflecting variations in the selection of cases for the decennial match. For example, all eligible SCP sample cases were included in the match study, permitting us to use the original sample design. Having combined the six surveys into a single dataset for analysis, we then created a "superstratum" for each survey. This weighting and variance estimation scheme was independently studied by Eltinge and Rope (1994) and found to have desirable properties relative to available alternatives.

Having determined an appropriate stratification and clustering scheme to permit variance estimation, all subsequent analyses were conducted using SUDAAN (Shah, Barnwell, and Bieler, 1996), a software system that uses Taylor series approximation to estimate variances and standard errors from complex sample designs. All results reported in this book are based on weighted estimates, with the standard errors reflecting the complexity of the design as specified in the definitions of strata and sampling error computing units in SUDAAN.

3.4.8   Analysis of the Decennial Census Match Data

We end with a discussion of the analytic uses of the decennial census match dataset. This section describes features of those uses that the reader should keep in mind when assessing the evidence from the match study.

First, the analyses we report in this book pool the six surveys into a single data set. This pooling is introduced in order to improve the ability of the analysis to detect reliably differences in contact or cooperation rates across subgroups. While statistical power is effectively increased with this strategy, the reader needs to question whether the pooling masks differences across the six surveys in the effects they manifest. Most often, key analyses were run pooled and unpooled, and when the unpooled results reveal differences across the surveys, those are discussed.

Second, in several places (especially Chapter 5) we examine a one-sixth subsample of the survey records—those households who completed the long form of the decennial census. This long-form subsample contains a richer set of independent variables to study influences on survey cooperation. When that subset is used, the statistical power to detect differences is reduced approximately to less than half its value (i.e., standard errors inflated by a factor of 2.45). The role of these analyses is to verify that large errors of conclusions are not made in simpler models using the leaner set of variables available on all cases (i.e., the short-form variables).

We again remind the reader that the data available to us from the decennial census match are generally household-level data. In other words, we do not have information on the characteristics of particular sample persons who may be selected into each sample survey. The past literature includes some variables measured at the level of the sample person, obtained from the sampling frame or through interviewer observation. For our part, we aggregate across all persons in the household, or use some other combined measure. Where this doesn't make sense, we rely on the characteristics of the reference person (loosely equivalent to the household head).

A final caution is that the surveys we examine have relatively high response rates. This is even more evident when restricting our analyses to cooperation rates (given contact), as we do in Chapter 5. It's useful to review a little arithmetic to understand the analytic challenge of empirical studies of cooperation rates versus overall response rates. The overall response rate is the product of the contact rate and the cooperation rate:

$$\text{Response Rate} = (\text{Contact Rate}) \cdot (\text{Cooperation Rate})$$

$$= \frac{\text{Number Contacted}}{\text{Number Eligible}} \cdot \frac{\text{Number Interviewed}}{\text{Number Contacted}}$$

Since the contact rate is always less than 1.0, the cooperation rate is always higher than the overall response rate.

Table 3.4 presents the overall response rate, the contact rate, and the cooperation rate for the six surveys in the decennial match study. Some of the government surveys we are using to test our theories of survey participation have response rates between 94 and 96%, with a range of 82–97%. Cooperation rates range from 87–98%, with the bulk of the surveys between 97 and 98%. We must address the question of how this affects definitions of what a *substantively important* difference in cooperation rates means, as well as what difficulties we might have in detecting *statistically significant* differences in cooperation rates.

Table 3.4. Contact, cooperation, and response rates for six surveys of the decennial census match study

Survey	Contact	Cooperation	Response
CEQ	98.6%	87.3%	85.7%
CPS	96.9%	97.9%	94.6%
NHIS	98.7%	97.5%	96.1%
NCS	98.7%	98.3%	96.9%
NHSDA	95.3%	88.7%	82.0%
SCP	96.4%	94.7%	90.1%

On the substantive side, if you are a field administrator of these surveys, and you learn of a method of increasing your response rate from 94% to 97%, you have reduced your nonresponse rate to one-half its former size. This can have substantial effects on changes in the nonresponse error component of the survey, depending on the relative characteristics of new respondents brought into the data set. It is most probably substantively important.

In such a case, will there be substantial decreases in the total error of survey statistics? It is unlikely that large nonresponse errors *in absolute value* arise from such low nonresponse rates, so the proportionate reduction in nonresponse error does not alone lead to large increases in the accuracy of statistics. However, many of these surveys have very large sample sizes (i.e., low sampling errors), and the major sources of error in their survey statistics are nonresponse and measurement errors. This again says, *of the error present in the statistics,* differences of 3 percentage points can be important improvements in the quality of the statistics.

On the statistical side, surveys that have high overall cooperation or contact rates pose challenges to detecting subgroups that exhibit large differences in such rates. Table 3.5 helps to illustrate this, using cooperation rates. Take, for example, the row of the table describing a comparison of two subclasses, each forming half of the sample (the fourth row of the table). This would apply to a two-category variable, with each category forming half of the sample (e.g., gender). For a survey with a relatively low cooperation rate, say 75%, it would be possible for the two subclasses to have cooperation rates that are 50 percentage points different. They would be a 50% cooperation rate for the low-cooperation subgroup and a 100% cooperation rate for the high-cooperation subgroup. In stark contrast, if we were examining a survey with an overall cooperation rate of 95%, the largest possible difference between the subgroups would be 10 percentage points, 90% and 100%. If it were a 98% cooperation-rate survey, the largest difference would only be 4 percentage points.

Surveys with high overall cooperation rates can contain small subclasses with low cooperation rates but not large subclasses with low cooperation rates. This is determined by comparing the rows of Table 3.5, showing that for a subclass forming a tenth of the sample, a minimal cooperation rate is 50% in a survey with an overall rate of 95% (yielding a 50 percentage point difference). In contrast, a sub-

Table 3.5. Maximum possible percentage point difference between two subclass cooperation rates, by proportion of total sample in the subclasses by overall cooperation rate of survey

Proportion of total sample in subclass		Overall Cooperation Rate of Survey					
Low Coop. Group	High Coop. Group	75%	80%	85%	90%	95%	98%
0.10	0.90	83	89	94	100	50	20
0.20	0.80	94	100	75	50	25	10
0.25	0.75	100	80	60	40	20	8
0.50	0.50	50	40	30	20	10	4
0.60	0.40	42	33	25	17	8	3
0.75	0.25	33	27	20	13	7	3
0.80	0.20	31	25	19	12	6	3
0.90	0.10	28	22	17	11	6	2

class forming three quarters of the sample in the same survey could not have a cooperation rate less than 93% (yielding a 7 percentage point difference).

The reader should use the discussion of this section in two different ways. First, one should keep in mind that bivariate tables showing cooperation or contact rate differences across subclasses from the decennial census match data will show small percentage-point differences, partly as a function of the high overall rates. Second, in examining multivariate models testing various impacts on cooperation or contact, one needs to be cautious of what level of risks of Type 1 errors are appropriate, given the nature of the phenomenon being studied.

While the match data form the core of the much of this book (especially Chapters 5 and 6), we supplemented these analyses both with additional measures (on interviewers and environment) and with other datasets that provided richer detail on the interaction between the interviewer and householder at the point of contact. We describe these additional data sources in the following sections.

3.5   DOCUMENTATION OF INTERACTION BETWEEN INTERVIEWERS AND HOUSEHOLDERS

As we note elsewhere in this book, a key ingredient of our theoretical perspective is the interaction between interviewer and householder at the time of the request. We believe that most householders do not have preformed judgements about participation in a survey, but rather make these decisions at the time of the initial request by the interviewer, using cues accessible during the interaction. Similarly, the interviewer is making decisions about the likely willingness of the householder to participate, and selecting among possible alternative strategies for the most effective means of persuasion. The measurement of these processes will, we believe, enhance our understanding of how and why survey nonresponse occurs.

Obviously, collecting household(er) socio-demographic characteristics provides little insight into the rich interplay between the two parties at the doorstep. Ideally, one would want to tape-record and code these interactions, as was done on a small scale by Morton-Williams and Young (1987) and later by Campanelli, Sturgis, and Purdon (1997). Not only is this expensive and time-consuming to do, but when we broached this idea, it was met with a great deal of opposition among field staff, primarily because of the concern about making interviewers' already difficult jobs even more so.

As an alternative, we developed a set of closed-ended measures to be completed by an interviewer, as soon after the contact interaction as was possible. The development of the items was guided by the series of focus groups described earlier in this chapter. The contact description measures emerged from loosely structured discussions with groups of interviewers about the kinds of things they said and did on the doorstep, and the kinds of responses they encountered from householders. Successive drafts of the form were shown to interviewer focus groups for comment and revision. Finally, the set of contact description questions has undergone repeated revision in use by several different surveys, in an attempt to distill the items down to what can reasonably be completed by interviewers. Most of the analyses in this book (especially Chapter 8) are from the first iteration of these measures, from the NSHS (see Section 3.6.3 below). While we measured a variety of contact-level events, our analysis focuses primarily on a set of householder comments and questions. The NSHS version of these items is reproduced in Figure 3.1 below.

The remaining questions on the contact description form included interviewer estimates of the time of the introductory interaction, the mode of contact (face-to-face, through a closed door, etc.), any difficulty with eye contact, a set of questions on interviewer behavior (what they said and did during the introductory interaction), and questions on distractions and mistaken identity. Initial analyses examined all these measures, but the most useful for our present purposes were the householder comments and questions reproduced in Figure 3.1.

The analyses in this book focus most attention on a collapsing of individual statements into thematic groups. Householder actions are classified into three groups, attractive theoretically and supported by factor analysis of the items:

(a) Negative statements include "I don't know anything about the survey topic," "I'm not interested," "Surveys are a waste of taxpayers' money," etc.

(b) Time-delay statements include "I'm too busy," "Let me think about it," etc.

(c) Questions include "How was I chosen?," "How long will this take?," etc.

A fourth group, positive statements, were rarely expressed, and are dropped from most of our analyses. While we lose some richness of detail with these collapsed measures, they offer a good deal of parsimony, and hence form the bulk of the analyses we report on in this book.

5. Did the HH informant/R make any of the following comments, whether or not these <u>exact</u> words were used? (CIRCLE ALL THAT APPLY.)
   a. "I'm TOO BUSY/I don't have time."
   b. "I ENJOY DOING SURVEYS."
   c. "I DON'T KNOW ANYTHING about the survey topic."
   d. "Let me THINK ABOUT IT."
   e. "I DON'T TRUST SURVEYS."
   f. "Surveys are a WASTE OF TIME."
   g. "I like to do things that will HELP THE COMMUNITY."
   h. "The PERSON that you want ISN'T HOME."
   i. "Surveys are a WASTE OF TAXPAYERS MONEY."
   j. "You ask too many PERSONAL QUESTIONS."
   k. "I'm NOT INTERESTED."
   l. "I/my spouse NEVER DO(ES) SURVEYS."
   m. "The GOVERNMENT KNOWS EVERYTHING about me already."
   n. Other (specify)
   o. NO COMMENTS

6. Did the HH informant/R ask questions about any of the following topics? (CIRCLE ALL THAT APPLY.)
   a. "What's the PURPOSE of the survey/What's this all about?"
   b. "Who is paying for this/Who is the SPONSOR?"
   c. "Why/how was I CHOSEN."
   d. "HOW LONG will the interview take?"
   e. "WHO'S GOING TO SEE my answers?"
   f. "Can I get a COPY OF THE RESULTS?"
   g. "Is there an INCENTIVE or gift for participating?"
   h. Other (specify)
   i. NO QUESTIONS

Figure 3.1. Extract of contact description questions, completed by interviewer, from NSHS.(*HH informant/R* is the person interviewer contacted).

3.5.1 Limitations of the Interviewer Observation Measures

The ideal data set for these analytic purposes would be measured at the level of each turn of the conversation between householders and interviewers randomly assigned to them. This would enable observation of how response propensities change based on what interviewers say in response to each householder behavior. The contact description data do not reveal the temporal order of utterances by the two actors, nor was there randomized assignment of interviewers to households. To address this weakness, statistical controls on householder characteristics are used when examining interviewer effects.

The second weakness of the contact description data is their limitation to a specific set of behaviors that were identified prior to the study. Only this set, derived from the theoretical interests of the investigators, is documented. Some important behaviors by either householders or interviewers may have been missed.

A third weakness is that the contact description data are collected by interviewers themselves. For some cases, these were completed after a lengthy interview was conducted. Whether interviewers are accurate and reliable recorders of their own behaviors and those of householders with whom they interact must be assessed. This issue is discussed in more detail below.

3.5.2 Quality Measures on the Interviewer Observations

A critical stance should be taken with any new measurement process, especially this first crude attempt to use a participant in an interaction to document *post hoc* characteristics of the interaction. There are three available approaches: attempts at validation, comparisons to prior studies of interaction, and empirical assessment of construct validity (i.e., do the measures exhibit interrelationships that make sense?).

In a coordinated study with European collaborators, McCrossan (1993) had trainers (supervisors) observing interviewers complete the same contact form as did interviewers. There were slight differences between this form and the one used in the NSHS, reflecting design differences across the two surveys, but the key items were identical across forms. Trainers completed the form *during* the interaction of the interviewer with the householder; interviewers completed the form, independently of the trainers, *after* the contact was completed. These data were graciously made available to us for analysis.

Unfortunately, the measurement circumstances are not identical because the British interviewers received less training in completing the form than did the NSHS interviewers. Measures of agreement between trainers and interviewers might be viewed as lower limits of accuracy of the NSHS data. For this purpose, trainers are assumed to produced error-free data and any deviation of interviewers from those data is evidence of errors on the part of the interviewer record.

Table 3.6 presents the results of this validation step for the 205 cases for which both trainer and interviewer observations were obtained. For negative statements, the observers and interviewers were in agreement 93% of the time. That is, 93% of all contacts had interviewers and trainers agreeing on whether one or more negative

Table 3.6. Agreement between trainers and interviewers on householder actions

	Negative statements	Time delay statements	Questions
Agree			
Both yes	7.3%	7.3%	23.9%
Both no	85.9%	78.0%	53.2%
Total agree	93.2%	85.4%	77.1%
Disagree			
Trainer yes, Interviewer no	3.4%	10.2%	9.8%
Interviewer yes, Trainer no	3.4%	4.4%	13.2%
Total disagree	6.8%	14.6%	22.9%
Total	100%	100%	100%
(*n*)	(205)	(205)	(205)

Source: Reanalysis of McCrossan (1993) data.

statements were made. For time-delay statements, they were in agreement 85% of the time. For questions, the agreement rate was 77%.

The most frequent occurrence for all three categories is that the householder does not say anything fitting in the category. Agreement between trainer and interviewer that nothing occurred is higher than agreement that a particular action occurred. For example, among those cases for which the trainer recorded a negative comment, interviewers also did for 68% of the cases. When the trainer recorded a time-delay statement, interviewers did so also for 42% of the cases. The corresponding number for questions is 71%.

The mismatch between interviewers and trainers is not merely a result of interviewers recording fewer householder behaviors than trainers. Trainers record negative comments in 68% of the contacts that interviewers record negative comments, agree with interviewer notes of time-delay statements 62% of the time and, with questions 64% of the time. Overall, interviewers record the same number of negative comments as trainers, fewer time-delay statements than trainers, but more questions than trainers.

The second evaluative step was comparison to counts of behaviors by householders coded from audio tapes of 149 first-contact doorstep introductions in Britain by Morton-Williams (1993). In both the NSHS and the British data, the most frequent first householder behavior is the question, "What is this survey about?" (Morton-Williams, 1993, Table 5.8, p. 90). The NSHS data show higher frequencies of questions about the length of the survey (NSHS was a longer interview (at least one hour) than the British (25 minutes)) and questions about how the household was chosen. NSHS shows lower percentages of the "I'm not interested" comment (note that the NSHS had a higher response rate, 86%, than the British survey cases, 79%). There are some differences in the classification of behaviors that make more detailed comparisons impossible, but the impression formed from the British data is not unlike that from the NSHS data.

Campanelli, Sturgis, and Purdon (1997) undertook a comprehensive study to examine these issues. They had interviewers on two different studies tape-record the doorstep introductions ($n = 258$ and $n = 218$, respectively), in addition to completing a form similar to ours. They report reliabilities between the two methods of 0.31 and 0.45 for detailed householder statements on the two studies. However, when the householder statements are aggregated to broad categories (negative, time-delay, questions, positive), the reliabilities improved to 0.61 and 0.68, respectively.

Lavrakas and Merkle (1991) attempted a narrower task, having interviewers record whether they provided additional information to householders either because it was requested or they judged the householders needed it. (The work thus focuses on householder questions, but it is somewhat confounded with interviewers' initiatives to provide more information.) In a telephone survey they found that 34% of the householders were provided more information about how they were sampled, the nature of the survey, confidentiality and uses of the data, length of the interview, that the interviewer was not attempting to sell them anything, or about how to obtain information to verify the source of the interviewer's call. We expect that the telephone mode would produce lower frequencies of householder questions; we also expect the statistic includes cases where interviewers provided more information without householder questions. We find some evidence for shorter interactions in general, and fewer householder questions in particular, from our own analyses of a small number of tape-recorded telephone interactions (see Couper and Groves, 1995).

Another indicator of the quality of these measures may be their reliability across surveys. We have used variations of these measures in three different Michigan Survey Research Center surveys to date. In addition to the NSHS, a set of contact-description questions was included on the 1990 National Election Study (NES) and the first wave of the Survey on Asset and Health Dynamics of the Oldest Old (AHEAD). The 1990 NES is a national survey of persons 18 or older, with one person randomly selected in each household. AHEAD is a national survey of persons 70 years old and older, with multiple persons selected in households (where eligible). The relative frequencies of these statements are consistent with the response rates obtained for the three surveys, and the type of samples used. For example, we expect those over 70 to have fewer time constraints, and this is reflected in the relative infrequency of these types of statements. More negative statements are reported for both the NES and AHEAD, reflecting the higher refusal rates for these surveys. These data suggest that these measures are relatively robust and stable across different surveys.

The final criterion used to evaluate the contact description data supplied by interviewers takes a construct validity perspective—do individual data items relate to one another in a sensible and theoretically consistent manner? This perspective will be taken repeatedly in Chapters 8 and 9, where we test the value of these data in predicting changes in householder behavior over successive contacts.

We use the contact description measures from two surveys in this book. These are the National Survey of Health and Stress (NSHS) and the University of Michigan Survey on Asset and Health Dynamics of the Oldest Old (AHEAD). The NSHS analyses form the bulk of Chapters 8 and 9. Chapter 11 presents postsurvey adjustments based on the data from AHEAD. We describe the two surveys briefly below.

3.5.3   The National Survey of Health and Stress

The National Survey of Health and Stress (NSHS) is a study of noninstitutionalized adults age 15–54 in the coterminous 48 United States (see Kessler *et al.,* 1994). The NSHS (referred to more formally as the National Comorbidity Study) was designed to study the comorbidity of substance-use disorders and nonsubstance psychiatric disorders in the United States. This was a face-to-face survey conducted by the University of Michigan Survey Research Center between September, 1990 and February, 1991. A multistage area probability sample was used (see Heeringa, 1992). The sample of households was designed to be self-weighting.

Within each housing unit, the interviewer attempted to obtain a complete listing of all household members. If no eligible adults age 15–54 were present in the housing unit, no interview was requested (these cases are treated as ineligible and excluded from our analyses). If one or more eligible adults were present, the interviewer proceeded to select a random respondent for the interview using the Kish procedure.

The NSHS interview consisted of two parts. The first part took an average of an hour to complete. This contained questions on a variety of physical and psychological health issues and related behaviors. This also served as a screener for the second part, which explored specific mental health experiences in more detail. Roughly half of the respondents received the second part of the questionnaire.

About 12,000 households were sampled for the NSHS, of which 9,863 yielded housing units with one or more eligible persons. The response dispositions of these eligible households are presented in Table 3.7. The overall response rate for the NSHS is 85.7%, while the cooperation rate (interviews/(interviews + refusals) is 89.2%.

We have records for 54,486 calls and 26,888 contacts for these cases. Part of the challenge is reducing these massive data files to yield analytically meaningful information. A variety of coding and aggregation activities were undertaken, which are described in more detail in the relevant chapters. In Chapter 4 we use the NSHS data to explore issues of noncontact, focusing primarily on information obtained from the call records and interviewer observations of the neighborhood. The contact description records from the NSHS form the core of the analyses in Chapter 8 and 9, focusing on the interaction between householder and interviewer at the time of contact.

Given that a stratified multistage design was used, the data from the NSHS are analyzed using Taylor series linearization to approximate variances and standard errors of the estimates. As with the match data, the SUDAAN software (Shah, Barnwell, and Bieler. 1996) was used.

3.5.4   Survey on Asset and Health Dynamics of the Oldest Old

We make use of one other source of primary data in this book. This study, more commonly known by its acronym AHEAD (*A*sset and *Health Dynamics*), is a longitudinal survey of older Americans (see Soldo *et al.,* 1997). We use only the first wave of AHEAD data collection here. The AHEAD data, together with that from the NSHS, are used to explore alternative postsurvey adjustments for nonresponse in Chapter 11.

The focus of the AHEAD survey is to understand the impacts and interrelationships of the changes and transitions for older Americans in three major domains:

Table 3.7. Response dispositions for the National Survey of Health and Stress

Disposition	Number	Percent
Interviewed	8,451	85.7
Refusal	1,026	10.4
Noncontacted	152	1.5
Other noninterview	234	2.4
Total	9,863	100.0%

health, finances, and family. The AHEAD survey is a national probability sample of persons 70 years old and older in 1993 (i.e., those born in 1923 or earlier), and their spouses, conducted by the University at Michigan Survey Research Center.

The sample for the survey comes from two sources. The majority of cases are from an area probability sample of households, and includes only noninstitutionalized persons in the first wave. Households containing eligible persons were identified in a massive screening effort (over 100,000 households) conducted a year earlier for the Health and Retirement Survey (HRS, a similar survey targeting the 1931–1941 birth cohort). This sample was supplemented with a list sample of Medicare recipients 80 years old or older, obtained from the Medicare Master Enrollment File of the Health Care Financing Administration (HCFA). The cases from the list sample were selected in the same PSUs as the area probability sample to reduce data collection costs. Oversamples of Florida residents (a state with a disproportionate number of retirees) and African-American and Hispanic areas (to provide sufficient sample sizes for subgroup analysis) were included. The analyses presented in this book reflect both the complexity of the design (stratification and clustering) and differential probabilities of selection.

Interviewing for the first wave of AHEAD began in October, 1993 and continued through July, 1994. A mixed-mode design was used, with attempts made to interview the younger group (those under 80) by telephone, while those 80 or older were targeted for face-to-face interviews. Mode switches were permitted at the respondents' request. All interviewing was done by the same set of interviewers using computer-assisted interviewing. Telephone interviews were conducted from interviewers homes using computer assisted telephone interviewing (CATI) while face-to-face interviewers were conducted in respondents' homes using computer assisted personal interviewing (CAPI). The interview lasted one hour. Approximately 130 interviewers worked on the AHEAD study.

The final dispositions of the eligible cases in the AHEAD dataset we use are found in Table 3.8. The response rate for the survey is 80.4% (interviews/eligible), while the cooperation rate is 82.6% [interviews/(interviews + refusals)].

As with the NSHS, we developed a contact description form for interviewers to complete. Given budget constraints, this was done for only the first contact with each household, and for the first contact (if any) with the selected respondent. Aside from this, the contact description form resembled that used for the NSHS, with minor variations to reflect the unique nature of the AHEAD sample.

Table 3.8. Response dispositions for the Survey on Asset and Health Dynamics of the Oldest Old (AHEAD)

Disposition	Number	Percent
Interviewed	8,224	80.4
Refusal	1,729	16.9
Noncontacted	109	1.1
Other noninterview	169	1.6
Total	10,231	100.0%

The use of the AHEAD dataset is described in more detail in Chapter 11, where we develop and evaluate alternative adjustment strategies for nonresponse in the AHEAD survey.

3.5.5   Documentation on Individual Calls on Sample Households

A final source of data closely related to the contact description measures is the collection of detailed call-record information. These are essentially a record of every call made on sample households, whether or not the call resulted in a contact. Information such as the date and time of call, the identity of the interviewer making the call, the mode (in person or telephone) and the outcome of the call provide useful information for studying survey nonresponse. With the computer assisted data collection, these data have become an automatic by-product of the collection process. We collected detailed call record data on both the NSHS and AHEAD to supplement the contact description measures. The call record data from NSHS are analyzed in Chapter 4.

3.6   SURVEYS OF INTERVIEWERS

Our theoretical framework suggests that interviewers are an important element in the likelihood of gaining participation from a household(er). For this reason, we attempted to measure stable interviewer attributes for all interviewers working on the studies described above.

The interviewer questionnaire was developed through several rounds of focus groups with interviewers from different organizations. The questions were designed to measure background characteristics (e.g., experiences, reasons for interviewing job, demographics), attitudes and expectations related to nonresponse, and interviewer reports of behaviors in the field.

However, some variation exists across the different studies. We were not permitted to ask for demographic information (e.g., race, sex) of Census Bureau interviewers. In addition, given the large number of interviewers working on the CPS (almost 2,000), we did not obtain support to administer the survey to this large a group, and so CPS interviewers are excluded. For the NSHS and NHSDA, we administered two versions of the survey: one during training, to elicit attitudes and expectation as well as obtain background information, and the second at the end of the field period to obtain reports of actual behavior on the NSHS. Finally, there are minor variations in the content of the questionnaires and wording of questions across organizations, reflecting different organizational practices.

In most cases, the questionnaires were administered to interviewers during training, prior to data collection. However, in the case of the Census Bureau interviewers working on ongoing surveys, the questionnaires were mailed to interviewers prior to the data collection period of the specific survey. In all cases, interviewers were either given time during training to complete the questionnaire, or were allowed to claim time taken for completion. This, together with extensive follow-up efforts, resulted in near-universal completion of the questionnaire by interviewers (over 97% for all surveys).

The counts of completed interviewer questionnaires for the five surveys included in the decennial match project (CPS excluded) are as follows:

CEQ	319
NHIS	204
NCS	490
SCP	113
NHSDA	280

In addition, completed questionnaires were also received from the following numbers of interviewers working on the two SRC surveys:

NSHS	165
AHEAD	138

The data from the interviewer questionnaires from the match surveys are primarily used in Chapter 7 to explore interviewer influences on nonresponse. For descriptions of interviewer characteristics and attitudes by organization, we present unweighted estimates. For the interviewer-level analyses of response rates, however, we weight the data by the denominator of the response rate, which is the number of eligible cases assigned to the interviewer. In this way, those interviewers who have small workloads, and hence more unstable response rate estimates, contribute less to the overall estimates.

In addition to the analyses in Chapter 7, variables from the interviewer questionnaire (e.g., experience, self-efficacy) are used in models throughout the book. Furthermore, the open-ended responses provided by the interviewers on the questionnaire have been a useful source of qualitative data to supplement the interviewer focus groups and observations of interviewers in the field.

While background characteristics and survey experience can be obtained from employment records of survey interviewers, we believe short interviewer surveys to obtain attitudinal and expectational measures are a useful adjunct to the interviewer data that are routinely available.

3.6.1 Interviewer Reports of Behavior

We discussed the validity and reliability of interviewer self-reports of behavior at the contact level. In the interviewer questionnaire we also included a series of questions on what interviewers usually or typically do on a survey.

In the NSHS we have reports of interviewer behavior both at the interviewer level and at the contact level. We can use this to evaluate whether interviewer self-reports of what they usually do or not do reflect their actual practices in the field.

In Table 3.9 we present the frequency with which interviewers report certain be-

Table 3.9. Percentage of NSHS-contacted households for which content descriptions recorded various behaviors, by frequency category chosen by same interviewer on general interviewer questionnaire after the survey

Behavior type	Report of typical frequency reported in interviewer questionnaire			
	Always	Sometimes	Rarely	Never
Compliment household	30.2%	19.6%	12.1%	4.3%
(n)	(1,258)	(6,115)	(1,515)	(798)
Mention interviewing others in neighborhood	25.4%	25.1%	23.4%	15.3%
(n)	(1,331)	(4,819)	(2,616)	(871)
Say that most people enjoy interview	33.0%	29.2%	18.2%	15.5%
(n)	(2,616)	(6,002)	(826)	(193)
Say that most choose to participate	35.3%	15.1%	10.7%	7.7%
(n)	(2,012)	(5,053)	(1,947)	(601)
Hand materials to householder	47.6%	43.9%	29.8%	21.6%
(n)	(3,560)	(5,213)	(701)	(171)

haviors for each household in the NSHS, by their more general reports of such behavior from the interviewer questionnaire. While the relationship between these two reports is well behaved ordinally, there appears to be gross overreporting on all items in the interviewer questionnaire. For example, among those who in the interviewer questionnaire reported *always* using compliments as a form of reciprocation, at only 30% of households was this behavior actually reported. At the other end of the scale, those interviewers who reported never using reciprocation actually reported doing so at 4% of households. It thus appears that while contact-level behaviors of interviewers may be more useful, general interviewer-level reports of behaviors can be used as weak proxies for what interviewers actually do in the field. We use these data from the interviewer questionnaires in exploring the role of the interviewer in survey participation in Chapter 7.

3.7 MEASURES OF SOCIAL AND ECONOMIC ECOLOGY OF SAMPLE HOUSEHOLDS

A final source of data that we use throughout this book are aggregate census data at the block, block-group (or tract), or county level. Unlike census microdata, these summary data are readily available to researchers, and should, we believe, be routinely used to supplement the usually sparse frame data available in most surveys.

Most large national area probability surveys within the United States use the decennial census geographical units and related data as a frame, so it is relatively easy to obtain this information for selected sample segments.

We adopted a similar approach for the decennial census match project. While the samples for the surveys included in the match study were based on 1990 census geographical boundaries, as part of the match to the household and address, higher orders of geography are also available. We used the census geocoding to extract block-level data from the Summary Tape File (STF) provided by the Census Bureau.

In the case of measures of crime and population density (see Chapter 6), this was

1. Does the building/property containing the housing unit have any of the following? (CIRCLE ALL LETTERS THAT APPLY.)
   a. Bars on the windows?
   b. Security doors, metal windowless doors?
   c. Crime watch or security system signs in window?
   d. Open lobby, then locked door?
   e. Locked lobby/Doorman?
   f. Locked gate?
   g. No trespassing sign?
   h. Beware of dog sign?
   i. No solicitors sign?
   j. None of the above

2. Are any of the following conditions of the building present? (CIRCLE ALL THAT APPLY.)
   a. Missing roofing material(s)?
   b. Boarded up window(s)?
   c. Missing/broken out window(s)?
   d. Missing bricks, siding, or other outside wall material?
   e. Punched out, torn screens on windows?
   f. Door(s) off hinges?
   g. None of the above

3. Which of the following are present within sight of the housing unit? (CIRCLE ALL THAT APPLY.)
   a. Boarded houses or abandoned buildings
   b. Abandoned cars
   c. Demolished houses
   d. Trash, litter, or junk in street/road
   e. Trash, litter, or junk around buildings in neighborhood
   f. Factories or warehouses
   g. Stores or other retail outlets
   h. None of the above

4. Relative to the other buildings on the same street/road, how well maintained is the sample HU? (CIRCLE ALL THAT APPLY.)
   a. Better than others
   b. Same as others
   c. Worse than others
   d. No other buildings

Figure 3.2. Neighborhood observation measures completed by NSHS interviewers before contact with sample units.

supplemented with county-level measures. The lowest level of detail for crime data are at the county level. These are from the Federal Bureau of Investigation's (FBI) Uniform Crime Reports (UCR) of incidents reported to the police. These data are from the 1988 County City Data book produced by the Census Bureau. These represent the latest data available at the time of the decennial match study.

For population density, a city-level measure of persons per square mile is used. For sample segments in unincorporated areas, we use county-level density measures. Both crime and density measures thus fail to reflect intracity variation in ecological characteristics.

The ecological measures used here were collected not only for the decennial match surveys, but also for the NSHS (based on a frame drawn from the 1980 decennial census but with links to 1990 census geography) and AHEAD (based on a frame from the 1990 census).

For the NSHS, the block-level census aggregate data were supplemented by a set of interviewer observations of the immediate surroundings of the sample unit, as well as characteristics of the building containing the sample unit. These neighborhood observations were made by the interviewer before any contact was made with the sample household. The observational measures are reproduced in Figure 3.2.

3.8   LIMITATIONS OF THE TESTS OF THE THEORETICAL PERSPECTIVE

Despite the efforts described above to assemble data to fully explore all aspects of survey nonresponse, we were still far from the ideal research design. Ideally, we would measure all sets of influences outlined in Chapter 2, on a wide variety of surveys. Such a design is not feasible. Hence we assemble evidence of the components of our theoretical framework from a variety of different sources.

As with other designs for studying nonresponse discussed in Section 3.1 and 3.2, the data we use in this book suffer from a number of limitations. We again review some of them briefly here.

A key limitation is that the decennial census match project yielded only demographic characteristics on responders and nonresponders. Such an approach cannot reveal *why* people make the decision to participate or not, and what social psychological factors underlie this decision making process. Second, by its very nature, a census tends not to yield rich data on people and households, and thus we have a limited set of variables available for all sample cases. Further, the surveys included in the match project represent a narrow range of possible survey types. Ideally, we would hope to include a sample of government, commercial, and academic surveys representing a variety of different data collection strategies, response rates, topics, and so on.

We have noted before that our studies have been limited to face-to-face surveys. We speculate in various chapters on how these findings may translate to telephone surveys, but we have little empirical information on possible differences. We do not address the issue of mail surveys directly, but suspect that much of what we discuss may be relevant to this mode of data collection.

The surveys we included in the match study all have relatively high response rates. This has two effects. First, it is harder to detect meaningful differences, given ceiling effects in the rates we examine. Second, it is not clear whether these findings translate to low response rate surveys. We believe that the mix of respondents and nonrespondents, and potentially the nonresponse bias, changes with varying response rates, but do not have the data to test this.

For the surveys in which we collected contact description information, we rely on the interviewers' reports of what transpired during the interaction. Ideally, we would want less-subjective measures of these interactions.

Finally, to evaluate the extent of nonresponse error, we would ideally have external measures of the key substantive measures. Such external data would also be helpful in evaluating the efficacy of alternative nonresponse adjustment strategies. Unfortunately, such data are extremely hard to come by, and we must infer the potential for error in substantive estimates from differences in demographic characteristics.

3.9 SUMMARY

The empirical analyses in this book are based on eight different household survey databases, on 1990 decennial census records matched to six surveys, on surveys of interviewers conducting seven of the surveys, and on interviewer observations about interaction with householders in two surveys.

Despite the wealth of data, they are limited in the breadth and depth of coverage of the survey participation decision. They do measure multiple parts of the influences we describe in Figure 2.1. The decennial census match project represents an unprecedented opportunity to explore a wide variety of demographic variables in a multivariate context across several surveys. Similarly, the measures of the householder–interviewer interaction provide us with a unique insight into this process on a large number of cases. In combination, the various data sources we have assembled provide an unrivaled opportunity to explore the many facets of the survey participation process. In the chapters that follow, we report on the analyses we have conducted on these data sources.

CHAPTER FOUR

Influences on the Likelihood of Contact

4.1 INTRODUCTION

Even survey interviews that a householder would be happy to provide will not occur if an interviewer never contacts the sample household. In most surveys, some sample households are never contacted by interviewers and hence never make a decision about their survey participation. This chapter describes what we know about the process of contacting a sample household.

Separate attention needs to be paid to noncontacts because they can have effects on nonresponse error that are distinctive from those of refusals. In the expression for error due to nonresponse, an additional term for noncontacts can be displayed to illustrate this point:

$$\bar{y}_r = \bar{y}_n + \frac{m_{nc}}{n}(\bar{y}_r - \bar{y}_{nc}) + \frac{m_{rf}}{n}(\bar{y}_r - \bar{y}_{rf})$$

where $\bar{y}_r =$ the sample estimate based on respondent cases, $r;$

$\bar{y}_n =$ the sample estimate based on all sample cases, $n;$

$\bar{y}_{nc} =$ the sample estimate based on the noncontacted cases, $nc;$

$\bar{y}_{rf} =$ the sample estimate based on the refusal cases, $rf;$

$m_{nc}, m_{rf} =$ the number of noncontacted and refusal cases, respectively.

Using this expression, appropriate for linear statistics, we can see that the deviation of the respondent-based statistic from the full sample statistic is a function of how those not contacted and those refusing differ from those who grant an interview. As we saw in Chapter 1, it is possible in the expression above that the two terms move in different directions. For example, imagine that y is a measure of frequency of using city recreational facilities (e.g., parks, tennis courts, playgrounds). It might be likely that those who are not contacted are disproportionately users of those facilities (and thus $(\bar{y}_r - \bar{y}_{nc})$ is a negative term), simply because using the facilities demands absence from the housing unit. Conversely, those refusing might be disproportionately nonusers of the facilities (thus uninterested in participating in a survey about them). This implies that $(\bar{y}_r - \bar{y}_{rf})$ is positive. In this example, whether the respondent mean is biased upward or downward depends on the mix between noncontacts and refusals.

The expression also forces the question of what happens when cases that are formerly not contacted are finally contacted and move either into the respondent group or the refusal group. In essence, the answer to this question rests on how the likelihood of being at home is related to the likelihood of granting the survey request. Using the example above, one can construct the argument that repeated efforts to contact those initially not at home may lead to higher response rates, not just the reclassification of a noncontact to a refusal category. That is, if noncontacts disproportionately use the recreational facility, once contacted they might be motivated to participate in a survey about those facilities. This implies that if field procedures dictated vigorous efforts to decrease noncontacts, response rates would effectively rise and estimated percentages of persons using the facilities would also rise, other things being equal.

This example illustrates the importance of separating noncontacts from refusals in understanding survey participation. If the two phenomena—the frequency of being at home and willingness to participate in the survey—are differentially related to the survey variables, survey estimates can be affected differently by reducing one source of nonresponse versus the other.

4.1.1 A Simple but Powerful Model of Contactability

Relative to the phenomenon of cooperation with a survey request, the process of contacting a sample household appears simple. By "contactability" we mean the propensity for a household to be contacted by an interviewer at any given moment in time. As we illustrate in Figure 4.1, contactability in fact must be a function of three factors: a) whether there are any physical impediments (locked gates, locked apartment entrances) that prevent visiting interviewers to alert the household to their presence, b) when household members are at home, and c) when and how many times the interviewer chooses to visit. (Equivalent impediments in a telephone survey include answering machines and caller ID.) In the absence of physical impediments to contacting the unit, if we knew when each household was at home, then interviewers could visit the housing unit at those times and contact would be made on the first visit in each case.

Because there is no such information available to interviewers at the time of sample assignment, they choose a time to call on a unit based on their work schedule and any guidelines given in interviewer training. For households in which at least one adult member is almost always home, they are likely to make contact on the first visit. Those rarely at home usually require more visits.

During the first visit in a face-to-face survey, the interviewer may observe attributes of the housing unit that help plan the timing of the next visit, if one is needed. For example, the first failed visit to a unit in a complex of small apartments, occu-

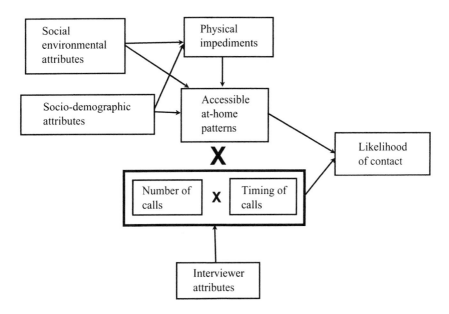

Figure 4.1. Influences on the likelihood of contacting a sample household.

pied mostly by young professionals, may lead the interviewer to call next in the evening or on a weekend day. By these behaviors, the interviewers attempt to identify successive times to call that match the at-home times of the sample household. They are, however, able to collect only weak proxy information on at-home times, and thus are forced to call on some units many different times.

In telephone surveys, the sampling frame typically reveals the name of the telephone exchange, which can be used to estimate the urbanicity of the household. Lower-level aggregates (e.g., zip code characteristics) can be linked to listed households. However, the telephone mode, even in the best of circumstances, cannot compete with the observational richness of face-to-face surveys.

If we were privy to both the at-home times of the sample household and the times at which the interviewer called on the household, we could predict with total certainty whether or not the unit would be contacted during the data collection period. The model diagramed in Figure 4.1 shows that the number and timing of calls, when combined with the accessible at-home patterns of sample households, are the proximate causes of contactability. The accessible at-home patterns, themselves, are affected by whether there are any physical impediments to gaining access to the unit, socio-demographic and lifestyle variables that determine when people are at home, as well as social environmental influences that affect at-home patterns (e.g., commuting times for workers in large versus small cities, nighttime crime incidence in neighborhoods).

We will study contactability with an imperfect tool—the observed number of calls required by an interviewer to make first contact with a sample household. It is important for the reader to understand why this is an imperfect measure of contactability. Ideally, the indicator would be an instantaneous likelihood of the difficulty of someone making contact with the household in some time period. In essence, it is a probability of contact, given a visit, during the time period. There exists an overall contactability for the total data collection period, estimated by the ratio of total time accessible at home over the total time of the data collection period. There are similar probabilities for evening times during the data collection period, for evenings in the first week of data collection, for the evening of the first day, etc.

We will estimate these probabilities by computing the ratio of households contacted to those visited for various combinations of time periods, visits, etc. The behavior of interviewers can bias our estimates of contactability for these time periods in a way that inference from our results to other surveys may be damaged. For example, imagine that we wish to estimate the probability that households can be contacted in the morning hours. To do this, we compute the ratio of contacts in the morning to all calls in the morning, using first calls made to households. If, however, the local interviewer, aware that a sample segment is located in an apartment complex, avoids calling on such units in the morning, and instead concentrates on other units, we may obtain a higher estimate of morning contactability than is true. In this case, we might overestimate the probability. We could also produce underestimates if interviewers, for example, chose to visit poorer urban neighborhoods first in the day, in order to assess their own physical safety there, despite the likelihood that such areas have very low contactability during the day. In short, the ideal data set would have fully randomized visit times for all sample units—a practical impossibility (although somewhat more feasible in a telephone survey).

To the extent that interviewers choose times of calling independent of the attributes of the sample units, the departures from the ideal data set may not be important. To the extent that we identify attributes of the sample households that are related to differential interviewer treatment, we can also purify our estimates of contactability. This we will attempt to do in the analyses that follow.

4.1.2   Measures of Contactability

To study in detail the process of contacting households, we need data on the outcomes of repeated calls to sample units. In this chapter we use data from the National Survey of Health and Stress (NSHS) conducted in 1990–1991 (see Chapter 3). These data contain interviewer observations about the housing unit, detailed records of call attempts on the sample unit, and administrative data on interviewers. The data are weak, however, on measured attributes of households that might be related to their patterns of being at home. Much of the analysis of the step-by-step process of contact will use these data. In addition, however, we employ data from the decennial census nonresponse match project. These data have richer measures of household attributes but very little data on the outcome of each call.

Figure 4.2 shows the distribution of number of calls in NSHS made to a sample unit to obtain the first contact with the unit. This shows that about half of the units (49%) are contacted on the very first call. By call 2, about 70% of the sample has been contacted; by call 3, over 80%; by call 4, nearly 90% have been contacted. The

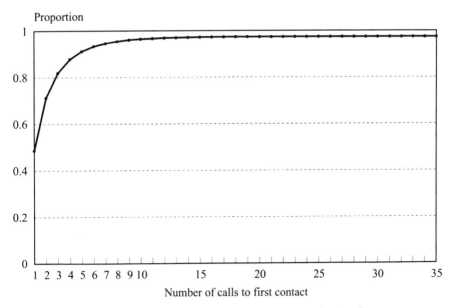

Number of calls to first contact

Figure 4.2. Cumulative distribution of sample units by number of calls to first contact.

distribution of calls required to first contact is thus highly positively skewed. The mean number of calls to first contact is 2.2, compared to the median of about 1.

Figure 4.3 provides another look at the same phenomenon. It presents the percentage of cases contacted at a specific call number, among all cases not previously contacted. For example, 44% of the sample units not contacted on call 1 are reached on call 2. About 37% of the units not reached on call 1 or call 2 are reached on call 3. The distribution is generally a monotonically declining one. With each successive failed attempt to contact a unit, the odds that the next call will result in a contact go down.

In essence, this chapter tries to discover why some units take many calls and why some units take few calls to obtain contact. That is, we try to explain the variation obvious in these distributions in order to predict difficulty of contact and extract principles of contactability.

In studying contactability, we note that there can be two types of patterns of influence. The first are transitory impediments to contact, which are easily repaired with information obtained on the first call. Examples of these might be a rural house with a note saying the farmer will return at 4 PM that day, a visit to an apartment building whose doorman reports the sample household is at home only in the evening, an evening visit to an urban high rise with a policy of locking external doors only in the evening, etc. Measures of attributes of these types of situations should be correlates of contact on the first call but not on later calls.

The second type of influence is a permanent impediment to contact. Examples of these are lifestyles of households that affect their at-home frequency and permanent impediments to contact (doormen who do not permit entrance to a multiunit structure). Detecting these statistically is a function of the relative size of the subgroup

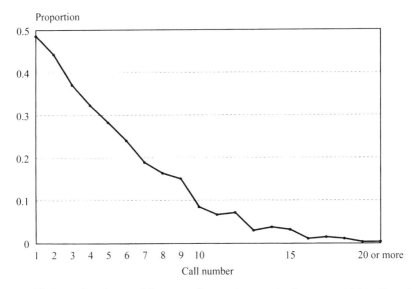

Figure 4.3. Proportion of successfully contacted among those previously uncontacted, by call number.

affected and the relative size of the group with temporary impediments to contact. On the first call, the lowered probability of contact with these "permanently difficult" units may not be vastly different than the probability of contact with the "temporarily difficult." Hence, no detectable differences will be found. Only in examining the later calls, for those cases never contacted, will the evidence of difficulty be discernable. Thus, we expect that relatively permanent influences on contactability will be of increasing statistical power as the number of calls rises.

4.1.3   Organization of this Chapter

This chapter is organized about the key blocks of variables presented in Figure 4.1. We begin by trying to identify types of households that differ in their at-home patterns. We start with characteristics of the sample neighborhoods to see if there are differences in contactability that are pervasive at that level. Then we go to the level of the housing unit and household itself.

Then we move to describing when interviewers call on sample housing units and what success they tend to have at different times of day and days of the week. We then step back to see if interviewers vary systematically in how they schedule their calls and what success they have contacting households.

Finally, we put these various levels of effects together, measuring the relative import of influences at different levels. In taking this step we compare the correlates of reaching easily contacted households to those of reaching difficult to reach households. We also combine data from several surveys to exploit a larger set of influences on contact.

4.2 SOCIAL ENVIRONMENTAL INDICATORS OF AT-HOME PATTERNS

In Chapter 6 we present evidence that over the years it is becoming increasingly difficult to gain the participation of sample households in the United States. One of the reasons for this, we believe, is the reduced accessibility of households, both because of efforts to control physical access to housing units and to have household compositions that lead to more hours of the day when no one is at home.

Some reasons for being away from home cluster spatially. Residents of large cities who work outside the home may spend more time each work day traveling to their place of employment than residents of smaller urban areas (see Robinson and Godbey, 1997). Further, they may require more time away from the home for grocery purchases, shopping, and other errands. They may also enjoy more entertainment options that take them away from home. We might expect this to be indicated both by the population density of the sample place, but also the urban size category of the area.

Residents of high-crime areas and multiunit structures may have special security features of their housing that limit access to the interviewer. Figure 4.4 shows that there has been a steady rise in the percentage of households that live in structures with five or more units, moving from about 14% in 1973 to about 17% in 1993.

Other environmental measures are aggregates of household or housing-unit char-

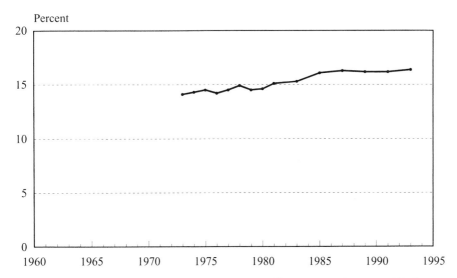

Figure 4.4. Percentage of households living in structures of five or more units, 1973–1993. *Source:* U.S. Bureau of the Census and U.S. Department of Housing and Urban Development, Annual Housing Survey, Current Housing Reports H-150-73,74,75,76,77,78,79,80,81,83, Table A-1; Current Housing Reports H-150-85, Table 1-1; Current Housing Reports H-150-87, Table 1A-1; and American Housing Survey Data Chart 1989, 1991, 1993, Table 2-1.

acteristics that may be related to ease of contact. Renters tend to be younger and less wealthy than home owners. Their lifestyles may involve more out-of-home activities than others. Similarly, there is some suggestion in earlier work (Juster and Stafford, 1985) that minority groups, perhaps because of the tendency to hold multiple jobs, spend more time out of the home.

Table 4.1 presents the percentage distribution of the call number on which first contact was made, by a variety of social environmental attributes. Since most of the

Table 4.1. Cumulative percentage distribution of number of calls to first contact by environmental attributes

Environmental attribute	Percentage contacted by call number							Percentage not contacted	Total
	1	2	3	4	5	6	>6		
Population density[a]									
1st quartile (<705)	52.6	76.4	87.7	93.1	95.5	97.1	99.3	0.7	100%
2nd quartile (705–2028)	52.0	74.1	84.7	90.4	93.4	95.3	98.5	1.5	100%
3rd quartile (2029–3687)	47.5	70.6	81.6	87.5	91.4	93.3	98.0	2.0	100%
4th quartile (>3687)	42.3	64.1	73.9	80.2	84.6	87.6	93.8	6.2	100%
Urban status									
In central city	42.1	62.8	72.7	79.1	83.2	86.6	92.5	7.5	100%
In balance of CMSA	46.8	68.8	80.1	85.9	89.8	92.2	96.8	3.2	100%
In other MSA	50.8	73.7	84.5	90.0	93.3	94.9	98.9	1.1	100%
In nonMSA	51.4	75.7	86.0	91.8	94.7	96.5	99.0	1.0	100%
% Multiunit structures									
1st quartile (<3%)	50.4	74.6	85.6	91.6	94.3	96.1	99.0	1.0	100%
2nd quartile (3–13%)	48.9	71.7	82.9	88.0	91.8	94.1	97.4	2.6	100%
3rd quartile (14–98%)	44.5	66.0	75.8	82.2	86.2	88.4	94.7	5.3	100%
4th quartile (>98%)	50.6	73.4	84.0	89.9	93.0	95.0	98.6	1.4	100%
% Homes owner-occupied									
1st quartile (<48.5%)	43.7	65.1	75.4	82.0	85.9	88.5	95.1	4.9	100%
2nd quartile (48.5–71.4%)	48.4	72.3	82.9	88.0	92.4	94.5	98.0	2.0	100%
3rd quartile (71.4–84.7%)	51.9	74.2	85.3	90.9	93.5	95.1	98.1	1.9	100%
4th quartile (>84.7%)	50.2	73.6	84.1	90.2	93.2	95.2	98.3	1.7	100%
Crime rate[b]									
1st quartile (<3459)	52.3	75.8	87.5	92.6	95.7	97.1	99.5	0.5	100%
2nd quartile (3459–5001)	46.7	70.9	82.3	88.4	92.0	94.4	98.6	1.4	100%
3rd quartile (5002–6878)	49.5	70.6	80.6	86.4	90.4	92.1	96.4	3.6	100%
4th quartile (>6878)	46.1	68.8	78.7	85.1	88.2	90.9	95.9	4.1	100%
% persons of minority race									
1st quartile (<2.5%)	50.4	73.1	84.5	90.6	93.8	95.7	98.4	1.6	100%
2nd quartile (2.5–7.9%)	47.6	73.4	84.2	89.6	92.9	94.6	98.7	1.3	100%
3rd quartile (8.0–24.0%)	49.2	69.5	79.8	85.4	89.0	91.5	96.4	3.6	100%
4th quartile (>24.0%)	47.2	69.3	79.3	85.6	89.3	91.6	96.1	3.9	100%

[a]Persons per square mile in county of the housing unit.
[b]Total personal crimes per 100,000 persons in county of housing unit.

cases are contacted by call 2 or 3, we can understand much of the variation in the table by comparing the percentages reached on call 1.

If an urban–rural categorization of sample places is used instead of population density, similar results are found. Central cities of the large metropolitan statistical areas (MSAs) have lower rates of call 1 contact (42%) versus small MSAs and non-MSAs (about 51%) ($\chi^2 = 25$, $df = 3$, $p = 0.0002$). In short, more calls are made in the large urban areas to make contact. We suspect this reflects true differences in at-home rates. The most plausible alternative hypothesis is that the times of calling on cases varies by the urbanicity of the area, in a way that harms the productivity of the urban interviewing. We will see later (Table 4.6), however, that there seems to be a tendency for interviewers to call on cases in these areas in the evening, the most productive time for contacts.

Table 4.1 also shows that the contact rate on call 1 is a rough monotonic function of population density of the area. The denser the area, the more calls made to contact a household. About 53% of the households in areas in the lowest quartile of density (up to 704 persons per square mile) are reached on the first call compared to about 42% of those in the top density quartile (3688 or more persons per square mile) ($\chi^2 = 36.9$, $df = 3$, $p < 0.0001$). We also note that the percentage of cases never contacted throughout the entire data collection period is higher for high-density areas (6.2%) than for the lowest density areas (0.7%) ($\chi^2 = 63.1$, $df = 3$, $p < 0.0001$).

The prevalence of multiunit structures tends to decrease the percentage of call 1 contacts (44% for areas almost completely multiunit versus 50% for areas in the lowest quartile of density of multiunit structures, $\chi^2 = 13.3$, $df = 2$, $p = 0.003$). Similarly, and related to this, we suspect, is the tendency for areas densely filled with owner-occupied units to generate higher percentages of call 1 contacts (44% for areas with less than 49% owner-occupied versus 50% for those over 85% owner-occupied).

There are smaller differences in contact rates for areas that differ in their crime rates and minority representation. The crime rate variable is measured at a high level of aggregation and is probably not a good indicator of the crime rate at the sample segment level. It is also possible that crime rates have counteracting influences on contact rates—encouraging some persons to stay at home more, but encouraging others to take protective measures that limit accessibility to their unit (e.g., locked gates). The minority representation of an area, by itself, may not be indicative of at-home patterns or may be counteracted by larger household sizes in minority areas, leading to greater ease of contact.

Similar results regarding ecological correlates of contact ability are obtained in the decennial census nonresponse match project. Here we can measure only whether or not the unit was contacted, not whether it was easy or difficult to contact. With these data, it is clear that central cities of large urban areas have higher non-contact rates (4.9%, s.e. = 0.53%) than do their suburban areas (3.0%, s.e. = 0.26%) or rural areas (1.6%, s.e. = 0.18%). Similarly, population density, crime rates, minority representation, and the prevalence of multiunit structures are correlates of contact rates. Despite having diverse designs and being conducted by different organizations, the surveys agree at this basic level.

4.3   HOUSEHOLD-LEVEL CORRELATES OF CONTACTABILITY

Following the simple model of contactability above, this section looks to the level of the household for correlates. We first discuss features of housing units that prevent easy access of a stranger to the unit. These are generally features that affect contactability at any time an interviewer would visit. We then review indicators of times when household members might be at home, using proxy measures of lifestyle and obligations that take the householder away from home.

4.3.1   Physical Impediments to Contacting Households

In face-to-face surveys, various structural aspects of sample housing units can affect the ability of an interviewer to contact the household. These appear at all levels of socioeconomic status of the population. In low-income, high-crime areas, residents with door grates, multiple dead-bolt locks, and alarm systems may not readily answer the door when a stranger calls, opting to have contact only with persons known to them already. Most of these cases, we believe, might be better labeled as reluctant respondents than noncontacts, but for those cases in which no conversation is had between interviewer and householder, some may be recorded as noncontacts.

Many apartment buildings have locked central entrances; some have doormen or security guards restricting access to units in the building. It is true that many of these have intercom systems, facilitating audio contact with householders, but not all have working systems.

Suburban and rural areas may have locked gates surrounding homes, which prevent access to houses. Others have posted signs that state that trespassers will be prosecuted or warning signs about dogs (or dogs themselves) guarding the property.

All of these features of a housing unit cause difficulties for the interviewer to make contact with the household. Some prevent the household from knowing that the interviewer is seeking contact; others prevent the interviewer from knowing whether the household is at home when contact is attempted. As part of the field administration, the NSHS asked interviewers to record observations about the sample unit. These are listed in Figure 3.2, and included observations about whether there was a locked gate or central entrance, bars on windows, etc. In addition, interviewers made judgements about the state of repair of structures in the neighborhood and of the sample unit itself.

In telephone surveys, we believe, answering machines, caller ID features, and call-blocking functions are the equivalent to the physical impediments present in face-to-face surveys. Unfortunately, these cannot be observed prior to the first call attempt. Further, we expect that the future will bring even more technological solutions to enhanced telephone privacy.

Table 4.2 presents the percentage distribution of the call numbers of first contacts with sample households by various indicators of physical impediments. The observation of physical impediments is a powerful indicator of how many calls the interviewers made before making contact with the sample household. About half of

Table 4.2. Cumulative percentage distribution of number of calls to first contact by housing-unit attributes

	Percentage contacted by call number							Percentage not	
Housing attribute	1	2	3	4	5	6	>6	contacted	Total
Physical impediment[a]									
None	50.4	73.3	84.0	89.6	92.9	94.8	98.2	1.8	100%
Some	39.1	60.8	71.5	78.6	82.7	86.0	93.3	6.7	100%
Structural disrepair[b]									
None	48.7	71.1	81.7	87.4	90.9	93.1	97.2	2.7	100%
Some	46.6	72.3	83.5	90.9	94.1	95.4	98.5	1.5	100%
Unit in worse repair than others									
Yes	45.8	72.9	83.9	90.4	93.7	95.0	98.8	1.2	100%
No	49.0	71.4	82.2	88.0	91.5	93.7	97.9	2.1	100%

[a]Observation by interviewer of building/property feature: bars on windows, security doors, crime watch or security system signs, locked entrances on apartment buildings, etc.
[b]Observations by interviewer of conditions of building: missing roofing materials, boarded-up windows, broken windows, etc.

those with no physical impediment were contacted on the first visit, much higher than the 39% of those with some physical impediment (χ^2 =29.5, df = 1, $p <$ 0.0001). Consistent with this, only about 2% of those with no physical impediment to access remained noncontacted at the end of the study, but 7% of those with impediments were never contacted (χ^2 = 17.0, df = 1, p = 0.0002). Indicators of the structural disrepair are not as powerful indicators of contactability, with the largest effects only on the percentage contacted on the first visit.

The decennial nonresponse match project provides some other measures of difficulty in accessing the housing unit. One strong correlate is whether the housing unit is a single-family structure or a mobile home in contrast to a multiunit structure. The single-family dwellings have lower noncontact rates (1.9%, s.e. = 0.12%) than other units (4.6%, s.e. = 0.35%). On the other end of the size dimension, households in large multiunit structures (with 10 or more units) tend to have higher noncontact rates (6.3%, s.e. = 0.71%) than others (2.2%, s.e. = 0.11%).

4.3.2   Reasons for Households to be Present in Their Housing Unit

If one person lives alone in a housing unit, contact is completely dependent on when he or she is at home. If two people form a household, contact can be made if either or both are there. If the at-home times of members of the same household were completely independent of one another, then the larger the number of persons in the household, the larger the probably that at any one call someone would be reached. Figure 4.5 shows a steady increase in the percentage of single-person households between 1960 and 1994, moving from about 13% to about 24% of all households. If

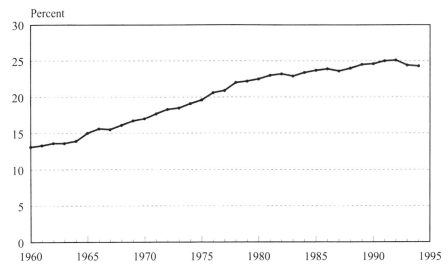

Figure 4.5. Percentage of single-person households among all occupied housing units, 1960–1994. *Source:* U.S. Bureau of the Census, *Historical Statistics of the United States, Colonial Times to 1970,* Bicentennial edition, Part 1, Washington, D.C., 1975, p. 42, Table A 335–349; U.S. Bureau of the Census, *Current Population Reports,* series P-20, No. 467 and earlier reports; unpublished data as cited in U.S. Bureau of the Census, *Statistical Abstract of the United States: 1993,* Washington, D.C., 1993, p. 56, Table 67 and earlier Abstracts; and Rawlings, Steve W. and Saluter, Arlene F., *Household and Family Characteristics: March 1994,* U.S. Bureau of the Census, *Current Population Reports,* P20-483, U.S. Government Printing Office, Washington, D.C., 1995, Table C.

the logic is correct, this trend may underlie the apparent increasing difficulty in gaining contact with households.

Households with young children more often have an adult caregiver at home than households without young children. This was the finding of Lievesley (1988) in the United Kingdom, for households with children less than 5 years of age. Figure 4.6 shows that the percentage of households with children 0–5 years in the United States has declined from about 30% in 1960 to about 23% in 1994. This trend is consistent with reported higher numbers of calls per contact in U.S. surveys over time.

Finally, we know that those households with someone outside the employed labor force will tend to have more time at home than those with all adults employed outside the home. One indicator of this is the proportion of households with persons who are over 70 years of age, as a proxy indicator for retirement from the labor force. This attribute, which should be a positive correlate of contactability, shows a modest increase between 1967 and 1994, from 19% to 22% of all householders who are 65 or older.

We do not know the household composition for the vast majority of the noncontacted units in the NSHS ($n = 256$) or for units contacted that did not supply a household roster ($n = 478$), but we do know roster characteristics for 9,384 house-

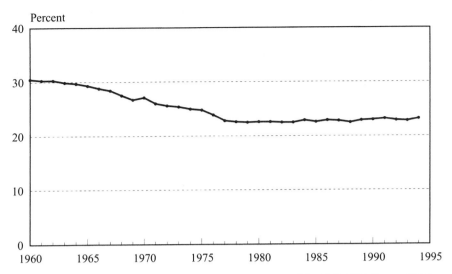

Figure 4.6. Percentage of all families with children under the age of six, 1960–1994. *Source:* U.S. Bureau of the Census, *Current Population Reports,* Series P-20, No. 477, and earlier reports as cited in U.S. Bureau of the Census, *Statistical Abstract of the United States: 1994,* Washington, D.C. 1994, p. 63, Table 74 and earlier Abstracts; and Rawlings, Steve W. and Saluter, Arlene F., *Household and Family Characteristics: March 1994,* U.S. Bureau of the Census, Current Population Reports, P20-483, U.S. Government Printing Office, Washington, D.C., 1995, p.1, Table 1.

holds. For example, if the hypothesis of larger households being easier to contact is not supported among those contacted, it could receive little more support looking at the full sample. If it is supported, we believe the effects we see would only be magnified if the full sample were examined. That is, among the noncontacted and roster-refused cases, smaller households would predominate (see Chapter 5 for evidence on the refusal cases.)

Table 4.3 shows the distribution of call of first contact by a variety of household-level attributes for those cases for which household roster data are available. Here the results are dramatic. Households with larger numbers of persons are much more likely to be contacted on early calls than those of smaller size. This is by far the most powerful correlate of contactability we have yet seen. Only 35% of the single-person households are contacted on the first call compared to 71% of those with five or more adults ($\chi^2 = 398$, $df = 4$, $p < 0.0001$). For the larger households over 90% are contacted in about 3 calls; for the single-person households over 6 calls are required to reach the same level.

The results for households with children less than 5 years old and those with elderly adults support the hypotheses in a similar way, but with smaller effects. About 55% of those with such young children are contacted in the first call relative to 48% of those without such children ($\chi^2 = 29.0$, $df = 1$, $p < 0.0001$). Over 63% of households with persons over 70 years old are contacted in one call, versus 49% of those without the elderly ($\chi^2 = 21.6$, $df = 1$, $p < 0.0001$). We conclude that the age compo-

Table 4.3. Cumulative percentage distribution of number of calls to first contact by number of persons in the household, among cases with roster information

Household attribute	Percentage contacted by call number							Percentage not contacted	Total
	1	2	3	4	5	6	>6		
Number of adults									
One	35.2	58.2	71.7	80.1	86.1	89.9	98.5	1.5	100%
Two	51.0	75.0	86.1	92.1	95.0	96.7	99.6	0.4	100%
Three	61.4	83.9	91.8	94.4	96.4	97.4	99.8	0.2	100%
Four	61.5	85.6	92.1	95.9	96.9	97.4	99.5	0.5	100%
Five +	71.1	86.8	91.7	95.0	96.7	97.5	99.2	0.8	100%
Missing	29.7	40.0	45.2	49.9	54.1	57.4	61.4	38.6	100%
Children <5 years in household									
No	47.8	71.5	82.6	88.7	92.5	94.8	99.2	0.8	100%
Yes	55.5	77.8	88.3	93.6	95.8	97.0	99.7	0.3	100%
Missing	29.7	40.0	45.2	49.9	54.1	57.4	61.4	38.6	100%
Adults over 70 years in household									
No	49.2	72.7	83.7	89.6	93.1	95.2	99.4	0.6	100%
Yes	63.2	85.7	92.1	96.6	97.4	98.1	98.9	1.1	100%
Missing	29.7	40.0	45.2	49.9	54.1	57.4	61.4	38.6	100%

sition is relevant to the at-home times of households. Households with the very young tend to have someone home to care for the children; households with the elderly tend to have someone home because of retirement from the labor force.

The decennial nonresponse match project generated data on correlates of contact at any time during the survey periods, but it did not include measures of call-by-call outcomes. Thus, we can see whether the household-composition measures are related to an ultimate contact. Those data show that relatively more single-person households remain uncontacted (4.1%, s.e. = 0.34%) than larger households (that average about 2.0% noncontact). Similarly, there are higher percentages of noncontacts among those without children under 5 (2.6% versus 1.3%), those with all members under 30 years of age (3.1% versus 1.7% for those with members all 70 or over).

There are other household-level indicators of at-home patterns for which we have little data to present. One common cause of at-home patterns is the employment activity of adults in a household. Figure 4.7 shows that the ratio of female civilian employment to the full population increased from about 35% in 1960 to about 55% in 1994. Adult women are more likely to be away from home during the day and are burdened with shopping and errands during their off-work hours than they were in earlier years. The view at the family level looks similar. Figure 4.8 shows that the percentage of married-couple families with two earners increased from about 40% in 1967 to about 51% in 1994. Both of these views point to housing units largely unoccupied during the weekdays and taken away from home for er-

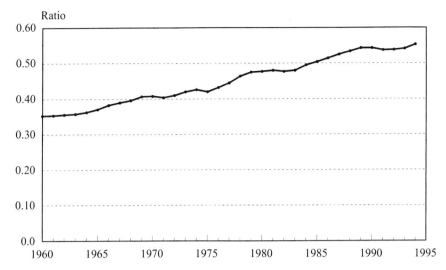

Figure 4.7. Ratio of female civilian employment to total population, 1960–1994. *Source:* U.S. Department of Labor, U.S. Bureau of Labor Statistics, *Handbook of Labor Statistics,* Bulletin 2340, August, 1989, pp. 8-9, Table 1 and U.S. Bureau of Labor Statistics, Bulletin 2307; and *Employment and Earnings,* monthly, January issues, as cited in U.S. Bureau of the Census, *Statistical Abstract of the United States: 1994,* Washington, D.C., 1994, p. 396, Table 616 and U.S. Department of Labor, Bureau of Labor Statistics, *Employment and Earnings,* Vol. 42, No. 1, 1995, p. 17, Table A-2.

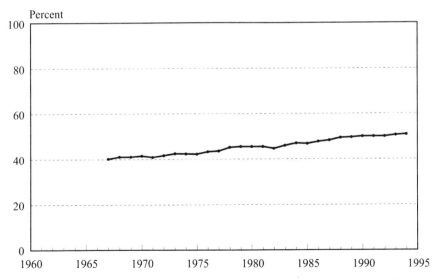

Figure 4.8. Percentage of married-couple families with two earners, 1967–1994. *Source:* U.S. Department of Labor, U.S. Bureau of Labor Statistics, *Handbook of Labor Statistics,* Bulletin 2340, August 1989, pp. 254-256, Table 61; U.S. Department of Labor, U.S. Bureau of Labor Statistics, *Marital and Family Characteristics of the Labor Force from the March 1994 Current Population Survey,* p. 6, Table 34, and earlier reports; and unpublished data provided by the Bureau of Labor Statistics.

rands during the weekends.

A fully specified model of household influences on contact would, therefore, include all attributes of households that prompt all adults to be out of the home at the same time. We suspect that for many households, members in the employed labor force is a strong predictor. For households out of the labor force, we must also consider consistently performed household, social, or community activities that take all persons out of the household. These would include volunteer work, church activities, exercise, social clubs, educational activities, and contact with extended-family members.

4.3.3   Summary of Environmental and Household Impacts on Contactability

Many of the population trends over the last few decades have conspired against easy contact with sample households. There are more households with all adults employed outside the home. There are more single-person households. There are fewer households with young children at home.

Examining bivariate relationships with contactability, the basic form of the model in Figure 4.1 is supported. Urban areas are more filled with households and housing units that pose problems for contact. They have densely populated areas of multiunit structures. They have larger proportions of single-person households.

We can also deduce from the above that households with any adults who are unemployed or out of the labor force tend to have someone at home more often than other households. Our indicators for this are households with young children and households with elderly adult members.

Some of the ecological indicators we examined in the NSHS data are aggregates of hypothesized causal mechanisms at the household level (e.g., percent multiunit structures in the block). For that reason, we would expect these measured relationships to be explained statistically by corresponding household-level attributes. For example, regardless of whether or not the neighborhood is filled with multiunit structures, it is more likely that an interviewer would have difficulty contacting households in a multiunit structure.

4.4   INTERVIEWER-LEVEL CORRELATES OF CONTACTABILITY

In many of the studies represented in this book, we gave a self-administered questionnaire to interviewers, asking them about their experience as an interviewer, their attitudes toward the job, and their ways of reacting to different field situations. (See Section 3.6)

Theoretically, we expect that interviewers' attributes can affect contactability of a case only through their choice of timing and number of visits to sample housing units. It is independently interesting, however, to question whether there are any net effects of interviewer experience and of two attitudinal states found predictive of gaining cooperation with sample households (see Chapter 7). Table 4.4 presents the distribution of number of call to first contact by three variables. This shows only

Table 4.4. Cumulative percentage distribution of number of calls to first contact by interviewer attributes

Interviewer attribute	Percentage contacted by call number							Percentage not contacted	Total
	1	2	3	4	5	6	>6		
Interviewer Tenure									
0–6 months	49.8	71.9	83.3	88.9	92.3	93.9	97.7	2.3	100%
6 months–1 year	47.8	71.4	82.0	88.4	91.3	93.4	97.2	2.8	100%
>1 year	48.4	71.0	81.4	87.1	90.8	93.1	97.4	2.6	100%
Rate versus quality									
Better to persuade	49.2	71.9	82.3	88.1	91.5	93.7	97.9	2.1	100%
Better to accept refusal	47.4	70.2	81.3	87.4	90.9	92.8	96.3	3.7	100%
Confidence									
Can convince the most reluctant	49.1	71.3	82.5	88.3	91.6	93.7	97.5	2.5	100%
Some will never participate	48.2	71.4	81.5	87.4	91.0	93.1	97.3	2.7	100%

small net effects of interviewer tenure—no important differences for experience, a very small advantage to interviewers with a devotion to persuading reluctant respondents to participate. We view all of these differences to be of only minor importance. We also asked interviewers if, upon failing to make contact with a sample household, they tended to leave handwritten notes, brochures describing the study, or their own business card. There were no meaningful relationships between answers to these questions and the number of calls to first contact.

We take this as evidence that, if these types of interviewers are assigned samples equivalent on contactability, the calling patterns of interviewers are relatively comparable. There appear to be no techniques of enhanced productivity that are learned through trial and error during the early months of an interviewer's tenure. Further, greater enthusiasm for the task of gaining cooperation doesn't translate into large differences in the ease of contacting sample households.

4.5   CALL-LEVEL INFLUENCES ON CONTACTING SAMPLE HOUSEHOLDS

Figure 4.1 asserts that one key ingredient in determining the likelihood of contact is the number and timing of calls on the sample unit. Before we begin to discuss the influences on the contactability of sample households, it is useful to present some descriptive information on the typical patterns of call attempts.

Figure 4.9 shows that a vast majority of first calls in NSHS occur during the day (between 9 AM and 6 PM). A relatively large number occur on Saturdays and Sundays (27% of all calls). Relatively few occur during evening hours (13%). The rate of evening calls is higher for weekdays than for Saturday and Sunday. Caution

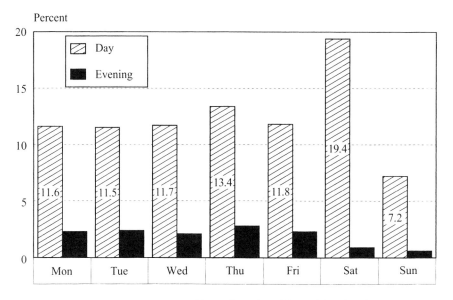

Figure 4.9. Distribution of first calls by time of day and day of week.

should be exercised in interpreting these percentages because of the relative number of hours available for calling in each of these time periods. If one considers each day divided into two parts, daytime (9 AM–6 PM or 9 hours) and evening (6 PM–9 PM or 3 hours), then there are many more hours to make calls on sample cases during the day than evening.

Based on our focus groups with interviewers, there appear to be two other reasons for first calls being disproportionately made during the day. Recall that many face-to-face surveys are based on clustered sample designs in which multiple houses in the same block or neighborhood are selected together into a sample segment. Interviewers report that, especially on surveys with long data collection periods and on surveys involving area frame listings, the first visit to the same segment is used to check listings of the sample for addresses missed in the listing process and/or to obtain some initial impressions of the neighborhood. The observations made include ease of identifying the sample housing units among others in the area, assessment of personal safety of the interviewer in the neighborhood, evidence of any locked gates or entrances, levels of visible activity in the area, places to park the car within observational distance of sample units, etc. Such visits are predominately made during daylight hours.

The typical interviewer on NSHS worked 18–23 hours per week. Some interviewers have other jobs. Weekends are likely to offer freedom to work more hours than usual. If interviewers were full-time employees, it is not clear that they would choose to work a disproportionate number of hours on the weekends.

How do the NSHS first call patterns compare to other surveys? Purdon, Campanelli, and Sturgis (1996) in the United Kingdom show a similar split between day and evening calls, but relatively fewer weekend day calls. This raises another possibility—that weekend calls may be differently accepted across cultures. For example, some government agencies, apparently out of sensitivity to religious practices, make few or no calls on Sundays. We might expect to see urban–rural or regional differences on this score in the United States, reflecting differing intensities of religious affiliations.

4.5.1   Patterns of Calling over Successive Attempts

Thus far we have seen tendencies for first calls to be made in the daytime. This is not true for later calls on units not successfully reached on the first call. Figure 4.10 shows that the proportion of day calls declines monotonically over successive calls, with more and more calls being made during evening hours. For example by call 4, about 30% of the calls are made in the evening, versus 14% of the first calls. The U.K. data of Campanelli, Sturgis, and Purdon (1997) show similar results. The biggest shift reflects moving calls from the weekday days to the evenings.

This is consistent with the notion above that interviewers attempt to resolve cases that can be reached both during the day and evening in early calls made during the day. This permits them to devote the more limited evening hours to those households at home only in the evenings. If such were the case, we might expect different timing of second and third calls on cases that differed on the timing of the earlier call.

As we saw in Figure 4.10, Table 4.5 shows that the proportion moving to evening calls increases successively regardless of what time the prior call was made. Relative to the 14% of the calls made in the evening for the first call, over a quarter of the second calls are made in the evening regardless of when the first call was made.

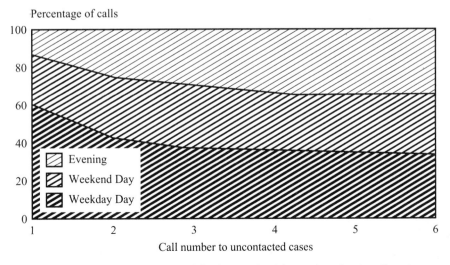

Figure 4.10. Percentage of calls on weekday days, weekend days, and evenings by call number.

Table 4.5. Percentage distribution of time of next call by time of prior call among those not yet contacted (standard error in parentheses)

1. Call 1 to call 2

	Time of call 2			
Time of call 1	M–F day	Sa–Su day	Evenings	Total
Call 1 M–F day	44.5	31.7	23.8	100%
	(1.6)	(1.6)	(1.4)	
Call 1 Sa–Su day	44.4	30.1	25.5	100%
	(2.1)	(1.9)	(1.7)	
Call 1 evening	34.6	30.8	34.6	100%
	(3.2)	(2.8)	(2.7)	

2. Call 2 to call 3

	Time of call 3			
Time of call 2	M–F day	Sa–Su day	Evenings	Total
Call 2 M–F day	41.4	32.9	25.6	100%
	(1.6)	(2.0)	(1.6)	
Call 2 Sa–Su day	36.5	32.7	30.8	100%
	(2.2)	(2.0)	(2.3)	
Call 2 evening	31.4	30.9	37.7	100%
	(2.0)	(2.6)	(2.8)	

Similar results pertain to call 3.

In addition, Table 4.5 shows that switches in call times do occur over successive attempts. There is a tendency for unsuccessful calls made during the day (either weekday or weekend) to be followed by weekend day calls. In contrast, unsuccessful calls in the evening tend to be followed by another evening call, relative to the other groups. In short, there is some evidence for stability of time of day of calling across calls, but a movement into the weekends for unsuccessful day calls.

These results alone do not suggest why this occurs. Why do units first visited in the evening (unsuccessfully) get repeat visits in the evenings? Why do units visited in the day continue getting day calls, albeit focused on weekends? One possibility is that the interviewer labels different areas as more well-suited to daytime versus evening visits. One way such labeling might occur concerns perceived personal safety. Another arises when one household in a segment is successfully contacted, and yields an appointment for a later visit. Other units in the same segment can be visited at that later time, at very low marginal costs to the interviewer.

The question that is stimulated by this is "what characteristics of sample segments or neighborhoods are related to interviewers choosing different times of day to call?"

4.5.2 Neighborhood Correlates of Calling Patterns

In telephone surveys, it is common to use software embedded in computer-assisted telephone interviewing (CATI) systems to direct the time and number of calls on sample telephone numbers. In the face-to-face mode, this protocol is generally left to interviewer discretion. Face-to-face interviewers may have a variety of reasons for choosing the time to call on sample units. In this section we examine whether there are systematic differences in calling times by various social environmental characteristics of the sample unit. In short, does the social context of the household affect when calls are made?

An important subquestion of this type is whether interviewers assigned to areas where there are likely to be fewer persons at home during the day make calls on those cases more frequently in the evening, or whether there appears to be no prior sorting of cases by interviewer knowledge of the sample segment. Table 4.6 presents

Table 4.6. Percentage of sample housing units with first call in the evening by environmental characteristics

Environmental attribute	Percentage with first call in the evening	
Urban status		
In central city	14.4%	
In balance of CMSA	11.9%	
In other MSA	16.0%	
In nonMSA	11.1%	$\chi^2 = 4.0, df = 3, p = 0.28$
Population density[a]		
Lowest quartile (<705)	11.8%	
Second quartile (705–2028)	14.2%	
Third quartile (2029–3687)	12.9%	
Highest quartile (>3687)	14.2%	$\chi^2 = 2.2, df = 3, p = 0.53$
Crime rate[b]		
1st quartile (<3459)	9.2%	
2nd quartile (3459–5001)	14.5%	
3rd quartile(5002–6878)	14.3%	
4th quartile (>6878)	15.1%	$\chi^2 = 6.2, df = 3, p = 0.12$
% Persons of minority race		
1st quartile (<2.5%)	9.1%	
2nd quartile (2.5–7.9%)	13.5%	
3rd quartile (8.0–24.0%)	16.5%	
4th quartile (>24.0)	14.1%	$\chi^2 = 8.2, df = 3, p = 0.05$
Units in disrepair in neighborhood		
No units	13.2%	
Some units	13.6%	$\chi^2 = 0.1, df = 1, p = 0.82$

[a]Persons per square mile in county of housing unit.
[b]Personal crimes per 100,000 persons in county of housing unit.

the percentage of first calls on cases that are made in the evening, at a time when households with employed members will more likely be at home. There is no evidence in the table for interviewers to avoid evening calls in difficult areas.

Almost all of these results suggest interviewers call at times when they think the household might be at home. Indeed, they argue against the focus group reports that when personal safety is an issue, first calls in the daytime are more likely.

4.5.3 Relative Success at Contact by Timing of Calls

Now we move from examining when calls are made to whether the relative success at contact varies by timing of calls. The information regarding at-home times of households is consistent in showing larger proportions of units with at least one person home in the evening hours and on the weekends (Weeks *et al.*, 1980; Weeks, Kulka, and Pierson, 1987). In the NSHS data as shown in Figure 4.11, about 49% of the first calls yield a contact with the sample household. However this varies from a low of 43% of the households called on weekday days (between 9 AM and 6 PM) to 53% of those called on weekend days, to 58% of those in the evenings. Evening calls are clearly more productive than calls at other times.

There is little variation over the days of the week on the productivity of daytime calling. Saturday and Sunday calls during the day show higher rates, 53% and 55%, respectively. The variation across days in evening contact rates is somewhat greater than for daytime calls, but this is likely to be a function of the smaller sample sizes producing the estimates. Small sample size also, we believe, lies behind the lower

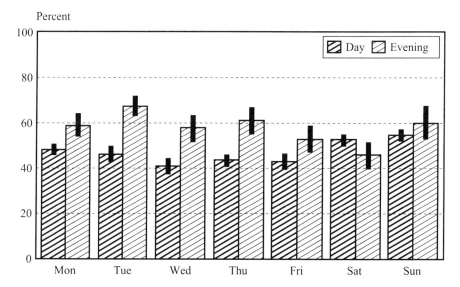

Figure 4.11. Percentage of first calls obtaining contact with sample household by day of week and time of day. *Note:* Lines on bars indicate one standard error of estimated percentage.

evening contact rate on Saturday evenings, although this might reflect true out-of-home higher rates on that day. (No such result occurs in the U.K. data, (Campanelli, Sturgis, and Purdon, 1997, Table 3.5).)

As we saw in Figure 4.2, for those not reached, each succeeding attempt at first contact achieves lower contact rates. Call 2 makes contact with 44% of those not previously contacted; call 3, 37%; call 4, 32%. Much of the effort in survey methodology to increase the efficiency of the survey-contact process has attempted to find combinations of days of week and times of day that obtain high contact rates. Table 4.7 presents data permitting us to measure whether the decline in contact over calls affects all calling times. We see that all time periods experience the downward productivity; there are no large changes in the relative efficiencies of the times of the day or days of the week.

An important question for interviewers to address, when the first call yields no contact, is what time they should choose to make the next call. In face-to-face surveys this decision is sometimes guided by observations they make on the first call.

Evenings appear to be more productive of contacts over all types of households, both the entire sample, and those that tend to be home less often. Weekend days are the next most productive. The simplest possibility, yielding the most parsimonious predictive model, is that the relative advantage of evening and weekend day calls over weekday day calls, remains true in all circumstances. We know from Table 4.7 that this is true on average at any given call number. Table 4.8 shows likelihood of contact on a later call, given no contact at a prior call, for different calling times on each of the calls. For example, the top panel describes the movement from call 1 to call 2. It shows (in the "Total" row) that the probability of a contact on call 2 is about 0.45, regardless of when call 1 was placed. In that panel, the contact rate on call 2 rises from weekday day to weekend day to evening in all cases, except for cases where call 1 was in the evening. For these cases it appears that call 2 on weekend day enjoys no advantage (0.40 contact probability) over those on a weekday days (0.41 contact probability).

The second panel of Table 4.8 describes the outcome at call 3 for cases not previously contacted. In this panel, there seem to be lower probabilities of call 3 contact when call 2 was unsuccessfully placed in the evening (0.35 contact probability), but because of fewer observations this is not a precise estimate ($\chi^2 = 5.6$, $df = 2$, $p = 0.07$). In general, call 3s yield contacts at higher rates in the evenings or weekend

Table 4.7. Proportion contacted among previously uncontacted by time of call and call number

	Call number					
	1	2	3	4	5	6
M–F day	0.44	0.41	0.33	0.32	0.26	0.25
Sa–Su day	0.54	0.45	0.39	0.34	0.29	0.28
Evenings	0.59	0.52	0.45	0.38	0.40	0.34

Table 4.8. Proportion contacted on call 2 or 3 among those not contacted on earlier calls, by time of previous call

1. Call 1 to call 2

Time of call 1	Time of call 2			Total
	M–F day	Sa–Su day	Evenings	
M–F day	0.43	0.45	0.53	0.46
Sa–Su day	0.36	0.47	0.48	0.42
Evenings	0.41	0.40	0.51	0.44
Total	0.41	0.45	0.52	0.45

2. Call 2 to call 3

Time of call 2	Time of call 3			Total
	M–F day	Sa–Su day	Evenings	
M–F day	0.32	0.42	0.47	0.39
Sa–Su day	0.33	0.43	0.48	0.41
Evenings	0.37	0.26	0.40	0.35
Total	0.33	0.39	0.45	0.39

days, except again for cases in which call 2 was in the evening. For these cases, following the evening call with a weekend day seems to yield lower contact rates (0.26) than weekday day (0.37) ($p < 0.05$).

Table 4.8 shows no support for a different model of call scheduling—one that sets the next call at a different time than the unsuccessful prior call. This is common to many call schedulers in CATI systems, attempting to minimize the number of calls prior to first contact. If this model were optimal, then interviewers would choose weekend days to call after unsuccessful evening calls to achieve higher contact rates. There seems no support for this from the data. (If one examines the call 3 outcomes, taking into account both call 1 and call 2, there is, however, some support that a call 3 makes more contacts on cases for which the first two calls were made during the same time of day.) We must remind ourselves in examining Table 4.8 that interviewers probably use information obtained at the time of the visit in order to choose a time to call.

4.6 JOINT EFFECTS OF MULTIPLE LEVELS ON CONTACTABILITY

Thus far we have observed that roughly half of the households are contacted in one visit, about 70% in two calls, and 80% in three calls. Overall, visits in the evening tend to yield higher contact rates than those at other times; weekend days are better than weekday days. The types of households that are contacted with least effort have consistent attributes. Those in urban, high-crime, high-density areas tend to require more calls than those in other areas. Housing units in locked apartment buildings, with intercom communication, locked gates, etc., tend to present difficulties in contact. Households with many adult members, those with children, and those with elderly members tend to be easier to contact.

Some of these influences on contactability are related to one another and thus the analyses above may mislead us with regard to the relative importance of individual attributes. To gain some sense of how these variables combine, we constructed multivariate models predicting contact at calls 1, 2, and 3. Then we compared these to models predicting contact at any call during the survey period, using both the NSHS data and the decennial match data.

We were motivated by the same conceptual model as described in Section 4.1.2. The indicators above are a) proxy measures of when the household is at home, b) observations of physical impediments to contact the household, and c) indicators of when interviewers visited the sample household. Since we have richer data on the households with completed rosters, we will present separately models for the complete sample base and those for rostered cases. For the rostered cases, we know household attributes that are theoretically better proxy indicators of the at-home characteristics of the unit. In order to replicate the findings and to link to other chapters in the book, we also present related models based on the decennial census nonresponse match project.

We expect three outcomes from the multivariate modeling. First, we expect some of the effects of the social environmental variables to be explained by the observations of physical impediments (especially multiunit structure). Others will be explained by characteristics of the householders; for example, the tendency for householders in areas with higher percentages of owned residences to be contacted easily may reflect age of householders and number of adults per household.

Second, we intend to compare coefficients in the three logistic models (those for call 1, call 2, and call 3) as a way to see whether the characteristics of those contacted changes over repeated calls. We expect that influences of a lasting and pervasive nature on the likelihood of contact will increase in their measured influence on contact success over successive calls. In contrast, those attributes that reflect temporary absences from the home will decline in their importance. These might arise for attributes related to being away from home in the first calls during the weekend but would disappear in later calls. With such variables, we would expect diminished effects in models including time of call.

Third, we intend to take two perspectives on successive calls to make contact on a household. The first concentrates on individual calls. The contrast of interest is between those successfully contacted on the call and those not contacted. We will restrict our attention to sample households not previously contacted. That is, we measure influences on successful contact on call 2 only among those cases not successfully contacted on call 1. Similarly, we will estimate the impact of predictors of call 3 contact only among cases not contacted on either call 1 or call 2. The discussion of these models will attempt to identify influences that increase their importance over calls. These will be the strong influences on contact that repeated calls cannot diminish. Ideally, in order to achieve contact rates that were constant over all subgroups, we would find that influences that are positive effects toward contact in early calls would be negative for later calls. That is, at the end of all efforts to contact the sample cases, there would be no measurable correlates of noncontact (the contact rate for all subgroups would be the same).

The second perspective focuses on a different contrast—how different are those contacted at call 1 from those first contacted at call 2, from those first contacted at call 3, and from those contacted at some later point? This will utilize a multinomial logistic model whose coefficients measure the impact on a successful contact at one call, relative to its impact on success at call 1. This perspective is more pertinent to assessing changes in the set of contacted households as a function of greater effort. We take this perspective later.

4.6.1   Models Predicting Contact for Previously Uncontacted Sample Units

The coefficients in Table 4.9 measure the influence of attributes on the probability of contact on a call relative to no contact. Three models are presented. The first measures the probability of contact on call 1, among all cases. The second predicts the probability of contact on call 2, among those not reached on call 1. The third predicts the probability of contact on call 3, among those not reached on calls 1 or 2. For most of the independent variables, the models show evidence of increasing influence as the number of calls increase, rather than declining influence over calls.

The models in Table 4.9 are the final form of a set of sequentially fitted models. The sequence moved from ecological models to household-level models. We first examined the gross effects of urbanicity and found that urban contact rates were progressively lower than nonurban with successive calls. Indeed, urban contact rates on each call attempt are lower, but even more so on call 3 versus call 1. The first analytic question was whether such contrasts would survive controls on various correlates of urbanicity that we examined above.

Urbanicity effects are brought to a negligible level when we control on the percentage of multiunit structures, crime rates, and density measures. This is true for call 1 and call 3 but not call 2, where the negative effects of urbanicity survives these controls. Note, however, the change of sign for the call 3 model, reflecting the tendency for urban dwellers not yet contacted by call 2 to come in at higher rates than their rural counterparts when call 3 is made.

Among the variables examined, it appears that the percentage of multiunit structures does much to explain the effects of urbanicity. In addition, population density acts to produce the urbanicity effects. In other words, one reason that urban areas have lower contact rates on each call effort is that urban areas tend to have many multiunit structures, which pose difficulties for interviewers accessing the sample units. Further, these areas tend to have high population densities, which may tend to be related to less at-home activities, because of travel times to work and other out-of-home activities. Another variable, crime rate, has the predicted negative effect on contact, but emerges as a significant contrast only on call 3 for previously uncontacted numbers.

The data permit us to move one more level closer to persons in households. We can add a measure at the housing unit level that measures physical impediments to

Table 4.9. Coefficients of logistic models predicting likelihood of contact among previously noncontacted households, on call 1, call 2, and call 3 (all sample cases) (Standard errors in parentheses)

Predictor	Call 1	Call 2	Call 3
Constant	0.60**	0.29**	0.43**
	(0.11)	(0.086)	(0.17)
Social environment			
Urbanicity			
Central city	−0.047	−0.29*	0.20
	(0.12)	(0.11)	(0.20)
Balance of CMSA	−0.081	−0.24**	0.22
	(0.083)	(0.066)	(0.17)
Other MSA	−0.0053	−0.16	0.17
	(0.085)	(0.094)	(0.16)
NonMSA	—	—	—
Percent multiunit structures	−0.0023	−0.0042*	−0.0083**
in block	(0.0016)	(0.0017)	(0.0016)
Population density[a]	−0.020	−0.016	−0.045*
	(0.011)	(0.0092)	(0.0020)
Crime rate[b]	0.00054	0.0020	−0.0075*
	(0.0025)	(0.0018)	(0.0028)
Housing unit			
Physical impediments to	−0.33**	−0.28**	−0.25*
household	(0.091)	(0.094)	(0.10)
Time of current call			
Weekday day	−0.65**	−0.43**	−0.58**
	(0.078)	(0.078)	(0.10)
Weekend day	−0.24*	−0.26**	−0.27*
	(0.099)	(0.080)	(0.12)
Evening	—	—	—

Note: Dependent variable coded 1 = contact, 0 = noncontact.
[a]Measured in thousands of persons per square mile.
[b]Measured in crimes per 1,000 persons.
*$p < 0.05$.
**$p < 0.01$.

access to the unit (see Section 4.3.1). Here we would expect some diminishing of the effect of multiunit structure neighborhoods (because they tend to have locked entrances), but also some of the urbanicity effects. There are, however, no powerful changes in the effects of the social environmental variables. Further, the effect of the physical impediments appears to be similar across the three calls, reducing contactability. (One predictor was dropped from the models because of lack of marginal effects controlling on the others. This was percentage of owner-occupied dwellings in the block.)

Finally, we add to the model measures of interviewers' choices of time to call on the sample unit. Here we do not expect large changes in coefficients of other variables, and basically find none. Instead, we see consistently lower contact rates for calls during the day and on weekends versus evening calls. Further, there is a suggestion of even greater contrasts in the later calls than on call 1. That is, the time of call is more important for contact after the portion easily contacted is removed. Evening calls become crucial for contacting many units.

The summary of the models in Table 4.9 is that the effects of urbanicity are only partially explained by structures that hamper interviewers' access to sample units, and to social environments that are not conducive to strangers calling on households or households feeling free to open their doors. The variables that are missing from the models are ones that more directly reflect the extent of time spent at home. For such attributes, we would expect that the higher commute times for larger cities among those who work out of the home would underlie some of the urbanicity effects.

4.6.2   Models Predicting Contact among Previously Uncontacted Numbers, with Household-Level Predictors

As Figure 4.2 shows, at the end of the NSHS survey, approximately 97.2% of the sample households were contacted. Of these, 9,384 provided a listing of the household, including the age and gender of each household member. From these units we can measure the effect on success at call 1, call 2, or call 3 of various indicators of household type that are motivated by the theory presented in Chapter 2.

These variables include the existence of young children (5 years old or less) in the household, whether all persons in the household are 70 years old or older, and whether the household is a single-person household. The first two attributes are associated with a higher proportion of time spent at home by some adult. The last indicator is negatively related to time spent at home.

When we fit the same models as shown in Table 4.9 using the subset of cases that provided household rosters, there are no important differences between coefficients. That is, the models as specified show no evidence of selection bias among those providing a roster. Given that check, we extended the models to include demographic variables related to household composition. Here we expected that some of the measured effects of multiunit structures and perhaps of the impediments to access of the housing unit might be masking effects of household compositional differences. We would thus expect a reduction of those predictors' effects in the extended model.

Table 4.10 presents these extended models. First, the effects seen earlier remain strong. Households with more adults tend to be contacted at higher rates, on each of the three calls. Households with young children are more easily contacted on each of the three calls. Finally, households with adults over 70 years of age are more easily contacted initially, but those not contacted by the second call show similar contact propensities as other households. The elderly persons more easily contacted early are probably not in the labor force and have more at-home time. Those not easily contacted probably have at-home patterns similar to younger adults.

These household composition variables are one step closer to indicators of the

Table 4.10. Coefficients of logistic models predicting likelihood of contact among previously noncontacted households, on call 1, call 2, and call 3 (cases with completed household rosters) (standard errors in parentheses)

Predictor	Call 1	Call 2	Call 3
Constant	−0.30**	−0.43**	−0.064
	(0.10)	(0.16)	(0.20)
Social environment			
Urbanicity			
Central city	0.063	−0.19	0.37
	(0.12)	(0.11)	(0.21)
Balance of CMSA	−0.052	−0.26**	0.25
	(0.086)	(0.069)	(0.17)
Other MSA	−0.0042	−0.19	0.14
	(0.091)	(0.097)	(0.16)
NonMSA	—	—	—
Percent multiunit structures in block	−0.00050	−0.0024	−0.0076**
	(0.0016)	(0.0020)	(0.0017)
Population density[a]	−0.027*	−0.017	−0.044*
	(0.011)	(0.010)	(0.020)
Crime rate[b]	−0.0014	0.0022	−0.0083**
	(0.0024)	(0.0018)	(0.0028)
Housing unit			
Physical impediments to household	−0.21*	−0.16	−0.17
	(0.087)	(0.097)	(0.11)
Household composition			
Number of adults	0.41**	0.37**	0.28**
	(0.025)	(0.052)	(0.064)
70+ year-old in household (1 = yes)	0.32*	0.40*	−0.018
	(0.13)	(0.19)	(0.35)
Children <6 years old in household (1 = yes)	0.33**	0.17*	0.25*
	(0.059)	(0.073)	(0.12)
Time of current call			
Weekday day	−0.65**	−0.49**	−0.59**
	(0.082)	(0.079)	(0.11)
Weekend day	−0.21*	−0.28**	−0.23
	(0.10)	(0.083)	(0.13)
Evening	—	—	—

Note: Dependent variable coded 1 = contact, 0 = noncontact.
[a]Measured in thousands of persons per square mile.
[b]Measured in crimes per 1,000 persons.
*$p < 0.05$.
**$p < 0.01$.

lifestyle and at-home habits of the sample households. In examining the change in measured effects of other variables, we see exactly what was expected—reduced effects of multiunit structure and indicators of physical impediments to contact. That is, some of the greater difficulty of contacting persons in multiunit structures is due to the fact that they are disproportionately small households, without children, and without older adults. Similarly, some of the effect of the indicator of physical impediments to access masks the fact that households in such units tend to be single-person households, without children, or without older adults. Finally, the effects of physical impediments appear larger in the first calls than in later calls. This probably reflects interviewer success in inventing ways to surmount the impediments.

The additional household composition variables do little to diminish the measured effect of the environmental influences. We suspected that the differences in the rates of single-person households in urban, high-density areas might have underlain some of the environmental contrasts. It appears, however, that the environmental effects are not explained by these household composition variables (number of adults, elderly householders, or child householders). Also note that, as in Table 4.9 without the household-level predictors, that the environmental variables have greater effects on contacts in call 3 than in call 1 or 2. In that sense, they emerge as important predictors for those most difficult to contact. We suspect arguments about higher frequencies of being away from home in urban areas for reasons of commuting to work, recreation, and other opportunities, are still attractive.

4.6.3   Models Contrasting First Contact on Later Calls with Contact on First Call

A different set of questions can be posed to the same data. This second perspective compares attributes of households with late contacts to those with early contacts. For example, we might wonder whether the characteristics of persons first reached on call 2 are different from those reached on call 1. This question and others like it focus on the nature of the respondent pool. They query the data with regard to how the composition of the respondent pool may be changed by continuing calls. These questions relate to whether the certain types of respondents tend to be first contacted early rather than late in the call sequence.

To answer such questions we have fit a multinomial logistic model, which compares first contact on calls 2, 3, or later to contact at call 1 (Table 4.11).

In comparing the influences on success in late calls to success at call 1, a different picture of contacting households emerges. First, regarding urbanicity, there are few differences between those contacted on calls 1 and 2, but those contacted on call 3 are more likely to come from urban areas than those on call 1. (This is consistent with the result in Table 4.9, as it must be, but focuses on a different contrast.) The later calls tend to offer a much smaller advantage to urban areas, one that falls below traditional statistical significance levels.

On the other hand, there is increasing ability to contact those with some physical impediment to their access. These are households for which, in some sense, repeated callbacks help the interviewer "discover" solutions for contact. Repeated calls

Table 4.11. Coefficients of multinomial logistic regression for first contact on call 2, call 3, or a later call, compared to contact on call 1 (standard errors in parentheses)

Predictor	Predicting success on		
	Call 2 versus call 1	Call 3 versus call 1	Later call versus call 1
Constant	−0.15	−0.56**	−0.49**
	(0.094)	(0.15)	(0.12)
Social environment			
Urbanicity			
Central city	−0.054	0.38*	0.18
	(0.12)	(0.16)	(0.20)
Balance of CMSA	−0.038	0.35*	0.16
	(0.092)	(0.13)	(0.13)
Other MSA	−0.12	0.10	−0.048
	(0.10)	(0.14)	(0.16)
NonMSA	—	—	—
Percent multiunit structures in block	−0.0014	−0.0032	0.0032
	(0.0012)	(0.0021)	(0.0024)
Population density[a]	0.0039	−0.011	0.020
	(0.0092)	(0.012)	(0.018)
Crime rate[b]	−0.00023	−0.0067	−0.0010
	(0.0020)	(0.0026)	(0.0030)
Housing unit			
Physical impediments to household	0.16	0.29*	0.48**
	(0.11)	(0.12)	(0.12)
Time of current call			
Weekday day	−0.87**	−1.28**	−1.41**
	(0.12)	(0.14)	(0.12)
Weekend day	−0.48**	−0.64**	−0.85**
	(0.12)	(0.16)	(0.13)
Evening	—	—	—

Note: Dependent variable coded 1 = contact, 0 = noncontact.
[a]Measured in thousands of persons per square mile.
[b]Measured in crimes per 1,000 persons.
*$p < 0.05$.
**$p < 0.01$.

permit them to find ways to gain access to the housing unit. Similarly, Table 4.11 shows that the relative disadvantage of daytime and weekend calls relative to evening calls becomes greater. In other words, evening calls are more successful for those contacted later than for those contacted on call 1.

In short, Table 4.11 shows that urban households and those with some physical barrier to access are disproportionately brought into the respondent pool with repeated callbacks. Further, evening calls become increasingly fruitful in later callbacks.

The results when we add household composition predictors (and thus limit our analysis to households with rosters) are quite similar (Table 4.12). Larger households tend to be contacted on call 1. Smaller households are added in later calls, with increasing contrasts to call 1. That is, added callbacks are increasingly successful in bringing in the smaller households. This is quite compatible with the model that at-home frequency is a simple function of number of persons in the household. Similarly, the later calls bring in households without young children. Finally, the indicator of elderly household members behaves somewhat differently. There is little tendency for those reached on the second call to be any different from those reached on call 1, on this attribute. Those contacted on call 3 or later, however, tend to be households disproportionately without elderly members, relative to those contacted on call 1. This implies that the nonelderly households tend to be brought into the contacted pool only in later calls.

4.6.4 Models Predicting whether the Household was Ever Contacted

This chapter utilized the NSHS data because they offered full information on the results of each call on the sample household, as well as observations on physical impediments to access. It is useful to see to what extent the same influences on early contact versus late contact are those that apply to whether the unit was ever contacted. When we examine the probability of ever making contact with the sample household, however, the number of available predictors is radically reduced to the environmental indicators (urbanicity, population density, crime rate, and percentage of multiunit structures) and the interviewers' observations of any physical impediments. The first column of Table 4.13 shows the model with these predictors.

The reader will recall that lowered contactability in urban areas appeared more likely for the first two call attempts but then among those not contacted in the first two attempts, there were no negative effects. In the overall contact propensity model in Table 4.13, the overall negative effects of urbanicity are evident. We take this as evidence that the initial underrepresentation of urban residents in contacts on early calls is never completely redressed. Among the other environmental indicators, population density was exhibiting negative effects on contactability in early call attempts. None of the other social environmental variables potentially related to at-home times reach standard levels of statistical significance, however, in the final contact model.

The housing unit indicator that was a strong negative influence on contacts in early calls, observations of physical impediments to access, remains a negative influence, but its power is diminished to a level just below that reliably detected with these data. We suspect that this reflects the ability of interviewers to overcome some physical impediments (e.g., locked apartment buildings). However, we also believe that continued negative coefficient represents the permanent effects of stronger physical impediments (e.g., high-security residential areas). The fact that there are few such units leads to a coefficient dipping below significant levels.

What do we make of these findings? In the call-by-call analyses earlier in this section, the household compositional variables (i.e., age and numbers of adults) and

Table 4.12. Coefficients of multinomial logistic regression for first contact on call 2, call 3, or a later call, compared to contact on call 1 (cases with household rosters) (standard errors in parentheses)

	Predicting success on		
Predictor	Call 2 versus call 1	Call 3 versus call 1	Later call versus call 1
Constant	0.30**	0.33	0.86**
	(0.12)	(0.18)	(0.20)
Social environment			
Urbanicity			
Central city	−0.079	0.34*	0.056
	(0.13)	(0.16)	(0.20)
Balance of CMSA	−0.069	0.34*	0.13
	(0.093)	(0.13)	(0.14)
Other MSA	−0.14	0.095	−0.059
	(0.10)	(0.14)	(0.16)
NonMSA	—	—	—
Percent multiunit structures in block	−0.0024*	−0.0050*	0.0011
	(0.0012)	(0.0022)	(0.0024)
Population density[a]	0.0080	−0.0029	0.031
	(0.010)	(0.013)	(0.018)
Crime rate[b]	0.00067	−0.0059*	0.00064
	(0.0020)	(0.0025)	(0.0030)
Housing unit			
Physical impediments to household	0.15	0.23	0.37**
	(0.11)	(0.12)	(0.12)
Household			
Number of adults	−0.20**	−0.43**	−0.69**
	(0.032)	(0.051)	(0.058)
70+ year-old in household (1 = yes)	−0.056	−0.43	−0.43*
	(0.14)	(0.29)	(0.20)
Children <6 years old in household (1 = yes)	−0.21*	−0.20*	−0.42**
	(0.076)	(0.081)	(0.088)
Time of current call			
Weekday day	−0.87**	−1.24**	−1.36**
	(0.12)	(0.14)	(0.12)
Weekend day	−0.48**	−0.62**	−0.82**
	(0.13)	(0.16)	(0.13)
Evening	—	—	—

Note: Dependent variable coded 1 = contact, 0 = noncontact.
[a]Measured in thousands of persons per square mile.
[b]Measured in crimes per 1,000 persons.
*p < 0.05.
**p < 0.01.

times of calling on the unit on the first call appeared to explain the lowered urban contact ability. In later calls, urbanicity effects sometimes emerged despite the household-level control variables. We cannot use these household-level variables as controls in this analysis of overall contact rates. Our judgement is that the tendency for urban areas to have lower contact rates in early calls is only partially explained by the household compositional variables.

One source of added evidence about the relative effects of environment and household on contactability is found in the decennial match data. By turning to the dependent variable of "ever contacted" versus "never contacted" we are also able to address the decennial match data with an expanded set of predictors reflecting the household composition of the sample unit. The pooled decennial match data set attained a contact rate of 97.6% compared to 97.2% for NSHS. In Table 4.13, we first make rough comparisons to the coefficients for similar variables in the NSHS data (compare the first two columns of Table 4.13). Here the two data sets produce similar results (with the percent of large multiunit structures in the block reaching significance in the match data).

Next we examine the same issue as above. Can household compositional variables explain the observed urbanicity effects on contact rates? With the decennial match data, we can use housing-unit structure attributes measured at the sample-unit level instead of the neighborhood or block level. We can also use additional indicators of at-home patterns from census data on sample units. These include indicators for single-person households, existence of children less than 5 years old, households where all members are less than 30 years old, and households where all are 70 or over. (Chapters 5–9 show that these variables predict cooperation likelihood in contacted households.) This expanded model is also in Table 4.13 (third column).

The results of the expanded model resemble those in the call-by-call models earlier in this section with regard to the estimated influence of the household variables. The indicators of at-home patterns perform as the NSHS data suggested in the call-by-call analysis. Households with young children and elderly persons tend to have higher contact rates, wherever they live. Single person households remain disproportionately not contacted at the end of the surveys. With regard to housing-unit structure, as expected, those in single-family homes tend to be contacted; those in large multiunit structures tend not to be.

However, despite the strong effects of the housing unit and household-level indicators, larger urban areas (both central cities and suburbs) show marginally lower contactability. Indeed, the urbanicity effects are on the same order of magnitude as those of the household variables. Clearly, in the decennial match data, the household compositional variables do not explain the observed negative effects on contact rates of urbanicity.

In short, our overall assessment is that the remaining effects of urbanicity reflect unmeasured variables in the predictive equations. The conceptual model in Figure 4.1 includes measures of the at-home attributes of the sample household, as well as pattern and frequency of call attempts on sample units. We have seen that there do not appear to be great distinctions in interviewer calling times in urban versus other

Table 4.13. Coefficients of logistic models for NSHS and decennial match surveys predicting whether the household was ever contacted during the survey period (standard errors in parentheses)

Predictor	NSHS	Decennial match data	
		Reduced model	Full model
Constant	5.83**	4.80**	4.51**
	(0.49)	(0.28)	(0.32)
Social environment			
Urbanicity			
Central city	−1.53**	−0.59**	−0.49**
	(0.45)	(0.17)	(0.16)
Balance of CMSA	−0.96**	−0.30*	−0.27*
	(0.33)	(0.13)	(0.13)
Other	—	—	—
Population density[a]	−0.011	−0.012	−0.013
	(0.020)	(0.013)	(0.013)
Crime rate[b]	−0.0080	−0.0035	−0.0059
	(0.0053)	(0.0024)	(0.021)
Percent multiunit structures in block	−0.012	−0.0068*	
	(0.0061)	(0.0031)	
Housing unit			
Physical impediments	−0.67		
to access	(0.35)		
Large multiunit structure			−0.41*
(10 or more units)			(0.17)
Single-family home			0.32*
			(0.15)
Household			
Single-person household			−0.57**
			(0.12)
Children <5 years old in household			0.50**
			(0.15)
Household age			
All household members <30 years old			−0.11
			(0.15)
Mixed ages			—
All household members >69 years old			0.59**
			(0.18)

Notes: Dependent variable coded 1 = contact, 0 = noncontact. Coefficients for dummy variables for individual surveys in decennial match model omitted from table.

[a]Measured in thousands of persons per square mile.

[b]Measured in crimes per 1,000 persons.

*p < 0.05.

**p < 0.01.

areas. We suspect, therefore, that the models above suffer from inadequate inclusion of measures of at-homeness for urban areas. Specifically, absent from the equations are measures of commute time and time away from home for reasons of shopping, entertainment, etc. We suspect that, for any particular household composition, urban dwellers may be out-of-home more frequently for commuting to work, for shopping trips, etc. The inclusion of these variables would, we speculate, eliminate the lowered contact rates of urban dwellers.

4.7 SUMMARY

Relative to the decision to participate in a survey, the process of contacting a sample household seems simple, at least on the surface. In this chapter, we have tested parts of a model that notes that contactability is a function of physical impediments to accessing the unit, at-home patterns of the household, and the timing and number of interviewer visits to the unit.

The results of the chapter largely support this model. We were hampered most by the absence of measures of when the household was actually at home. Instead we examined socio-demographic attributes of the household logically correlated to at-home patterns to test the model indirectly. Most of the household composition indicators behave as the theory would assert. Households with many members are easier to contact, presumably because the chances are higher that at least one member is at home at any given moment. Households with elderly persons or young children are easier to contact because they tend to have someone at home at more hours in the day.

The lower rate of contacting sample households in urban areas was not fully explained by variables in the models. It is the case that some of the differences are due to the preponderance of multiunit structures, single-person units, and units with some physical barrier to access. However, it appears that other, unmeasured variables underlie the urban–rural contrasts on contactability. We suspect that true differences in the at-home rates predominate in these explanatory attributes. We have noted that urban dwellers are likely to have more time-consuming commutes to and from work, sehool, and other routine out-of-home activities. Time requirements for shopping and household errands may be higher in urban areas. Opportunities for out-of-home recreational activities may be larger and more highly valued, as a way to escape the density of urban life.

The measure of physical impediments to the sample housing unit was in many models a strong and pervasive influence on contactability. In the overall contact models (Table 4.13), there is evidence that some units with physical barriers to access offer temporary hurdles for the interviewer. Repeated callbacks on units can obtain information or invent protocols to conquer the barriers, permitting the interviewer to make contact with the unit. These no doubt include contacting building managers and security offices, seeking the assistance of neighbors, as well as less-sanctioned behavior like slipping through a locked apartment-house door by following a resident of the building. Informal conversations with interviewers suggest these and even more ingenious schemes are commonly used, once traditional contact attempts fail.

The measures of timing of calls proved simpler in their effects than earlier research implied. We found a rather strong, consistently positive effect of making calls in the evening. Weekend day calls were generally more productive than weekday day calls. There were few interpretable conditional effects of timing of calls. That is, there was no consistent result that pointed toward the efficiency of calling in the morning among cases that were not previously reached in the evening. Finally, there was evidence that the advantage of evening calling increases in the later calls on cases not yet contacted. It appears that cases not easily contacted lie disproportionately among those not at home during the day.

In sum, we have learned the following lessons in the analyses presented in this chapter:

- Locked apartment buildings and high-security housing developments complicate and slow down the process of contacting sample households. Some physical impediments cannot be overcome with repeated attempts at contact, but many seem to be overcome in two call attempts.
- Households with young children or elderly adults tend to be at home more often and thus are more easily contacted.
- Households with more adults are easier to contact that those with fewer adults, probably because of the increased times during which at least one person is at home.
- Evening calls are consistently more productive than calls during the daytime.
- The findings above (physical impediments, household composition, and times of calling) partially explain the prior observations of lower contactability in urban versus rural areas. Full explanation of urbanicity effects await better measures of at-home times of the sample households.

Later chapters of this book examine the process of cooperation with a survey request, given contact. This other source of nonresponse needs to be contrasted with that of noncontact. Many survey researchers retain hopes that the characteristics of the noncontacted cases will, in some sense, be different from those of the refusal cases. The result they seek is that the net nonresponse error will be negligible, despite the fact that noncontacts and refusals, singly, might have quite different values on the survey variables than do the respondents. In Chapter 5 the reader will be able to compare the nature of predictive models of cooperation, given contact, with those of contact.

4.8   PRACTICAL IMPLICATIONS FOR SURVEY IMPLEMENTATION

Household surveys begin by sampling (either through a telephone number or an address) housing units occupied by persons who may be eligible for the sample survey. If some of the attributes of households related to difficulty of contact were observable, then survey designs might be altered to improve efficiency and reduce noncontact rates.

Which attributes are impediments to contacting a household are dependent on the mode of data collection. In face-to-face surveys these include locked apartment entrances, units with bars on the windows, security gates at subdivision entrances, or units with "no soliciting" signs. Area probability designs offer the chance for listers of the sample segments to observe some of those attributes. When those residential areas with high security issues are sampled, interviewers assigned the cases might be asked to call on those units first during the field period, in order to allow the maximum amount of time for negotiation of access to the units. For dangerous areas, interviewers can arrange to work in teams. It might be wise in cases of large developments with centralized access controlled by security personnel, to begin negotiation with the complex's authority as soon as listings are complete, even prior to sample selection of individual units in the development. Indeed, direct contact with the survey organization's leadership might be wise, in order to establish the importance of the access to the survey's goals. Sample units in such structures also might be candidates for switching from face-to-face to telephone mode. This would require efforts by interviewers to obtain home numbers for sample units. Finally, advance mailings might be tailored to such large developments, to emphasize the legitimacy of the survey request.

With telephone surveys, many of the physical impediments are neutralized, but technological analogues take their place. For example, answering machines can be used to avoid telephone contact with strangers. "Caller-ID" and "call-blocking" features of phone systems allow householders to restrict their telephone contact to those with whom contact is desired. Such cases require more effort at contact than others. Upon first sign of an answering machine, cases might be moved to a different calling procedure, one that places more calls in the evening (when machines are more likely to be turned off). Unfortunately, caller ID is invisible to the caller (although the caller can usually prevent their number from being displayed by entering special "marking" codes).

In random-digit dialed (RDD) samples, unlisted telephone numbers have a higher chance of being nonhousehold numbers (e.g., ringing without answer). They might be called earlier in the survey period in order to give interviewers time to resolve their status.

The other attribute that is strongly predictive of contactability is the age of persons in the household. Again, at the time of sample listing, some proxy evidence can sometimes be gleaned from observation. Small bicycles and toys are evidence of child household members and therefore of higher contactability for the household. Visible physical assistive devices (e.g., ramps for wheelchairs) might indicate the presence of an elderly person, or another likely to be at home more hours of the day. These observations can be recorded and used to schedule calls on these sample units. Such units could be scheduled for first visits later in the survey period.

Collecting call record data and observations on the neighborhood can help survey managers guide interviewers in their calling strategies and make cost–error tradeoffs about additional contact efforts in the latter part of a data collection period. With a computer-assisted interviewing (CAI) system, these data can be cheaply obtained in almost real-time, permitting more informed decision making during the data collection process.

CHAPTER FIVE

Influences of Household Characteristics on Survey Cooperation

5.1 INTRODUCTION

This chapter shifts attention to another stage of survey participation. It examines only those sample households the interviewer successfully contacts, and seeks to understand why people do or do not cooperate with the face-to-face interviewer request. The majority of the rest of the book concentrates on this cooperation step. We will see that it is the most complex theoretically, involving the interaction of socio-economic and demographic, social and cognitive psychological, and interactional influences. It is also the component of survey participation that is increasingly problematic in many societies.

As Figure 5.1 reveals, our theoretical perspective asserts that effects on a sample person's behavior arise from multiple levels of aggregation of psychological and sociological phenomena. This chapter, however, focuses on only one of the blocks of hypothesized influences on cooperation—that associated with relatively fixed attributes of the sample household.

We begin with household-level attributes because the survey-methodological literature contains more analysis and commentary on that level of measurement than on any other type. Many times these characteristics of sample units are recorded on the sampling frame or are observable by interviewers. This permits easy comparison of respondents and nonrespondents on these measures. For that reason, the household level will be familiar terrain for many readers.

By starting at the householder block of influences we can also set the stage for later chapters. There we will address how much of the influence of the household level remains, controlling for effects of other variables at different levels of aggregation (e.g., the social environment, the interaction between householder and interviewer).

Although the survey literature is replete with socio-demographic correlates of nonresponse, the literature suffers from varying definitions of nonresponse, from reliance on bivariate results, and from an overemphasis on case studies. These all

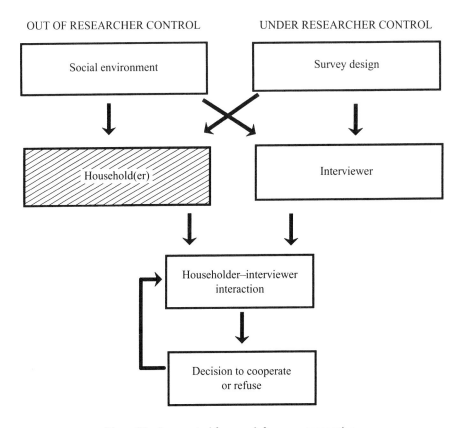

Figure 5.1. A conceptual framework for survey cooperation.

contribute to a lack of conceptual integration of these findings. This chapter uses the theoretical framework outlined in Figure 5.1 as a tool for organizing the findings, both from our own data and from past research on household and householder correlates of cooperation.

We do not hypothesize that many of the socio-demographic variables are direct *causal* influences on cooperation. Rather they are indirect measures of what are essentially social psychological constructs. Furthermore, the mapping of measures into relevant concepts is imperfect. For example, older persons might exhibit *higher* rates of cooperation because of greater perceived civic duty to respond. At the same time, they might exhibit *lower* rates because of increased fear of crime, relative to younger persons. Age thus maps onto two constructs with different hypothesized effects on cooperation. The decision to participate is based on the combined influence of interacting factors (not all considered at the time of the request), some facilitating cooperation, others constraining or mitigating against it.

We view these variables as setting the context within which survey requests are interpreted. Variables like age, race, and socioeconomic status are useful to us in understanding survey cooperation, we believe, to the extent that they are related to certain shared life experiences. The life experiences of import are those shaping the interpretation of a request from a government agency, an educational institution, or a commercial firm to provide private information about persons in a sample household. The shared experiences of socio-demographic groups may produce various predispositions to those requests and reflect features of their current lifestyles that affect how they react to such a request.

We expect the effects of these attributes to be specified by characteristics of the request (e.g., survey topic, agency of the request). For example, we might expect that those who are largely dissatisfied with the service of a firm will be reluctant to respond to a customer satisfaction survey from that firm (see Trice and Layman, 1984), but would have no such reluctance for some other survey. These attributes should not be totally causal of the outcome of the request. Our theory says their effects can be modified by the behavior of the interviewer and by which parts of the survey request are made most salient to the householder.

This chapter introduces the reader to a new set of data to test the theoretical notions discussed. For orientation purposes, a review of Chapter 3, Section 3.4, describing the decennial census match project, is recommended.

To remind the reader, we pooled data across six surveys in analyzing the decennial census match project. Further, at various points in this chapter we examine data from the one-sixth sample (or census long form). Even with the pooled data, this severely restricts the statistical power to detect differences. In addition, we again note that the data are household-level rather than person-level data. Finally, as discussed in greater detail in Chapter 3, the six surveys we examine all have relatively high cooperation rates. This means that we focus on relatively small percentage-point differences.

The chapter is organized about a set of theoretical constructs that help describe the influences on householders' decisions. They range from constructs endemic to rational-choice models of survey participation decisions to those that are much more social psychological in nature. We start by using the notion of opportunity costs that householders must weigh in agreeing to spend their time responding to a survey interview (Section 5.2). We then move to perspectives involving social exchange concepts, which focus on perceived obligations owed to the interviewer, the survey organization, the sponsor of the survey, or the beneficiaries of the survey (Section 5.3). Then we examine a related construct, that those most connected to social institutions in the society would tend to cooperate with surveys, especially those viewed as information collections in support of those institutions (Section 5.4). To complete the theoretical discussion, the chapter proceeds to examine effects of authority (Section 5.6) and other influences (Section 5.7). The chapter ends with multivariate models combining indicators of several concepts in order to measure their marginal impact on cooperation.

5.2 OPPORTUNITY COST HYPOTHESES

A fully "rational" view of the decision making of a prospective survey respondent would have him or her weigh all the costs of participation against the benefits of participation, with the outcome of this calculus being a decision one way or the other. The costs of the participation would, in this perspective, include the time required to complete the interview, the lost opportunity to perform other activities, the cognitive burden incurred in comprehending and answering the survey questions, and the potential embarrassment of self-revelations that the questions require. The benefits of participation might include the avoidance of alternative, more onerous tasks, the satisfaction of contributing to a socially useful enterprise, the enjoyment of thinking about novel topics, the pleasure of interacting with the interviewer, the gratification that one's opinion was sought by those in authority, and the satisfaction of fulfilling a perceived civic duty, among others.

This perspective is common to a rational choice theory of decision making and to a "central route" protocol for assessing the validity of the interviewer's arguments for participation (Petty and Cacioppo, 1986). It is *not* compatible with the notion that much decision making about survey participation is likely to be based on temporary features of the home situation, peripheral aspects of the interviewer request, and minor components of the survey that become disproportionately salient in specific survey interactions. This latter viewpoint leads to hypotheses that much of the variation in likelihood of participation is explained by what particular features of the request situation become most salient to the householder.

While we do not subscribe fully to this theory of a deliberate and considered view of decision making, it is important to inquire whether it receives any empirical support. This section concentrates on one component of this perspective—the effects of the amount of discretionary time for the sample household on survey cooperation.

All other things being equal, the burden of providing the interview is larger for those who have little discretionary time. Time limitations of the household should affect both contact and cooperation. Those with less discretionary time are less likely to be found at home and, when they are found, less likely to feel free to participate in a survey. Not only are there societal level changes (such as increased labor-force participation of females) over the last few decades that may be contributing to such phenomena (see Chapters 4 and 6), but there is also individual variation in discretionary time available that may be revealed in survey data. In fact, many of the survey households we approach tell us this—witness the relatively large proportion of time constraint reasons provided by householders in initial interactions with interviewers (see Chapter 8).

Large households have increased likelihood of finding someone at home (thereby reducing the noncontact portion of nonresponse). For surveys using a household informant (as opposed to the selection of a random respondent within the household), or permitting proxy reporting, larger households should also present a larger substitution pool, increasing the likelihood that at least one person will have time for the interview. This hypothesis stems from the observation that the tasks required to maintain a household (cooking, cleaning, bill paying, and so on) do not increase proportionally to the size of a household. Larger households share the duties among household members, freeing each for other pursuits.

Is there empirical support for this aspect of a rational choice approach to survey cooperation? A number of studies (e.g., Kemsley, 1975; Paul and Lawes, 1982; Rauta, 1985; Redpath and Elliot, 1988) report higher response rates among larger households, as we hypothesize. However, many of these report bivariate results (i.e., they do not control for age, presence of children, and other factors), and do not distinguish between noncontact and refusal components of nonresponse. In examining the effects of household size, a distinction should also be made between the number of adults (substitutability) and presence of children (social isolation, discussed in Section 5.4). However, Barnes and Birch (1975), Comstock and Helsing (1973), and Smith (1983) all report a positive relationship between cooperation rates (given contact) and household size, both for surveys with a randomly selected respondent (e.g., Smith, 1983) and for those with a household respondent (e.g., Barnes and Birch, 1975).

One exception to this trend is reported by Foster and Bushnell (1994), who find decreasing cooperation by household size for the British Family Expenditure Survey. However, this survey requires participation by *all* adult members of the household to be considered a responding unit. Thus, this contrary result may be explained by the increased burden associated with larger households, rather than the substitutability of household members.

For three of the surveys in the match study (the Current Population Survey (CPS), the National Health Interview Survey (NHIS), and the National Crime Survey (NCS)), a household respondent was sought for the initial household information, after which person-level data may be sought from individuals within the household. In such cases, households with larger numbers of adults would have greater substitution pools for the initial informant, reducing the potential burden on individual household members. However, Table 5.1 with the pooled data shows no monotonic increase in cooperation as the number of adults gets larger. However, we do find, consistent with the past literature, a larger contrast between single-person households and other household types. Single-person households have lower cooperation rates.

Thus far, we see little support for the hypothesis of increasing cooperation as the number of adults increases. Perhaps the indicator of the total number of adults is too weak an indicator of the discretionary time hypothesis. If we use, instead, measures of the time commitments of the adults, we may get closer to the concept of opportunity costs for the survey interview.

One measure we have on households from the census long form is the number of

Table 5.1. Cooperation rate by number of adults in household

Number of adults	Cooperation rate	(Standard error)
One	94.3%	(0.42)
Two	95.5%	(0.25)
Three	94.5%	(0.68)
Four or more	96.0%	(0.76)
$\chi^2 = 7.65$, $df = 3$, $p = 0.05$		

Table 5.2. Cooperation rate by presence of nonworking adults in household

Presence of nonworking adults	Cooperation rate	(Standard error)
One or more nonworking adults	95.0%	(0.73)
No nonworking adults	95.0%	(0.64)
$\chi^2 = 0.005$, $df = 1$, $p = 0.94$		

adults who do not have a job outside of the home. Households where no one works would, presumably, have higher time availability for the survey interviews. Based on census match data from the United Kingdom, Redpath and Elliot (1988) and Kemsley (1975, 1976) report increasing response rates with more employed adults in the household (not controlling for household size). However, the matched sample data once again show little support for this (see Table 5.2), with both households with and without a working adult achieving about a 95% cooperation rate.

Finally, we have available other indicators of the amount of discretionary time for the household—the number of minutes of commute time and number of hours at work. These variables are available only on the sample cases that received the long form of the 1990 census. These are even more direct indicators of the amount of time at home and thus the amount of discretionary time to give to surveys. We created three different measures of the amount of time away from home, presented in Table 5.3. For each of these measures, we expect that the mean time away from home (measured in hours per week) would be higher for noncontacted than contacted households, and higher for households producing refusals than those leading to interviews. While we find significant effects in the expected direction for contacts, we find no differences for cooperation. Kennickell (1997) found similar effects of average commute time on contact and cooperation in the Survey of Consumer Finances.

Table 5.3. Means of three discretionary time indicators by contact versus noncontact and interview versus refusal (given contact) (standard error in parentheses)

Hours away per week	Contact	Noncontact	Cooperation, given contact	
			Interview	Refusal
Mean hours away from home for	42.6	45.2	42.7	42.2
householders who work	(0.56)	(1.27)	(0.58)	(2.17)
One-tailed t-test	$p = 0.03$		$p = 0.42$	
Least hours away among	38.5	42.2	38.5	39.2
those who work	(0.63)	(1.64)	(0.66)	(2.22)
One-tailed t-test	$p = 0.02$		$p = 0.38$	
Least hours away for those who	39.7	43.0	39.6	42.1
work outside the home	(0.63)	(1.57)	(0.65)	(1.97)
One-tailed t-test	$p = 0.03$		$p = 0.12$	

After viewing empirical results on several indicators of discretionary time, we find little support for the hypothesis that reduced time at home leads to reluctance to cooperate with surveys. Further, when we combined these indicators in a multivariate model predicting cooperation, no further insights were gained. What better indicators could we suggest for tests of the hypothesis? We have examined no measures of the household's obligations of time away from employment tasks. Some persons have work obligations, even at their home site; some have commitments to friends and relatives that might raise the opportunity costs of a survey interview; and others, primarily the self-employed, may lose income producing opportunities by participating in interviews. Lindström (1983), Rauta (1985) and Redpath (1986) all report lower response rates among the self-employed. We have thus omitted in our tests of the discretionary time hypothesis a set of measures of what alternative activities the household might pursue, in the absence of providing a survey interview.

We suspect that the impact of this measurement weakness is an attenuation of the effects of the measures. When we compare households who are home often to those who are absent more often (but who were contacted by an interviewer), we have combined those with many other attractive time-use options at home with those having few other attractive time-use options at home.

Going beyond the data, we speculate that empirical support would be stronger if we introduced measures of interviewer behavior as well as householder behavior. Limited householder time availability should increase refusal rates, if interviewers do not effectively communicate their willingness to conduct the interview at any time the respondent might be available. Interviewers are trained to offer such flexibility to sample persons, in order to reduce nonresponse. One interpretation of these results is that most interviewers successfully do this. Indeed, examining tape recorded telephone interactions, it appears that interviewers are more adept at reacting to householder pleas of time pressures than other sources of reluctance (see Chapter 8). (Note this implies that for surveys with limited callbacks, those with reduced discretionary time are disproportionately nonrespondent.)

5.3 EXCHANGE HYPOTHESES

In the next two sections we are going to entertain two extensions of the rational decision-making approach, each adding sociological or social psychological content to the perspective. The first is heavily dependent on social exchange theories; the second, on social isolation theories (treated in Section 5.4). Social exchange considers the perceived value of equity of long-term associations between persons, or between a person and societal institutions (Blau, 1964).

Social exchange hypotheses have been popular in the discussion of survey participation in the literature (see Dillman, 1978; Goyder, 1987). Social exchange may operate at any number of levels, ranging from exchange in dyadic interactions (e.g., reciprocation in social interaction; see Chapter 2) to exchange relationships between an individual and the larger society (e.g., civic duty in return for personal welfare). Dillman (1978) focused on social exchange within a relatively closed system (survey organization and householder), with relatively small gestures on the part of the survey organization (personalized letters, token incentives, reminder letters) hypothesized to evoke a reciprocating response from the householder. On the other hand, Goyder's (1987) discussion of exchange evokes a wide array of obligations and expectations over an extended period of time between an individual and various institutions of society.

Central to all conceptualizations of social exchange is the notion that, unlike economic exchange, all social "commodities" (ranging from measurable entities, such as time and information, to less tangible socio-emotional goods, such as approval) are part of an intuitive bookkeeping system in which debts (obligations) and credits (demands, expectations) are documented. Thus, virtually any relationship can be described in exchange terms.

Whether the social exchange perspective applies to face-to-face survey requests may depend on whether the householders connect the request to some "relationship" they have or will have with another person or institution. For single-contact surveys, conducted by organizations with no prior connection to the household, the exchange perspective may have little value. For surveys conducted by organizations with an ongoing relationship with the households, however, the exchange notions may be useful.

When given a request for a government survey, householders may consider the cumulative effect of multiple government contacts. This might include the full weight of past relationships with the agency or organization making the survey request (or the broader class of institutions it represents, e.g., government, academia, commercial interests). Those receiving fewer services from the government may feel less need to reciprocate. Since government services are differentially provided across economic groups, indicators of socioeconomic status should broadly reflect exchange influences on survey participation.

Unfortunately, social exchange theory can lead to two different hypotheses about differences in cooperation by different socioeconomic status (SES) groups. First, one can argue that lower SES groups may have the greatest indebtedness to the government for the public assistance they (or others in their community) may receive. Surveys funded by the government might seem like another encounter with an institution to whom they are indebted. In contrast, those at the high end of the SES scale may have the least need for government services, and least sense that they owe the government any sort of repayment. In fact, they may resent government intrusion into their lives, and feel that the balance of exchange lies in their favor. This suggests a monotonically negative relationship between socioeconomic status and cooperation propensity.

Alternatively, those in the lower SES groups may believe that in relationships with those more fortunate they are routinely unjustly disadvantaged. Survey interviewers, to the extent they are viewed as agents of those more fortunate, may evoke memories of that exchange history, and the householder may tend to refuse the survey request. Those in the highest SES groups might perceive similar long-run feelings of inequity, that large-scale social institutions repeatedly target them for contributions of time and money, despite having over the years contributed little to their achievement of the SES status enjoyed. Both groups feel relatively deprived in the relationships and tend to refuse survey data collectors. This suggests a curvilinear relationship between SES and cooperation.

The two hypotheses are distinguished by what relationship is cognitively most accessible and judged relevant by the low SES householder faced with a survey request. Is this a request from a person better off than themselves, again wanting something from them, but offering relatively little in return? Is this a request from the government?

Of the two alternative hypotheses, the more popular in the existing literature may be the latter, positing curvilinear effects. However, the only evidence in the literature supporting a curvilinear hypothesis for income finds effects in the opposite direction to that expected based on social exchange theories. Smith (1983) found that refusers were more likely than respondents to be in the middle-income category, and less likely to be in the low or high categories. However, these data are based on interviewer estimation of gross income categories of nonrespondents, on which there is 31% missing data among refusers. There is support for higher cooperation among low SES groups. DeMaio (1980), in a government survey, found significant differences in refusal rates by income, with middle-income households being most likely to refuse, while low-income households were least likely to refuse. Weaver, Holmes, and Glenn (1975) also found highest refusal rates among the middle-income group in a telephone survey of city employees in San Antonio, and the lowest rates of refusal among the lowest income group. However, there are also contradictory findings. Benus and Ackerman (1971) found the lowest nonresponse among the highest income group. A number of studies report refusal rates decreasing with other indicators of socio-economic status, such as social class (Lindström, 1983; Redpath and Elliot, 1988; Redpath, 1986) and property value (Goyder, 1987; Goyder, Lock, and McNair, 1992). Analyzing data from the 1972 Swedish Household Income Survey, Lindström (1983, p. 44) found that "there is a tendency toward higher income and less social assistance benefits among the respondents than among the non-respondents." Foster and Bushnell (1994) find significant bivariate effects of socioeconomic status (class) on cooperation for three of the five surveys they matched to the 1991 U.K. census, but these effects largely disappear in multivariate analyses. Ekholm and Laaksonen (1990) also report no effect of income on response to the Finnish Household Budget Survey in multivariate analyses.

The most appealing indicator of SES would probably be a combination of education, occupation, and personal and family income. All of these variables are measured only for the one-sixth sample sent the long form of the decennial census. Further, family income and occupation suffer severe item missing data problems. For example, income on the decennial form is missing for over 30% of the households. Hence, we are forced to deal with imperfect measures of SES, available on the short form of the census.

Economic indicators.   One gross measure of SES status available on all cases is the nature of housing costs faced by the household. For renters, this is an estimate of the

Table 5.4. Coefficients from a logistic model predicting cooperation versus refusal rate by housing tenure and costs

Predictors	Coefficient	(Standard error)
Constant	3.38**	(0.13)
Owner occupied (1 = yes)	−0.24	(0.56)
Monthly rent for renters[a]	−0.070**	(0.025)
House value for owners[b]	−0.017**	(0.0049)

Notes: Dependent variable coded 1 = interview, 0 = refusal. Coefficients for dummy variables for individual surveys omitted from table.
[a]Measured in units of $100.
[b]Measured in units of $10,000.
**$p < 0.01$.

mean monthly rent, while for home owners it is a self-estimate of the house value. We thus use three variables in a logistic regression to measure the combined effects of housing costs on cooperation. Table 5.4 shows lower cooperation rates among those renting or owning more expensive housing units. (No changes in conclusions occur if the model includes dummy variables for the different surveys.) Using various transformations of these variables, we fail to find support for the curvilinear effect of SES with these data. We thus find a general decline in cooperation with increasing SES, in support of the first hypothesis forwarded above. One argument for the higher cooperation found among lower SES households (as measured by housing costs) may be the government sponsorship of the surveys in our sample, and the greater perceived reliance of this group on government largesse, leading to greater felt social debts.

Education. Another traditional SES indicator is education, available only on the subsample of cases completing the long form of the census. To the extent that surveys are viewed as research/information gathering, those with higher educational achievement might have benefited from such efforts, might appreciate the utility of such efforts, and thus tend to cooperate. (Note that this is a hypothesis in a different direction from that made above for higher economic groups.) This hypothesis should apply both to government-sponsored surveys and those conducted by academic organizations, and even to some surveys conducted for commercial purposes.

The past literature tends to show that lower education groups disproportionately fail to participate in surveys (Benson, Booman, and Clark, 1951; Dohrenwend and Dohrenwend, 1968; Foster and Bushnell, 1994; Kemsley, 1975, 1976; O'Neil, 1979; Wilcox, 1977). Using a measure of education of reference person (from the census long form), our data tend to show somewhat higher cooperation among lower-education groups, particularly those with less than high school education (see Table 5.5) (statistically significant only at $p = 0.10$). This is consistent with the finding on SES as measured by housing costs. Given that the surveys we include in the match are conducted or sponsored by the federal government, this again may suggest support for the notion of indebtedness to government among those of lower SES.

Table 5.5. Cooperation rate by education of reference person

Education	Cooperation rate	(Standard error)
Less than high school	97.0%	(0.78)
High school	94.2%	(1.03)
Some college	95.1%	(0.87)
Completed college	93.9%	(1.21)
$\chi^2 = 6.2$, $df = 3$, $p = 0.10$		

Government Transfer Payments. A more direct measure of exchange "debts" owed to the government may be whether the household currently receives any public assistance from a governmental agency. From an exchange perspective, we would expect that those households currently receiving benefits would be more willing than others to comply with a survey request from a government agency. We are again restricted to a question from the census long form for this analysis, and find no support for this hypothesis in the data (see Table 5.6).

We end this section with a test of the combined effects of those socioeconomic measures that have some marginal bivariate effect on cooperation. For the exchange hypotheses, this is housing costs and education. Given that the latter is measured only on the long form, whereas the former comes from the short form, we do this in three stages. First we model the effect of the housing cost variables only, using full data (decennial short form data). This is the first model in Table 5.7. We see that housing costs have a significant negative relationship with cooperation, as shown earlier. This relationship generally holds when restricted to long-form cases only (second model in Table 5.7). The third model in Table 5.7 shows the combined effect of housing costs and education on cooperation. The effect of housing costs does not change much with the inclusion of the education measure, while the latter set of indicators does not reach statistical significance ($p = 0.10$).

What can we conclude from these various tests of social exchange concepts? We have comments both on the theory and our indicators of its constructs. First, with regard to the theory, we initially note that it implied two different empirical relationships with SES. This alone portended low falsifiability of the theory in this case. (If low SES households had high or low relative cooperation, the theory provides an explanation.) Blau (1964, p. 93) distinguishes social exchange from economic exchange by stressing that the former entails "diffuse" and "unspecified" obligations (see also Goyder, 1987, p. 170). But therein may lie a key drawback in applying so-

Table 5.6. Cooperation rate by receipt of public assistance

Receipt of public assistance	Cooperation rate	(Standard error)
One or more household members receive assistance	95.6%	(0.90)
No household members receive assistance	94.7%	(0.58)
$\chi^2 = 0.65$, $df = 1$, $p = 0.42$		

Table 5.7. Coefficients for logistic models predicting cooperation versus refusal using social exchange indicators, simple model using short form variables, expanded model using short and long form variables, by sample type (standard error in parentheses)

Predictor	Simple model		Expanded model (long-form cases)
	Total sample	Long-form cases	
Constant	3.38**	3.56**	3.66**
	(0.14)	(0.33)	(0.46)
Owner occupied	−0.24	−0.28	−0.25
	(0.16)	(0.39)	(0.42)
Monthly rent for renters[a]	−0.070**	−0.031	−0.056
	(0.025)	(0.059)	(0.069)
House value for owners[b]	−0.017**	−0.036**	−0.035**
	(0.0050)	(0.011)	(0.013)
Education			
Less than high school			0.29
			(0.36)
High school			−0.43
			(0.30)
Some college			−0.089
			(0.27)
Completed college			—

Notes: Dependent variable coded 1 = interview, 0 = refusal. Coefficients for dummy variables for individual surveys omitted from table.
[a]Measured in units of $100.
[b]Measured in units of $10,000.
**$p < 0.01$.

cial exchange theory to survey participation: it may be *too* diffuse or unspecified to lead to testable hypotheses of specific behavior. Almost every behavior can be interpreted *post hoc* in an exchange context, but to be able to predict behavior would require knowledge of the accumulation of exchanges over an undefined period of time, and using exchange criteria (valuation of acts) that may differ from one individual to another. Social exchange theory thus too infrequently leads to testable hypotheses that can be refuted.

Second, with regard to the indicators, we find support in our data for the notion that those in high SES households cooperate less with surveys than those in the low SES groups. The indicators available are housing costs and education. SES, as a concept, can at most be an indirect indicator of perceived exchange relationships. Direct indicators would include perceptions by the householder of previous or future obligations to the interviewer, the agency collecting the data, or those potentially aided by the survey results.

The appeal of the social component to cost–benefit analyses remains. However, a narrowing of the scope of the candidates exchanges may be more appropriate to survey requests. We believe the more narrow concept of reciprocation is useful, especially for understanding the effects of survey design decisions such as the use of incentives or questionnaire length (see Chapter 10) and in the context of the interaction between interviewer and respondent (see Chapters 8 and 9).

5.4 SOCIAL ISOLATION HYPOTHESES

Theories of social connectedness, isolation, or disengagement are related to those of social exchange. Social exchange influences arise from ongoing relations between two actors, or more broadly between an actor and a social group. The influence arises because of the ubiquitous norm that benefits to such relationships must be roughly equally shared over long periods of time. On the other hand, social isolates are out of touch with the mainstream culture of a society. They tend not to feel the influence of norms of that dominant culture, but behave in accordance with subcultural norms or in explicit rejection of those of the dominant culture.

The approaches are related conceptually. For example, a long history of inequitable social exchange relationships between a subgroup and the larger society may lead to the development of a subculture that explicitly fails to include the norms of the larger culture. If a person feels cheated by larger society because of their membership in a group, he or she might tend to ignore the norms of the larger society. This logic has been applied to findings of lower response rates among racial and ethnic subgroups. Similarly, the absence of ongoing relationships between one group and the larger society will lead to the absence of shared norms. This logic has been applied to findings of lower response rates among the elderly (Glenn, 1969; Mercer and Butler, 1967).

Those with strong feelings of social isolation will not be guided by norms of the dominant culture. Survey researchers sometimes have noted that feelings of "civic duty" prompt survey participation, especially when the agency collecting the data represents an important social institution (e.g., government or academia). Those who are alienated or isolated from the broader society/polity would be less likely to cooperate with survey requests that represent such interests. To the extent that large-scale national surveys are a tool of the central institutions of society, those more at the periphery of society would feel less obligation to participate.

Another rationale behind this set of hypotheses is the view that surveys are inherently "social" events because they contribute to knowledge about the full society. To the extent that sample persons feel cohesion with the defined population, cooperation will be enhanced. This is one explanation for the relatively higher rates of cooperation in organization membership surveys. Cooperation in household surveys may be seen both as an obligation of membership in a society or group, and indeed as an affirmation of one's membership.

There are both structural and social psychological aspects of alienation or social isolation. Some groups (e.g., the "underclass"), by virtue of their position in society, may not be bound to the larger society to the same extent as others. This may be reflected both in "input alienation" (e.g., powerlessness, lack of political efficacy) and "output alienation" (lack of trust in government or in the responsiveness of government institutions) (see Southwell, 1985; Weatherford, 1991). Both would lead, we assert, to lower levels of cooperation with surveys representing those agents of government. This view equates survey participation with other acts of political or social participation such as voting (see Couper, Singer, and Kulka, 1997; Mathiowetz, DeMaio, and Martin, 1991). The attachment of alienated groups (often defined in terms of race and/or socioeconomic status in the United States and by class in other countries) to society is believed to be weak, and such groups are posited to be less likely to participate in a variety of social and political activities, including responding to surveys.

In the previous section, we examined housing costs and education as indicators of social-exchange influences. They could also be used to provide insight into the social isolation hypothesis. Social isolation theory suggests the opposite effects of SES than do social exchange theories. The lower SES groups should be likely to be alienated from the central institutions of society, and resentful of their dependence on the government. The higher SES group may perceive themselves to hold an important place in society, and may as a consequence have a greater sense of civic obligation or recognize the value of survey data for the common good. This suggests a positive relationship between SES and cooperation propensity. The reader will recall that a negative relationship was found, and thus the data refute isolation theory applicability to survey cooperation, at least as indicated by SES.

5.4.1   Demographic Indicators of Social Isolation

As with many social psychological theories of survey cooperation, our tests of concepts of social isolation will depend on proxy indicators that are socio-demographic characteristics of sample households.

Race and ethnicity.   Social isolation has been a popular hypothesis with regard to the behavior of racial and ethnic subgroups in surveys. However, there is little evidence in the literature that nonWhites cooperate with survey requests at different rates than Whites. O'Neil (1979) found lower rates of resistance among Blacks to a telephone survey in Chicago, while Hawkins (1975) reports similar results for a face-to-face survey in Detroit (see also Brehm, 1993). Although Weaver, Holmes, and Glenn (1975) found more problems of accessibility among Blacks; they obtained refusal rates of 4% for Blacks, 10% for Mexican-Americans and 15% for White Anglos. Both DeMaio (1980) and Smith (1983) fail to find effects of race on cooperation rates. As we see in Table 5.8, we find, if anything, higher rates of coop-

Table 5.8. Cooperation rate by race/ethnicity of reference person

Race/ethnicity	Cooperation rate	(Standard error)
Hispanic	96.9%	(0.59)
Black nonHispanic	95.8%	(0.46)
Other	94.8%	(0.68)
$\chi^2 = 11.72, df = 2, p < 0.01$		

eration among these minority groups. We will examine the influence of race/ethnicity later, in a multivariate model.

Age. Social isolation has also been used to explain the behavior of elderly persons. The disengagement hypothesis, first articulated by Cummings and Henry (1961; see also Glenn, 1969; Krause, 1993; Mercer and Butler, 1967) has been used to explain lower cooperation rates among the elderly. For example, in the political context, both Abramson (1983) and Jennings and Markus (1988) find declines in political involvement and efficacy among the elderly. A mitigating factor, however, may be that the current cohort of elderly in the United States have a greater sense of civic duty, and are more likely to perceive government as making legitimate demands on its citizenry. Thus, lower levels of political participation among the elderly may be attributable to such factors as encroaching infirmities and the likelihood that older cohorts have less education and lower SES, rather than higher levels of disengagement (see Bennett and Bennett, 1986; Rosenstone and Hansen, 1993).

As we've seen in Chapter 4, elderly persons tend to be at home more frequently than other age groups, because of their lower employment rates and, at advanced ages, reduced mobility. However, the group also is disproportionately classified as "other noninterviews" because of health problems preventing their survey participation. These two nonresponse categories thus counteract one another. It is important, therefore, whenever possible in citing past research findings to distinguish between overall nonresponse rates, contact rates, cooperation rates, and other noninterview rates.

First, let us examine what the literature finds regarding overall response rates by age. We would expect this literature to show mixed age differences, as a function of noncontact rates. That is, in surveys with few callbacks, relatively more of the contacted cases would be elderly, generating higher response rates for elderly than for nonelderly, other things being equal. A mixed picture is exactly what the literature shows.

The most striking demonstration of the age effect is from British data starting with Kemsley, for the 1971 Family Expenditure Survey (FES) (Kemsley, 1975) and 1971 National Food Survey (NFS) (Kemsley, 1976) (see Figure 5.2). Similar declines in overall response rates by age are found by Norris (1987), Rauta (1985), Redpath (1986), and Redpath and Elliot (1988), all on U.K. government surveys.

In a census match study much like ours, Foster and Bushnell (1994) find much higher nonresponse rates for older heads of households on NFS and FES in 1991, but not for the Labor Force Survey, General Household Survey, and National Travel Survey. In fact, the lowest response rate for the LFS is among those 16–24. This may reflect a survey design effect. Both NFS and FES are fairly burdensome surveys involving both interviews and diaries (2-week expenditure diary and 1-week food consumption diary).

Christianson (1991) found response rates to be highest among those 65–79 and lowest among those 15–24 in Swedish TV audience surveys conducted by telephone. However, nonresponse to these surveys appears to be dominated by noncontacts (5/6ths of all nonresponse). Similarly, in summarizing the results from a num-

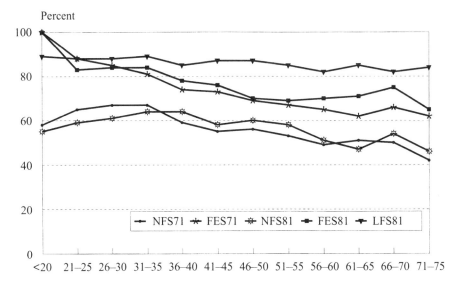

Figure 5.2. Response rates by age of household head for five British surveys

ber of face-to-face surveys in Japan, Sugiyama (1991) found that response rates tend to increase with age. In a study of nonresponse to the Swedish Labor Force Survey, Kristiansson (1980) found no tendency for the elderly to have lower response rates. Paul and Lawes (1982) demonstrate the importance of controlling for household size when examining the impact of age on nonresponse. They find that for each household size group considered separately, individuals 65 or older exhibited the highest response rates, while this age group had the lowest response rates overall. This results from the fact that the majority of persons 65 or older live in households of size 1 or 2, where the response rate was the lowest (Paul and Lawes, 1982, p. 62).

What does the literature suggest if we focus exclusively on refusal rates? Refusal rate studies produce fewer mixed findings on age differences. There are a set of studies showing higher refusals among older persons (Brown and Bishop, 1982; Dohrenwend and Dohrenwend, 1968; Goyder, 1987; Hawkins, 1975; Herzog and Rodgers, 1988; Smith, 1983; and Weaver, Holmes, and Glenn, 1975). Comstock and Helsing (1973) found that refusals increased with age in bivariate analyses; however, when controlling for other variables in multivariate analyses, a reversal of this trend was found, with refusal rates of 16% for those 25–39, 11% for those 40–54 and 10% for those over 55. Benson, Booman, and Clark (1951) also found no effect of age on refusals in a face-to-face survey in Minneapolis.

In short, there appears to be more consistent support for higher rates of refusals among the elderly than for overall lower response rates. The literature also suggests that we might expect different results in multivariate models (especially controlling on household size) than in bivariate models.

Given that we have household-level data from the match, we can examine the effect of age in different ways. One is to look at age of reference person, as was done in the U.K. match studies. The cooperation and contact rates by age of reference person are presented in Figure 5.3. Tests of these effects show a significant relationship ($\chi^2 = 54.37$, $df = 7$, $p < 0.01$) between age and contact, with the likelihood of contact increasing with increasing age, with the most pronounced effect being for those households where the reference person (head of household) is under 25. The relationship with cooperation is not statistically significant, however ($\chi^2 = 12.0$, $df = 7$, $p = 0.10$), nor can any clear trend be discerned in the figure.

Another operationalization of the age variable would distinguish "young" households and "old" households from other types of households. In a survey seeking cooperation from any household members (as do most of those included in the match), the effect of age is likely to manifest itself through the person contacted in the household, rather than the reference person. Table 5.9 shows that "young" households have cooperation rates higher than other households (in which at least one person is between the ages of 30 and 69). However, social isolation hypotheses would have led us to expect lower cooperation rates for the old households versus the other two groups. Thus, with neither of these measures do we find support for the contention that elderly persons (or households) have lower cooperation rates.

A number of hypotheses can be advanced for why we don't find the expected effect of age. One is that older persons in the United States have a higher level of civic duty that mitigates the potential effects of age for government surveys. Another is that the increased legitimacy of government surveys (relative to those conducted by other organizations) may decrease the fear of criminal victimization disproportionately experienced by the elderly (see Chapter 6 and Section 5.5).

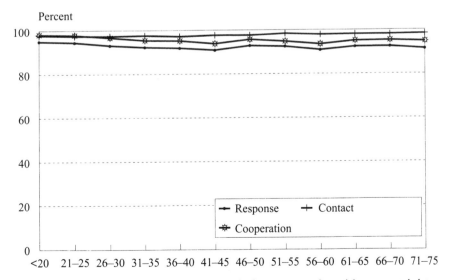

Figure 5.3. Cooperation and contact rates by age of reference person, decennial census match data.

Table 5.9. Cooperation rate by age composition of household

Household age composition	Cooperation rate	(Standard error)
All household members under 30	97.5%	(0.31)
Some household members between 30 and 69	94.6%	(0.25)
All household members 70 or older	95.1%	(0.66)
$\chi^2 = 45.12$, $df = 2$, $p < 0.01$		

In an attempt to explore this issue further, we have examined a number of interaction effects. For example, is fear of crime a factor; are older persons in inner-city areas more likely to refuse than their rural counterparts? Are elderly cooperation rates related to the composition of the interviewer labor pool (e.g., older female interviewers may be less threatening to elderly respondents)? Does topic salience play a role; that is, do older persons have relatively higher refusal rates on topics not directly affecting them (e.g., labor force participation) than those of more direct interest (e.g., health and retirement issues)? We tested these and other interactions, but failed to find any effects that provide plausible explanations for our findings on age. The finding also appears robust to other manipulations of the age variables. However, we will return to the age question again when examining multivariate models of household-level effects on cooperation (Section 5.6).

Gender. Another demographic variable commonly examined in nonresponse studies is the gender of the sample householder. Most all studies have found either no gender effect on cooperation or the tendency for males to have lower cooperation rates (Smith, 1983; Lindström, 1983). Some of the findings confound noncontact with refusals, and there are clear tendencies for males to be at home less frequently than females, who more often accept larger responsibilities for child care and other household duties.

There are several possible theoretical explanations for lower cooperation among males. Many of them are related to social isolation and role differentiation of males and females on that dimension. Females are more frequently telephone answerers in mixed sex households (Groves, 1990); they more often take on the role of maintaining social relations with friends and neighbors. Hence, when a request for information comes from outside the home, they are more accustomed to interacting with nonhousehold members from the home setting. As we mentioned above, we do not have measures at the person level in these data, so we cannot estimate the marginal effect of householder gender on cooperation. However, we can and did examine cooperation rates in all-female versus all-male households. Such a test is a purer effect of gender rather than gender-based roles. There was no evidence for different cooperation rates in these two types of households.

We suspect that gender-based roles lead to lower cooperation rates in surveys of randomly selected adult respondents when males are the sampled persons. We thus speculate that respondent rules disproportionately targeting males lead to lower response rates versus rules that permit any adult to respond.

Table 5.10. Coefficients of a logistic model predicting cooperation versus refusal using socio-demographic indicators of social isolation

Predictor	Coefficient	(Standard error)
Constant	2.72**	(0.057)
Race/ethnicity		
Black reference person	0.26	(0.14)
Hispanic reference person	0.52*	(0.23)
Other	—	
Age composition		
All < 30 yrs.	0.89**	(0.15)
Mixed ages	—	
All > 69 yrs.	0.088	(0.16)

Notes: Dependent variable coded 1 = interview, 0 = refusal. Coefficients for dummy variables for individual surveys omitted from table.
*$p < 0.05$.
**$p < 0.01$.

Combined effects of demographic variables. Table 5.10 presents a logistic regression model measuring the joint effects of race, Hispanic status, and household age composition. The table shows that Hispanics and young households (with everyone under 30 years of age) tend to exhibit higher cooperation rates. Elderly households are similar to households with mixed age groups; Black households resemble those of other races (dominantly White). To summarize the results of the demographic indicators of social cohesion, we find generally the opposite of those hypothesized. If minority ethnic or racial groups and the elderly are relatively socially isolated from those institutions connected with the survey request, there is no evidence that this alone affects their rates of survey cooperation. (If dummy variables reflecting the different surveys are included in the model, the coefficient for Hispanic households is reduced sufficiently to lose statistical significance at the traditional 0.05 level.)

5.4.2 Household Status Indicators of Social Isolation

The social isolation hypothesis also has both macro and micro features to it: at a macro level, the connections involve the polity/society as whole; at a micro level, "community" ties might generate civic duty that has impacts on requests affecting the locality. Thus, groups considered isolated at a macro level (e.g., Southeast Asian immigrants, Blacks) may be part of highly cohesive local communities, in which a local survey focusing on community needs may receive a very different reaction than a national survey on topics deemed important by the central institutions of society.

The disengagement hypothesis, on the other hand, refers to an individual's withdrawal from society at both macro and micro levels. Some indirect indicators of social isolation might be derived from various properties of households. These might include whether the household is a single-person household (those living alone tending to be less socially integrated), whether there are children in the household (those with children having higher social integration through schools and friends networks), whether the household had moved recently (those more transient having fewer community roots), and whether the household lives in a large multiunit structure (greater transiency, less contact with neighbors). Glorioux (1993, p. 171) notes that "having children seems above all to have a strong positive effect on defining activities in terms of duty." He continues, "the more people are alone, the less they define their activities in terms of 'obligation,' 'social alliance,' 'duty' and 'instrumentality.' The family, on the other hand, seems to be the most fundamental social tie in our society." He thus links household structure (single-family home and presence of children) not only to social isolation, but also to social obligation. Atchley (1969) found that refusers to a study of retired women tended to be loners, that is, people with few contacts with friends, and those who prefer to do things alone.

Single-Person Households. Findings from the series of survey census match studies in Great Britain all point to lower overall response rates for single-person households (e.g., Kemsley, 1975, 1976; Norris, 1987). However, Barnes and Birch (1975) found that single-person households have the highest noncontact rates, but their refusal rates are not much different from 2–3 person households. Foster and Bushnell (1994) found higher refusal rates to the National Food Survey for single-person households, but lower refusal rates to the Family Expenditure Survey. The latter survey requires interviews with all adult members of the household, thus substantially increasing the burden of participation for large households. They fail to find significant effects of single-person households on cooperation rates in multivariate analyses of the remaining three surveys they examined. Paul and Lawes (1982) find higher rates of nonresponse among single-person households (but as reported above, note that these households disproportionately consist of older persons). Smith (1983) reports higher refusal rates among single-person households, as do Brown and Bishop (1982), Ekholm and Laaksonen (1990), and Wilcox (1977).

In our data (see Table 5.11), we also find lower cooperation rates for single-person households than for multiperson households.

Presence of Children. Without exception, every study that has examined response or cooperation finds positive effects of the presence of children in the household (e.g., Cartwright, 1959; Kemsley, 1975; Norris, 1987; Rauta, 1985; Redpath, 1986; Lievesley, 1988; Ekholm and Laaksonen, 1991; Lindström and Dean, 1986; Foster and Bushnell, 1994). While Kemsley found that the presence of children has a posi-

Table 5.11. Cooperation rate by household size

Household size	Cooperation rate	(Standard error)
Single-person household	93.7%	(0.52)
Other	95.4%	(0.22)
$\chi^2 = 10.28$, $df = 1$, $p < 0.01$		

Table 5.12. Cooperation rate by presence of children in household

Presence of children under 5	Cooperation rate	(Standard error)
Yes	97.6%	(0.29)
No	94.6%	(0.24)
$\chi^2 = 51.44$, $df = 1$, $p < 0.01$		

tive effect on response, the *number* of children has no effect. Our data support these findings, whether for young children under 5 (Table 5.12) or for all children under 18 years of age (not shown).

Household Mobility. Contrary to expectation, the findings on residential mobility suggest that higher refusal rates are found among nonmovers than among movers. Barnes and Birch (1975) found that this difference persists when comparing mobility in the past year, or the past 5 years, or whether examining mobility of the head of household or all members of the household. Redpath (1986) found that those who had moved in last 12 months were more likely to respond to the Family Expenditure Survey. Comstock and Helsing (1973) also found more refusals among nonmovers. However, while Foster and Bushnell (1994) found higher noncontact rates among movers for 4 of the 5 surveys they examined, they found no apparent effect of mobility on cooperation, given contact.

From the decennial census long-form data at our disposal, we have a measure of household mobility. As with the other studies cited above, our data run counter to the social isolation hypothesis. We find significantly higher rates of cooperation among households that have moved in the last 5 years (Table 5.13). However, this does not include appropriate controls for correlates of residential mobility such as age, household size, and socioeconomic status.

Previous residential mobility appears not to be a threat to cooperation in cross-sectional surveys (or for the first wave of panel studies). This could be because more mobile persons tend to have young children that facilitate integration into their new communities, that they are of lower socioeconomic status (see Section 5.3), or some combination of these factors. We need multivariate analysis to unravel these effects.

Type of Housing Structure. Another indicator of social isolation at the community level is the type of structure the household occupies. It could be argued that those who

Table 5.13. Cooperation rate by household mobility in last 5 years

Household mobility	Cooperation rate	(Standard error)
All household members moved	96.4%	(0.60)
Some or none moved	94.4%	(0.64)
$\chi^2 = 5.39$, $df = 1$, $p = 0.02$		

live in single-family homes have more contact with neighbors and are more integrated into their community than those who live in large multiunit structures. A common theme in the literature on urban architecture (see, e.g., Aiello and Baum, 1979; Baum and Valins, 1977) is the impersonality of such large structures relative to single-family dwellings that are the stereotypical neighborhoods of middle America.

Gower (1979) reports higher nonresponse rates (both noncontacts and refusals) for apartment dwellers (in buildings at least 5 stories high and with 30 or more units) relative to other urban dwellers for the Canadian Labor Force Survey. Goyder, Lock, and McNair (1992) similarly find apartment dwelling in Canada to be negatively associated with response. Fitzgerald and Fuller (1982, p. 9) found that apartment dwellers are hard to contact, and also yielded high refusal after contact, but Hawkins (1975, p. 479) found a small positive effect of apartment dwelling on response.

We also find significantly higher rates of nonresponse for residents of large multiunit structures (10 or more units). However, as shown in Table 5.14, this is largely due to differences in contact rates rather than cooperation. We find that residents of multiunit structures have lower contact rates than others. This supports interviewer reports that gaining access to such structures and finding their residents at home are the biggest problems. Once contacted, however, such persons appear no less likely to cooperate with the survey request than other households.

Combined Effects of Variables. The indicators reviewed above are correlated. For example, a common occurrence with the birth of a new child is a residential move, to quarters of a size suitable for the larger household. We expect that the tendency for mobile households to cooperate may reflect the fact that mobile households tend to have young children in them. Theoretically, we believe that the presence of the children induces the social connections leading to cooperativeness.

Table 5.15 presents coefficients from models estimated on the short- and long-form sample cases, examining several of the social isolation correlates. Using the short-form cases, only the presence of young children has positive effects strong enough to be reliably detected with samples of this size. Controlling on that variable, the observed effect of single-person households is diminished to trivial levels. Similarly, there are no stable effects of living in a large multiunit structure.

Table 5.14. Contact and cooperation rates by housing type

	Contact rate	(Standard error)	Cooperation rate	(Standard error)
Single-family homes	98.1%	(0.12)	95.1%	(0.24)
Multiunit (10+) structures	93.7%	(0.71)	94.0%	(0.70)
Other	96.6%	(0.33)	95.1%	(0.54)
	$\chi^2 = 51.96, df = 2, p < 0.01$		$\chi^2 = 2.77, df = 2, p = 0.25$	

Table 5.15. Coefficients of logistic model predicting cooperation versus refusal by household status indicators of social isolation, short-form and long-form predictors

Predictor	Reduced model short-form cases		Expanded model long-form cases	
	Coefficient	(Standard error)	Coefficient	(Standard error)
Constant	2.51**	(0.16)	2.83**	(0.15)
Large multiunit structure (1 = yes)	−0.17	(0.15)	−0.32	(0.30)
Single-person household (1=yes)	−0.17	(0.12)	−0.28	(0.23)
Presence of children < 5 (1 = yes)	0.86**	(0.16)	0.80**	(0.27)
Residential change within last 5 years (1 = yes)			0.46*	(0.20)

Notes: Dependent variable coded 1 = interview, 0 = refusal. Coefficients for dummy variables for individual surveys omitted from table.
$*p < 0.05$.
$**p < 0.01$.

Using the long-form cases, we gain access to the residential mobility indicator. The expanded model shows that even the reduced sample size can detect positive effects on cooperation from the presence of children and the experience of residential mobility in the last 5 years.

Table 5.15 shows that the initial finding that households with young children are disproportionately cooperative survives initial multivariate controls. However, the puzzle about the underlying causes of more transient households providing higher cooperation rates remains. We will return to that finding at the end of the chapter with a larger multivariate model, containing controls on socioeconomic status. We are especially interested in observing whether this finding is really masking effects of housing tenure and other socioeconomic indicators. That is, do movers have higher cooperation rates because movers tend to be poorer?

5.5   THE CONCEPT OF AUTHORITY AND SURVEY COOPERATION

A concept related to social isolation is that of "authority," the influence of legitimized power of a person or institution over the behavior of others. Cialdini (1984) and others have noted that when persons or institutions with authority over the lives of the requestees seek assistance, decisions might be made with less attention to the costs and benefits of the task. In a sense, this is a subset of the influences over those who feel connected to a larger society and applies to their behavior with regard to specific instruments of power in the society.

We believe that influences of authority are important in understanding compliance to government survey requests but probably less so to requests from academic or commercial survey organizations. Many large scale national surveys in the United States are conducted for or by government agencies or using government funds. Such surveys can be viewed as representing the interests of the central power structures in society. To the extent that the authority or power of such institutions over the members of a society are weakened, cooperation with survey requests may be negatively impacted. This could be viewed at a societal level, as part of the "social climate" for surveys (see Chapter 6).

Here we are concerned with individual or household-level indicators of authority. An authority hypothesis may explain why we did not find the expected effect of SES on cooperation (see Section 5.3). It could be argued that the government exercises greater control over those at the lower end of the SES scale, thus producing higher levels of compliance among this group to survey requests from government institutions.

Using the decennial census (long form) data, we examine two indicators of authority—whether a household member was in the military (as a behavioral commitment to government authority), and whether all members of the household were citizens (with those who are not citizens more likely to be sensitive to government authority than others). Households with members in the military do indeed exhibit higher cooperation rates (98% to 95%), but there are no direct effects of citizenship on cooperation. When the two variables are combined, however, reflecting the fact that they are highly correlated, more information about the nature of the effect emerges (see Table 5.16). The lowest cooperation occurs among households with nonmilitary citizens (as the authority hypothesis would suggest), while those with some noncitizens or those with military members have the highest cooperation.

Another variable we have available from the census long form is the language spoken by members of the household. This could be used as a proxy for immigrant status, the expectation being that those with greater dependence on government or under greater threat of sanction may be more likely to cooperate with requests from government institutions or their agents. In Table 5.17 we see that cooperation rates *are* higher in those households where some or all members do not speak English.

This finding is consistent with that reported for race/ethnicity in Table 5.8, in which Hispanics have relatively high cooperation rates. Non-English speaking

Table 5.16. Cooperation rate by citizenship and active military duty of household members

Citizenship/active duty	Cooperation rate	(Standard error)
Some or all noncitizens	97.4%	(1.53)
All citizens, no one on active duty	94.8%	(0.51)
All citizens, someone on active duty	97.9%	(1.65)
$\chi^2 = 4.57, df = 2, p = 0.10$		

Table 5.17. Cooperation rate by language spoken in household

Language spoken	Cooperation rate	(Standard error)
All household members speak English	94.8%	(0.54)
Some or none speak English	97.0%	(0.92)
$\chi^2 = 4.05$, $df = 1$, $p = 0.04$		

households in the United States are likely to be dominated by Spanish-speakers. An alternative explanation could be the reciprocation inducing effects of extra efforts made to include Spanish-speaking residents in the surveys (e.g., Spanish versions of questionnaires, Spanish-speaking interviewers, etc.). We do not have enough sample cases to separate out the effect for this group relative to other immigrant or non-English speaking groups in the United States.

Noncitizens are more likely to use some other language than English at home. Hence, the two indicators of the authority concept are correlated. A multivariate model might yield a different interpretation than the bivariate tables above. When the indicators of citizenship, military service, and language are combined into a single logistic regression model, the effects of language are diminished. In short, these indicators of the authority hypothesis do not yield much insight into the process of cooperation with the surveys covered in the decennial census match data.

5.6 JOINT EFFECTS OF INDICATORS OF SOCIAL ISOLATION AND AUTHORITY

We could posit interaction effects for the impact of social cohesion on cooperation by survey type. For example, the impact of social integration should be greater for studies impacting the local community (e.g., drugs and crime) than on studies focussing on broader societal issues or of more personal concern (unemployment, cost of living, etc.). Unfortunately, we have insufficient cases in each survey to explore this further, and these remain untested hypotheses.

Finally, we combine the indicators of social isolation and authority in a single multivariate analysis to examine the relative contributions of the different variables. As we did previously, Table 5.18 presents (1) the short-form variables only, modeled on the full sample, (2) the short-form variables modeled on long-form cases only, and (3) short- and long-form variables together. When we combine these indicators into a combined multivariate model, people living in multiunit structures ($p = 0.08$ for the third model), multiperson households, households with children, younger or older households, as well as Blacks, Hispanics, and people experiencing some move in the last 5 years are more likely to be cooperative. We retain these variables for further analysis.

Table 5.18. Coefficients of logistic models predicting cooperation versus refusal by social isolation and authority indicators

Predictor	Simple model		Expanded model (long-form cases)
	Total sample	Long-form cases	
Constant	2.82**	2.89**	3.37**
	(0.076)	(0.19)	(0.72)
Social-isolation indicators			
Race/ethnicity			
Black	0.22	−0.0050	0.0084
	(0.13)	(0.34)	(0.34)
Hispanic	0.50*	1.17**	0.94
	(0.20)	(0.40)	(0.48)
Other	—	—	—
Household age			
All household members <30	0.86**	0.99**	0.87**
	(0.14)	(0.28)	(0.30)
Mixed ages	—	—	—
All household members >69	0.30	0.036	0.12
	(0.16)	(0.36)	(0.36)
Large multiunit structure (1 = yes)	−0.22	−0.44	−0.60
	(0.13)	(0.35)	(0.34)
Single-person household (1 = yes)	−0.24*	−0.30	−0.31
	(0.11)	(0.25)	(0.25)
Children < 5 years in household (1 = yes)	0.24*	0.17	0.11
	(0.10)	(0.25)	(0.25)
Residential change in last 5 years (1=yes)			0.44*
			(0.21)
Authority indicators			
Citizenship/military duty			
Some or all noncitizens			—
All citizens, no active duty			−0.32
			(0.62)
All citizens, some on active duty			0.45
			(1.03)
All household members speak English			−0.29
(1 = yes)			(0.34)

Notes: Dependent variable coded 1 = interview, 0 = refusal. Coefficients for dummy variables for individual surveys omitted from table.
*$p < 0.05$.
**$p < 0.01$.

5.7   OTHER HOUSEHOLD-LEVEL INFLUENCES ON COOPERATION

Fear of Crime. Suspicions that some persons are reluctant to cooperate with surveys because they fear criminal victimization is a specific form of a script error— the misunderstanding of the survey request as one that might in fact be an attempt to gain entry and physical access to the householder for purposes of theft or assault (see Chapter 8). We would expect that fear of crime might produce reluctance of householders to respond to an unexpected knock on the door, but also to be a stronger influence on behavior when the person attempting contact with the household appears threatening in any way. We note that while fear of physical safety may be a concern in household surveys, the fear of being victimized through telemarketing scams (often targeted at the elderly) may be no less a concern in telephone surveys. The common use of female interviewers in surveys might dampen the effect of fear of crime in face-to-face surveys.

We have very imperfect indicators from the short form to test this hypothesis, namely women living alone, and elderly (those over 69) living alone (see Table 5.19). However, there is little evidence from our data that women living alone (who might have greater fears of victimization) or elderly respondents (who do exhibit more fear of crime; see for example, Miethe and Lee, 1984; Rucker, 1990) have lower cooperation rates.

Topic Saliency. A common metaphor for survey interviews is a "conversation with a purpose" (see Kahn and Cannell, 1957; Schaeffer, 1990). There is much speculation that when the purposes of the conversation are goals shared by the sampled persons, they tend to cooperate. This speculation is based on hypotheses that surveys on salient topics may offer some chance of personal gain to the respondents because their group might be advantaged by the survey information, and also that the chance to exhibit one's knowledge on the topic would be gratifying. When the topic of the interview is used as an important attribute by the interview in persuading the householder, then prior knowledge about the topic and personal relevance to the householder can affect response propensity.

Couper (1997) finds that those who express little interest in politics are more likely to decline participation in electoral behavior surveys. The decennial census data offer few indicators to test hypotheses concerning topic saliency. We would ex-

Table 5.19. Cooperation rate by indicators of fear of crime

Fear of crime indicators	Cooperation rate	(Standard error)
Women living alone	94.4%	(0.44)
Other	95.3%	(0.24)
$\chi^2 = 2.80$, $df = 1$, $p = 0.09$		
Person over 69 living alone	94.3%	(0.97)
Other	95.2%	(0.22)
$\chi^2 = 0.83$, $df = 1$, $p = 0.36$		

pect saliency to be enhanced especially in those cases where the survey purpose is to construct benefit programs affecting sample person (e.g., the CEQ for Social Security beneficiaries, the CPS for Unemployment Insurance beneficiaries).

5.8   MULTIVARIATE MODELS OF COOPERATION INVOLVING HOUSEHOLD-LEVEL PREDICTORS

Many of the hypotheses reviewed above are not independent of one another. For example, those socially isolated in the society are less likely to have the normative guidance of social exchange relationships with survey takers. Hence, deeper understanding of the nature of the process of survey participation might be gained by combined multivariate analysis of the different indicators. In combining these household-level variables we also need to control for survey design and social-environmental variables when possible, as we will do in later chapters.

5.8.1   Combining the Household-Level Predictors

The first step is the examination of a model that combines the indicators of the household-level hypotheses that found some support in the data. Of the social exchange indicators, the housing cost variables and housing tenure variables seemed most powerful. For the social isolation indicators, race, ethnicity, age composition of the household, whether the household is a single-adult household, whether there were children in the household, and whether the household had moved in the last 5 years appeared useful.

As we noted earlier, the concepts of social exchange and social isolation are related to one another. Two socially isolated groups cannot develop ongoing exchange relationships. Further, some of the indicators for the various concepts themselves are correlated. For example, minority racial and ethnic groups tend to live in housing of lower cost than majority groups. For that reason, some of the effects we appear to be measuring in the models examining each hypothesis may be spurious effects of other concepts. A larger multivariate model can be specified to check on that possibility.

The first column of Table 5.20 shows a multivariate logistic regression based on the full sample, predicting the likelihood of an interview relative to a refusal, among sample households contacted for the six surveys. The model controls for base response rate differences among the six different surveys (to control on omitted design differences among the surveys).

The socioeconomic indicators that we use as proxy measures of social exchange influences (tenure and housing costs) retain their impact. The finding is that those incurring lower housing costs tend to cooperate with the survey requests. This is consistent with the notion that requests coming from a government source, with implications for government policy and social services, tend to be viewed as potentially beneficial to those in lower socioeconomic groups.

In the presence of these socioeconomic measures, the race/ethnicity indicators

Table 5.20. Coefficients from logistic models predicting cooperation versus refusal; model using social exchange, social isolation, and authority indicators; model adding environmental indicators (standard errors in parentheses)

	Household level		
Predictor	Short-form household	Expanded (long-form) household	Environment + household
Constant	2.78**	3.03**	2.72**
	(0.22)	(0.52)	(0.29)
Environment			
Urbanicity			
Central city			−0.27
			(0.17)
Balance of CMSA			−0.14
			(0.13)
Other			—
Population density[a]			−0.022*
			(0.011)
Crime rate[b]			−0.0053
			(0.019)
Percent under 20 years old			0.0096
			(0.0047)
Household			
Social exchange			
Owner occupied	−0.10	−0.30	−0.20
	(0.18)	(0.43)	(0.18)
Monthly rent[c]	−0.062*	−0.031	−0.044
	(0.029)	(0.068)	(0.033)
House value[d]	−0.016**	−0.040**	−0.011
	(0.0052)	(0.013)	(0.0060)
Social isolation			
Race/ethnicity:			
Black reference person	0.24	0.092	
	(0.15)	(0.29)	
Hispanic reference person	0.39	0.92*	
	(0.22)	(0.38)	
Other			
Household age			
All < 30 years	0.70**	0.52	0.67**
	(0.15)	(0.35)	(0.14)
Mixed ages	—	—	—
All > 69 years	0.40*	−0.16	0.42*
	(0.19)	(0.45)	(0.19)
Single person household	−0.37**	−0.70*	−0.36**
	(0.12)	(0.32)	(0.13)

(*continued*)

Table 5.20. Continued

	Household level		
Predictor	Short-form household	Expanded (long-form) household	Environment + household
Household			
Social isolation			
Children <5 in household	0.65**	−0.040	0.63**
	(0.15)	(0.28)	(0.15)
Residential change in last 5 years (1 = yes)		0.16	
		(0.26)	

Notes: Dependent variable coded 1 = interview, 0 = refusal. Coefficients for dummy variables for individual surveys omitted from table.
[a]Measured in thousands of persons per square mile.
[b]Measured in crimes per 1,000 persons.
[c]Measured in $100 units.
[d]Measured in $10,000 units.
*$p < 0.05$.
**$p < 0.01$.

have negligible influence on cooperation. In models addressing only the social isolation hypotheses, these showed higher response rates among Hispanics. We interpret this as evidence of the more pervasive effects of socioeconomic status than any subcultural influences associated with racial and ethnic groups. In other analyses (not presented) we find that the estimated effects of race/ethnicity are also sensitive to what environmental predictors are included in the model. Given these empirical results and the mixed findings about race/ethnicity in the past literature on nonresponse, we excluded these measures from the final model.

The curvilinear effects of age remain in the first model in Table 5.20. Young and old households tend to exhibit higher cooperation rates than do households with persons exclusively in the 30–69 year old range. We'd expect that employment rates might differ across these three types of households. A careful reader will note that we have lost controls on the employment status of householders. We had included such variables under the hypothesis of opportunity costs, with the result that there were no differences by employment status. We suspect that the higher rates among young households reflects an interest in social participation because of greater social engagement in general. The higher rates among the elderly may reflect stronger norms of civic duty among the current cohort of elderly.

There were two other indicators of social connectedness we discussed earlier. Those who live alone tend not to cooperate with the survey request, other things being equal. Conversely, those households with young children tend to participate. These two groups, one assuming all the responsibilities of household maintenance with a single individual, the other sharing it among several persons, are sharp contrasts. While one can easily avoid contact with large numbers of strangers in their day-to-day life, the other interacts with the community in an ongoing manner. This includes health care and education workers, as well as neighbors or others involved in child care. The adults in these households are well acquainted with requests for their time. With full multivariate controls, younger person households and those with children show higher cooperation.

The second column of Table 5.20 presents model coefficients for a long-form model, on a reduced sample size. This allows us to add to the model an indicator of residential mobility, the variable indicating whether the household moved in the last 5 years. We found earlier, in a model testing effects of social isolation indicators, that movers tended to cooperate at higher rates, in the presence of controls for other social isolation indicators. Absent from those models, however, were indicators of the housing tenure of the household. Those living in rental housing tend to move more frequently than those in owned housing. Hence, we expected that the measured effect might mask the effects of housing tenure. As shown in Table 5.20, when we control on housing costs, the effect of residential mobility diminishes to a negligible level.

5.8.2   Combining Household Predictors with Predictors from Other Levels

The third column of Table 5.20 examines whether the entire set of household-level predictors continues to exhibit hypothesized effects in the presence of controls at the social environmental level (the influences to be discussed in Chapter 6). A comparison of the first column and the third column of the table shows that only the housing cost variables appear to be substantially affected by controls on social environmental variables. When we control on urbanicity, crime rates, and population density, the negative effects of the housing cost indicators decline to just below traditional levels of statistical significance ($p = 0.08$, for the three predictors combined; the coefficient for house value among owners is closest to the chosen level of statistical significance). We interpret this as reflecting the relatively higher cost of housing in urban areas. Thus, the overlapping influences of urbanicity and housing value weakens the marginal impact of the latter coefficient. It appears that the effects of higher living costs we observed in Section 5.3 were in part reflecting the negative influences of urbanicity and urban living conditions on cooperation.

One way to quantify how the addition of environmental predictors affects the estimates of household-level effects is to compare the change in the two likelihood ratio statistics associated with the household-level variables, one without the environmental level controls, and one with the environmental level controls. (We use as the base model one with a constant and the survey dummy variable predictors.) The change due to household predictors is 66.80 without the household controls, and 17.04 with the controls. In short, about $(66.80 - 44.00)/66.80$ or 34% of the measure of fit of the model associated with the household variables seems to overlap with the environmental predictors.

The last controls we added to the model concern the interviewer (discussed in Chapter 7). These are not shown in Table 5.20. Here we were testing whether the household-level predictors maintain their effects in the presence of characteristics of interviewers found to influence cooperation. The household-level effects remain substantially the same, and the same conclusions would be made as from the model in the third column of Table 5.20.

5.9   SUMMARY

Earlier chapters in this book have noted that our understanding of survey participation has been heavily shaped by studies examining bivariate relationships between some variable and participation. These have often been case studies, not studies of sets of surveys. They have tended to study demographic indicators, not underlying social psychological concepts. They have emphasized overall response rates without separately studying noncontacts and refusals.

This chapter no doubt has results surprising to some survey professionals. When it is combined with the results of Chapter 4, on the process of contact, a different picture on survey participation is formed for the six surveys in the decennial match study than we assumed prior to examining the data.

Most survey researchers might assume lower cooperation rates among lower socioeconomic groups, among racial/ethnic minority groups, and among the elderly. However, once contacted, poorer groups (as indicated by the proxy measures of housing costs) appear no different from other groups. This result becomes clear only when controls on social environmental influences are applied to the equation. That is, the result that those in more expensive housing do not respond is partially explained by the fact that more of those persons live in urban areas, the sites of low cooperation in general. There is some tendency for owners (versus renters) in expensive housing to not cooperate, but the effect is a weak one, deserving of further attempts to replicate.

A resolution of this chapter's findings regarding age is more complex. We find a curvilinear effect of age (using a proxy indicator of age grouping at the household level). It shows young and old households cooperating at higher rates than the middle-age households. In Chapter 4 we found that the elderly were easier to contact than others. These two findings combine to suggest the largest overrepresentation of the elderly will occur in surveys with low contact rates and no efforts to convert initial refusals (which should convert disproportionately younger persons).

The young and old households may have higher cooperation rates for different reasons. Younger persons may tend to exhibit more curiosity about efforts to seek information from them. They most recently experienced standardized information-seeking associated with school and jobs. The elderly cohort, in contrast, may not have the curiosity but may maintain norms of civic duty regarding requests from the government not shared by younger cohorts. The households in the middle-age category or with mixed ages, on the other hand, as a group do not share as much of the history of standardized measurement nor the civic norms of the elderly. Clearly, we have no direct tests of these notions with these data.

Of further relevance to age effects, our multivariate models control on household size. Most of the past findings of lower response rates among the elderly, for example, did not control on the fact that many elderly live alone. (Controlling on household size, we found older householders more willing to respond.) We have consistently found that those who live alone are a distinctively difficult group to interview in surveys. Chapter 4 shows that, other things being equal, they tend to be difficult to contact. In this chapter, we've learned that once contacted, they tend to refuse, relative to others. This is strong evidence, we believe, of the underlying tendency for those who live alone either to have an outward orientation, nurturing social relations by being away from their home often, or to be relatively socially disengaged, self-oriented in a way that leads to avoidance of acceptance of stranger requests at the door.

The preponderance of the evidence thus far points to support for a two-prong view of household-level correlates of nonresponse. First, the process of contacting some households affects nonresponse error differently than the cooperation step. At the contact step, household size, age of householders, and other household compositional variables can exert their influence on the nature of the nonrespondents. How important the correlates of noncontact remain in the final data set depends on the efforts to reduce noncontacts.

Second, the nature of cooperation is most fully explained by concepts that describe the perceived relationship between the survey requester and the sample household. Notions of social isolation and social exchange appear consistent with the multivariate findings. In this perspective, the request for a survey interview is seen in the context of the prior set of dealings with the sponsor of the survey over time (or with others perceived to be similar to the sponsor). When the government is involved, those lower in socioeconomic status or with stronger senses of civic duty may be more willing to participate because of potential rewards of providing information to the government. We might speculate that when universities are the sponsors, those in higher education groups might be more cooperative.

It is important to note that the data used in this chapter come from surveys with response rates that are higher than those typical of academic or commercial surveys. They result from designs that employ unusual efforts to contact all sample households; thus, their noncontact rates tend to be unusually low. Their refusal rates are also low, but have some parallels in other surveys. In this book, surveys with few noncontacts allow a study of the reaction to the survey request of a more diverse pool of householders. They thus are more valuable for studying the correlates of cooperation.

Since the analyses concentrate on cooperation of sample persons once they are contacted, the inferential limitations of the work depend on whether the influences on cooperation are different for lower response rate surveys than for these high response rate surveys. We note that there remain large net differences in response rates across surveys, controlling on these demographic factors. These arise from different design features (such as survey topic or length of interview). These design features do not necessarily affect household-level influences because they are part of the survey request for all the demographic and social subgroups. By pooling data over surveys that vary on some of these design features, we measure the effects of household-level attributes that are robust to those design features.

Only if some subgroups of the population react to design features differently than do others will the kinds of models discussed in this chapter be inapplicable to surveys with lower response rates. One example of this might be the tendency for persons uninterested or uninformed about a topic to refuse a survey request (see Couper, 1997) or for those who are busy to disproportionately be nonrespondents to surveys that are conducted over just a few days. For a survey with very low levels of effort to contact sample persons, the refusal cases will consist of those who are easily contacted. In such cases, the magnitude of predictive power for some of the household-level variables may be larger, because their effects have not been attenuated by repeated persuasive efforts of interviewers. For the most part, however, we speculate that the general form of the models will remain the same at lower response rate levels.

In short, we have learned the following key lessons from the empirical analyses in this chapter:

- The tendency for those in high cost housing to refuse survey requests is partially explained by their residence in urban, high density areas. We interpret this as stronger support for hypotheses about the influence of social isolation and disintegration than for social exchange hypotheses concerning cooperation.

- Tendencies for those in military service and those in non-English speaking households to cooperate in surveys appear to be explained by their having household compositions favoring survey participation. We interpret this as partial refutation of the hypothesized role of government authority in survey cooperation.

- The apparent tendency of racial and ethnic minorities to cooperate with surveys when asked is largely explained by their lower socioeconomic status (as measured by housing costs). Once socioeconomic status is analytically controlled, minority cooperation rates are much closer to those of the majority group.

- Multiple indicators of social integration (or isolation) share strong marginal effects on cooperation. Households with young children or young adults tend to cooperate; single-adult households tend not to cooperate.

- In examining the participatory behavior of elderly households, it is important to consider the tendency for the elderly to live alone. After analytic controls on household size, elderly households tend to cooperate.

There are many features that define the psychological and behavioral context of a conversation between the interviewer and a sample person, in which the interviewer requests participation in the survey. These conversations comprise the proximate causes of survey response rates. This chapter has shown that even without measurement of these proximate causes, systematic and measurable variations exist across persons and households in their tendency to comply with survey requests. Adding measurement of these proximate causes, through observation of the conversations between interviewers and respondents, provides further insight into the process of survey participation (see Chapters 8 and 9). These are necessary for a complete understanding of the phenomenon.

5.10 PRACTICAL IMPLICATIONS FOR SURVEY IMPLEMENTATION

Now that we have studied both the process of contacting households and the process of gaining cooperation, we can for the first time consider design strategies that jointly consider noncontacts and refusals. We learned in Chapter 4 that single-person households are difficult to contact. In this chapter we learned that they also tend to refuse survey requests. We found that elderly households and households with children are easy to contact; here we found they also tend to be cooperative once household size is controlled. (We do know from other work, however, that elderly persons tend to be nonrespondent for health and other physical reasons more often than younger persons.)

Many of the findings of this chapter concern the influence of attributes that are not usually knowable from the sampling frame itself. They are socio-demographic attributes that we believe are imperfect indicators of social psychological states affecting reactions to survey requests.

The most important practical implication of these results is to urge interviewers to learn these attributes of a sample household as early in the survey process as possible. That is, in early face-to-face calls on the household or in the first contact with the household, some effort should be made to observe these demographic characteristics.

For a limited number of these, proxy indicators might be useful to alert the face-to-face interviewer to possible challenges. For example, single-bedroom apartment units are sometimes systematically dispersed in a complex. Knowing the configuration of those in a area segment might give useful, albeit imperfect indicators of single-person households. Similarly, complexes that are devoted to young singles might indicate difficulty with contact and need for flexibility in negotiating the time of the interview once contact is made. The observation of evidence of young children mentioned in Chapter 4 applies here also.

Only if interviewers learned of these attributes can they use knowledge to customize approaches to the households. To younger single-person households, who might have relatively less time at home, flexibility in doing the interview in several shorter segments may be important in gaining cooperation. In urban areas, persistence and explicit argument about the unequal value of the householders' participation might be useful. We have elaborated in Chapter 2 the notion that tailoring of interviewer approaches to real concerns of the householder can make a difference in cooperation rates. This chapter is our first note that much of this tailoring must begin after initial contacts with householders.

There are many features that define the psychological and behavioral context of a conversation between the interviewer and a sample person, in which the interviewer requests participation in the survey. These conversations comprise the proximate causes of survey response rates. This chapter has shown that even without measurement of these proximate causes, systematic and measurable variations exist across persons and households in their tendency to comply with survey requests. Adding measurement of these proximate causes, through observation of the conversations between interviewers and respondents, provides further insight into the process of survey participation (see Chapter 8). These are necessary for a complete understanding of the phenomenon.

CHAPTER SIX

Social Environmental Influences on Survey Participation

6.1 INTRODUCTION

As we have noted earlier, the request for participation in a survey is inherently a social activity. The interaction between interviewer and householder does not take place in a vacuum. In this chapter we examine the broader context or social environment in which the interaction between the two actors takes place. The role of ecological influences in affecting survey participation is illustrated in Figure 6.1.

While we believe the more proximate causes of the decision to participate in a survey lie at the level of the householder and his or her interaction with the interviewer, the environment affects this decision in shaping the context in which the decision is made, both in influencing the prior dispositions and reactions of the householder and in influencing the expectations and behaviors of the interviewer. Studying nonresponse at the higher level of aggregation also helps us in another respect. We have noted that many of the variables we need to understand survey cooperation are social psychological in nature (e.g., social isolation, fear of crime). We are generally unable to directly measure such attitudes or dispositions among nonrespondents, a key factor underlying the intractability of the nonresponse problem. Examining aggregate societal-level attitudinal variables and how they may covary with nonresponse, either over time or across societies, may give us insight into the effect of such variables on nonresponse.

The social environment of the survey request can be conceptualized at two broad levels. The first are societal-level conditions that facilitate or mitigate survey participation in a particular society, and the second are more local variations in context at the community or neighborhood level that shape the decision to participate or refuse. We examine these in turn.

Lyberg and Dean (1992) introduced the notion of a "survey-taking climate," in which societal-level differences or trends in survey participation can be understood. However, these macro-level influences are well-nigh impossible to detect with a cross-sectional study such as we have in the decennial census match study. We need

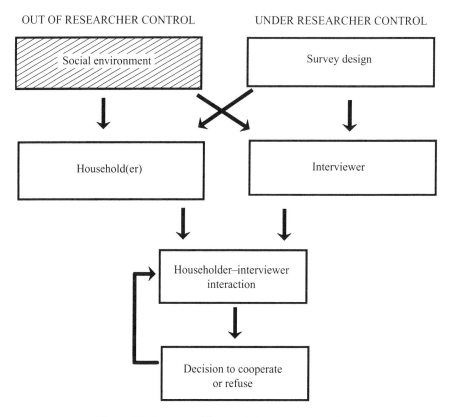

Figure 6.1. A conceptual framework for survey cooperation.

to rely on other data sources to examine the effects of the survey climate on participation. There are three related approaches to examining these effects. The first is to
examine response rate trends over time within a single country. A second approach
is to examine cross-national differences in survey participation that may reveal differences across countries in the societal context for surveys. Finally, we can examine "natural experiments" that may have occurred. These are events or changes in a
society that may lead to marked perturbations in response rates. The problem with
all of these approaches is that survey design features change over time (or differ
across countries), making it difficult to disentangle true societal impacts from design differences, and leading to counter-hypotheses for the trends or differences that
may be detected.

6.2 TRENDS IN RESPONSE RATES OVER TIME

In order to measure the effects of societal-level factors on cooperation within a single society, we need to examine trends on survey cooperation over time along with

indicators of hypothesized correlates of cooperation. To the extent that these covary, it is suggestive of a relationship between these macro-level features of a society and the levels of cooperation obtained within that society over time.

Most of the concepts we discuss here (e.g., alienation, discretionary time availability) have their parallel at the household level and are also discussed in Chapter 5. Whereas in that chapter we were concerned with understanding differences among householders in their propensity to respond, here we focus more on aggregate-level effects on overall levels of cooperation as additional evidence of the effect of these factors. The societal level factors may have an added effect of changing the perceived value of surveys in society, over and above that of the individual's calculus of the worth of a particular survey. As social beings, people take their cues from others in society, and to the extent that surveys as a "public good" are devalued, this could influence the behavior of individual members of that society in response to a survey request. In other words, normative responses to survey requests are, we believe, heavily shaped by the survey-taking climate extant in a particular society at a particular time.

Before we explore these influences in more detail, we first turn to the issue of tracking response rate trends over time. Are things getting better or worse for survey research in terms of response rates? Are people more or less inclined to participate in surveys than 10, 20, or 30 years ago? While these are clearly important questions for the survey research industry, obtaining clear answers to these questions are not that easy, despite several efforts to assemble such data, and clear recognition that such issues should be explored.

In 1973 the American Statistical Association (ASA) convened a conference to explore the possible effect on surveys of changes in society or public acceptance of this research tool. This group found that little hard data were available for evaluating changing acceptance by respondents or other survey difficulties. Despite the paucity of empirical evidence, the group concluded that "there are a significant number of reports of completion rates declining, or where achieving a satisfactory completion rate is becoming increasingly more difficult" (American Statistical Association, 1974, p. 31).

The Committee on National Statistics (CNSTAT) organized a Panel on Incomplete Data in 1977 to explore all aspects of survey incompleteness (including noncoverage and item-missing data, as well as unit nonresponse). This work culminated in a three-volume report (Madow, Nisselson, and Olkin, 1983). The committee recommended that nonresponse should be measured and reported, including the reasons for nonresponse. This would facilitate (it was argued) the study of trends in nonresponse rates and biases among surveys and over time (Madow, Nisselson, and Olkin, 1983, Vol. I, pp. 5–6).

Similar efforts, albeit on a less grand scale, have been undertaken by other organizations. Bradburn (1992, p. 392) notes that the AAPOR (American Association for Public Opinion Research) council "appointed a small task force to look into the response rate problem, but they abandoned their effort because they couldn't get comparable data." The British Market Research Society has convened two working parties to investigate issues of respondent cooperation (Market Research Society,

1976, 1981). The first concluded that "there is some evidence . . . of an increasing reluctance on the part of the general public to agree to cooperate in market research surveys." Their report continued, "On balance, we feel it is at least harder to obtain co-operation from the general public than it was, say, ten years ago" (Market Research Society, 1976, p. 133). The second working party's conclusions were somewhat more equivocal, noting that the evidence of declines in response rates is not consistently reported. However, they note further that "the continuation of high response rates could have been achieved at the cost of considerable increase in effort to maintain high levels, and there is evidence of such efforts" (Market Research Society, 1981, p. 19).

We both participated in a more recent effort to assemble evidence on response rate trends in government surveys in the United States (see Shettle *et al.*, 1994; Gonzalez, Kasprzyk, and Scheuren, 1995). This group assembled trend data for 26 federally sponsored demographic surveys for a 10-year period. They echoed the recommendations of the earlier Panel on Incomplete Data. For example, Gonzalez, Kasprzyk, and Scheuren (1995, p. 613) write:

> It is true that every survey program examined by the Subcommittee calculated response rates in some fashion and had some auxiliary information about nonresponse. It is also true that most survey programs did not have *readily* available what the Subcommittee viewed as "basic" data on nonresponse; nor did repeated surveys have the time series easily available of nonresponse rates and nonresponse components.

Despite the limited success reported by these committees and groups, almost all these reports end with the call for continuing and intensifying the effort to collect trend data on nonresponse and its components. What have we learned from these efforts? We can summarize some of the lessons from this work as follows:

1. It is oftentimes difficult to assemble time trend data, particularly for earlier decades of long-running surveys.

2. Many important design changes that may help explicate trend effects are difficult to uncover without relying on the institutional memory of those who worked on the surveys in the past.

3. Trend data are generally only available for long-running surveys, which are often of a particular character that makes generalization to other surveys difficult. Many of these are conducted by government organizations (such as those included in our match study), or are flagship surveys (e.g., the National Election Study or the General Social Survey) that may experience different pressures on response rates than other surveys conducted by the same organizations. More specifically, surveys conducted by the commercial sector are chronically underrepresented in comparisons across time.

4. Many of the surveys for which trend data do exist do not distinguish between different sources of nonresponse (e.g., refusals and noncontacts), or have only begun to do so more recently. Alternatively, different definitions of response rate components make comparisons across surveys difficult. Recent efforts to

standardize the reporting of response rate components for all surveys at Statistics Canada (see Hidiroglou, Drew, and Gray, 1993) are noteworthy, and we hope that this approach is adopted by other agencies and organizations.

5. We believe that the effort expended to obtain a desired response rate is not constant over time; however, information on effort or costs associated with data collection are rarely made available. This will change, we hope, with the advent of computer assisted interviewing, which makes the automated collection of effort measures from call record data routine.

6. The examination of response rate trends needs a relatively long time series to reduce noise, such as seasonal variation on relatively frequent surveys or minor fluctuations in response rates on others. We believe trends in response rates may more appropriately be measured in terms of decades than years.

Despite these difficulties, these efforts have not been entirely unsuccessful and a large body of data has been assembled representing a wide diversity of surveys and organizations. It is not our intention to present or even review all of these trends here. Instead, we refer the interested reader to the many studies that review these trends (e.g., Baim, 1991; de Heer and Israëls, 1992; Smith, 1995; Steeh, 1981). We present here a few exemplary cases, focusing on those that have a relatively long time series and distinguish (at least) refusals from overall nonresponse.

The prevailing view, both based on these trend data, and from practitioners' knowledge of the situation, is that response rates are declining, and have been doing so for some time (see, e.g., Baim, 1991; Bradburn, 1992; Brehm, 1994; Brown, 1994). Mervin Field echoed this view in an interview for the *Washington Post* (April 14th, 1992, p. A1), in saying "One of the problems facing the research industry today is the continuing and alarming decline in rates of participation."

However, this view has been contested by Smith (1995, p. 158) who writes that "Despite the overwhelming consensus that response rates have been falling and continue that decline into the future, the empirical evidence from the U.S. is both more equivocal and less uniform." Gonzalez, Kasprzyk, and Scheuren (1995) reach similar conclusions: "In summary, despite the prior beliefs of many in the survey community, there was little evidence of declining response rates over time among either the establishment or demographic surveys included in the study."

Why do these beliefs persist in the face of apparent evidence to the contrary? Or, are response rates really declining, but it is simply too difficult to assemble appropriate data to demonstrate this? This debate reminds us of the story of the blind men and the elephant: it really depends on what perspective one takes, and what data one looks at. In this sense, it resembles the discussion of the literature on household-level influences in Chapter 5, and again alerts us to the danger of generalizing from case studies and to the need to differentiate between various sources of nonresponse.

The Current Population Survey (CPS) is one of the longest continuous-running surveys in the United States, and it is also regarded as a flagship survey for the federal statistical system. Nonresponse trends for the last 35 years of the CPS are presented in Figure 6.2. Aside from drawing attention to the low overall nonresponse rate (an average of 4.3% over the 35-year period), the CPS shows remarkably little change in the overall nonresponse rate during this time. Fitting a regression line to the nonresponse rate produces a slope coefficient of 0.0004 (s.e. = 0.0056), a trend not significantly different from zero. However, the refusal rate produces a different picture. Here it is clear that the general trend is toward increasing refusal rates in the CPS. Fitting a regression line to these data produces a positive slope coefficient of 0.065 that is significantly different from zero (s.e. = 0.0036). We speculate from this figure that the overall nonresponse rate has been maintained at the expense of increasing effort to reduce the noncontact portion of nonresponse (the major portion of the difference between overall nonresponse and refusals). If these trends were to continue, the overall nonresponse rate must eventually be pushed up. Another thing to note from Figure 6.2 is that the rate of increase in refusals is not constant. If one were to look only at the period 1985–1990, one might conclude that refusal rates are remaining steady or even declining. Conversely, looking only at the period 1970–1975 would lead to a much less positive conclusion.

Figures 6.3 to 6.5 present trends for three other ongoing surveys included in the decennial census match study: the National Health Interview Surveys (NHIS), the Consumer Expenditure (Quarterly Interview) Survey (CEQ), and the National Crime Survey (NCS) respectively. Not only do these surveys not cover as long a time period as the CPS, but the trends are also less clear. However, both the NHIS and the CEQ suggest increasing refusal rates over time. The NCS is slightly different in that a household is regarded as having participated in the survey if *any* adult

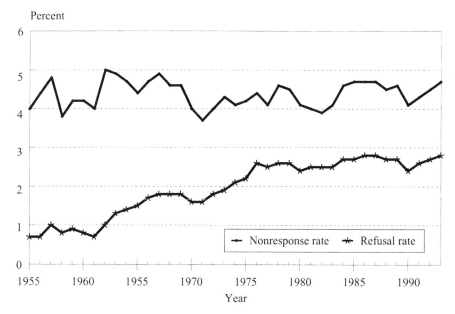

Figure 6.2. Nonresponse trends for the Current Population Survey.

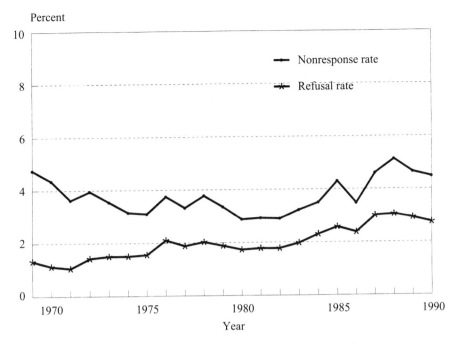

Figure 6.3. Nonresponse trends for the National Health Interview Survey.

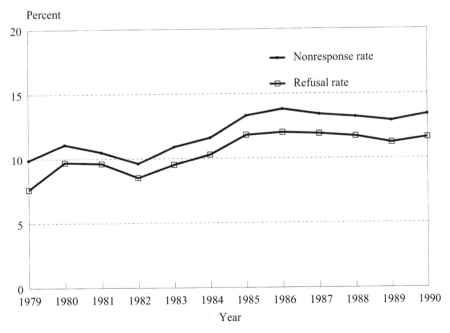

Figure 6.4. Nonresponse trends for the Consumer Expenditure (Quarterly Interview) Survey.

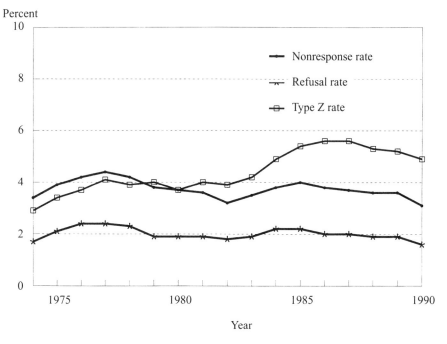

Figure 6.5. Nonresponse trends for the National Crime Survey.

member responded to the household portion. The household-level nonresponse rates and refusal rates suggest a decline in nonresponse, if anything. However, the NCS is designed to obtain self-reports from all eligible persons within each household. The person-level nonresponse rates suggest more of an increase over time, with the exception of the latter portion of the 1990s.

Two other sets of trends from U.S. surveys can be examined. The National Election Studies (Figure 6.6) are conducted periodically (currently every two years) by the Survey Research Center at the University of Michigan (see Luevano, 1994). In contrast to the CPS, refusals appear to dominate the nonresponse (especially in later years), reflecting the longer field period for NES. The General Social Survey (GSS), conducted annually by the National Opinion Research Center (NORC), reveals a trend that appears to run counter to those we have seen (see Figure 6.7), supporting the contention of Smith (1995) that response rates are not declining. The GSS and NES trends show the difficulty of tracking response rate trends in a survey where many other factors (e.g., length of questionnaire, field procedures, etc.) are not constant.

The relatively few long-run trends that are available on surveys conducted abroad lead to more mixed conclusions. Nonresponse rates to the Canadian Labor Force Survey (the equivalent of the CPS in the United States) are remarkably flat over a 24-year period (see Figure 6.8) (Hidiroglou, Drew, and Gray, 1993).

Nonresponse rates for the Swedish Labor Force Survey and Survey of Living

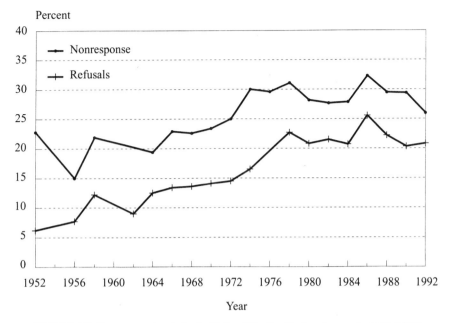

FIGURE 6.6. Nonresponse trends for the National Election Studies. *Source:* Luevano (1994)

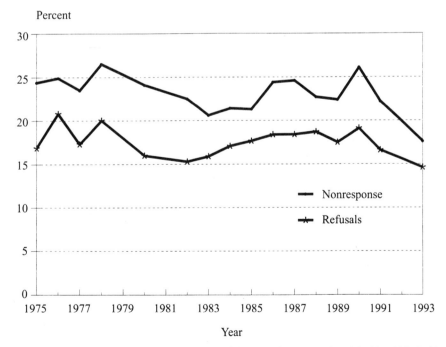

Figure 6.7. Nonresponse trends for the General Social Survey. *Source:* Davis and Smith (1990), Smith (1995).

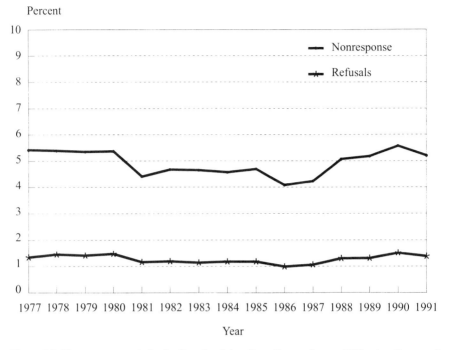

Figure 6.8. Nonresponse trends for the Canadian Labor Force Survey. *Source:* Hidiroglou, Drew, and Gray (1993).

Conditions are presented later in Section 6.3 (see Figures 6.15 and 6.16) (De Heer and Israëls, 1992). On both surveys, refusal rates (and overall nonresponse) has been increasing over a 10-year period. There is also evidence that response rates were declining in the previous decade (Lindström and Dean, 1986).

Finally, we examine two examples from Japan. The first is the "Survey Concerning Social Consciousness" which has been conducted annually since 1974 among a sample of 10,000 persons nationwide (Sugiyama, 1991). Both the refusal rate and nonresponse rate have been increasing steadily over this time frame (Figure 6.9). A similar trend is also observed for the second, the "Survey Concerning National Life" (Sugiyama, 1991).

In summary, depending on which surveys one examines, and for what time period, one could reach different conclusions about response rate trends. Why are these trends not as clear as the claims made by survey practitioners suggest? We have already alluded to some key reasons above, which we shall return to now.

First, efforts to maintain response rates at relatively constant levels may have increased over time. This is especially evident in the CPS trend, which suggests that increased calling efforts have contributed to the reduction of the noncontact portion of nonresponse. (An alternative explanation is that it is easier to contact people today than in past decades, an argument that would not be taken seriously by many

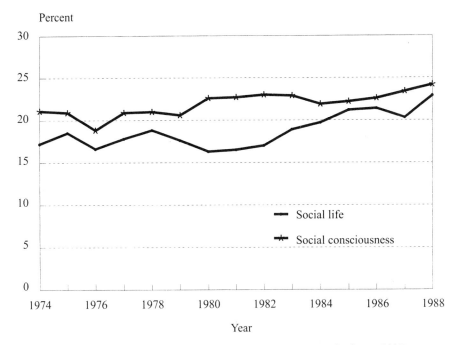

Figure 6.9. Nonresponse trends for two Japanese surveys. *Source:* Sugiyama (1991).

survey researchers.) Documented evidence of increased effort is hard to come by. Botman and Thornberry (1992) report that the average number of call attempts per respondent has increased from 2.6 to 3.2 from 1981 to 1991 on the NHIS. Similarly, Meier (1991) reports increases in the average number of calls per address from 2.4 in 1983 to 3.4 in 1989 on the British National Readership Survey.

A second reason may be that survey organizations are often committed to reaching or maintaining a certain response rate for a survey (whether as public demonstration of survey quality for government surveys, or because of promises to clients or funding agencies in the case of academic surveys). Given such commitments, organizations will allocate or redistribute resources to meet the stated response rate goal. This is also true of continuing surveys. For example, the Census Bureau prides itself on the 95% response rates for the CPS, and this expectation is probably communicated through all levels of the organization.

These arguments may suggest two types of patterns in response rate trends. Firstly, a survey may exhibit a gradual decline that is hardly noticeable from year to year, but is revealed through longer time trends. This may lead to a realization of the decline and concerted action to rectify the situation, leading to a modest upswing in response rates for a period of time. This cycle may make the detection of trends more difficult. This may be one reason for the decline in nonresponse to the NES following the sudden upswing in 1985 (see Figure 6.6).

Another pattern that may occur results from a change in procedures that allows slippage in a previously established response rate to be more readily accepted, or at least attributed to changes in data collection procedures. An example of this is the CPS before and after the introduction of the redesigned survey in January 1994 (see Tucker and Kojetin, 1994). The monthly nonresponse and refusal rates before and after the transition are shown in Figure 6.10. The nonresponse rate in the two years prior to the transition averaged 4.57%, whereas it is 6.42% in the two years following the transition. Similarly, the average refusal rate has increased from 2.77% to 3.71% for the same time periods. A similar effect was found for the Canadian Labour Force Survey (Dufour, Simard, and Mayda, 1995). We expect that response rates to the CPS will not return to the pre-1994 level, and that the new level has become accepted as the target for future data collection efforts. An earlier example is cited by Scott (1978), in which a sudden drop in SRC response rates in 1963 was precipitated by an experiment on limiting the number of calls. Despite this being a one-time experiment, response rates never regained their former level. Scott (1978, p. 2) concludes: "This suggests that interviewer and supervisory expectations may have been lowered by the experience of having two studies on which low response was acceptable."

In sum, we believe that cooperation with surveys is declining in the United States, or at least it is becoming harder to maintain the same levels of cooperation as were evidenced previously. If we assume that changes in design or data collection procedures are not primarily responsible for this trend, we must search for reasons

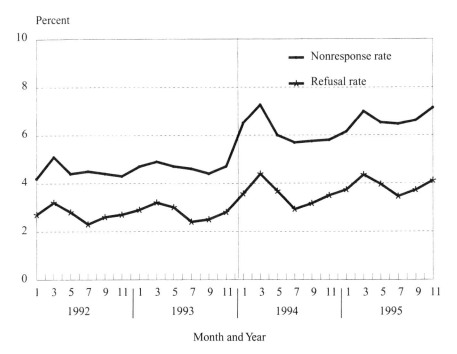

Figure 6.10. Nonresponse rates before and after the January 1994 redesign of the Current Population Survey. *Source:* Tucker and Kojetin (1994).

outside the survey research community. To what do we attribute these declines in survey cooperation? In other words, what societal trends help us understand the phenomenon of survey nonresponse and why it occurs?

A number of hypotheses have been advanced for the impact of secular changes in U.S. society on survey cooperation. While some of these hypotheses have been discussed elsewhere (Chapter 5) to explain differences in cooperation rates among subgroups, here we are interested in changes over time at an aggregate level and what that may mean for response rates in general.

One set of hypotheses relate to the changing demography of U.S. society. Arguments relating to demographic changes in society are really not causal at the societal level. Rather, the changing mix of U.S. society is hypothesized to affect nonresponse through changes in the relative proportions of various subgroups, and their different levels of likely cooperation. Nonetheless, changing demographics are a popular hypothesis for declines in survey participation. What are the characteristic demographic trends in U.S. society that could explain such declines? In Chapter 4 we discuss a number of trends that may impact contact rates; in this chapter, we focus on some of the trends relevant to cooperation.

One crucial trend relevant to social environmental correlates of survey cooperation is urbanicity. The increasing urbanization of the U.S. population over the last several decades is well known, and is illustrated in Figures 6.11 and 6.12. Figure 6.11 shows the decline in the rural population over the last five decades. While rela-

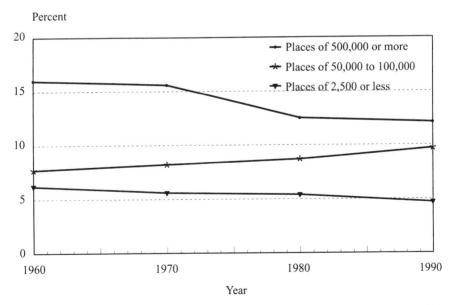

Figure 6.11. Percentage of persons living in large cities, other urban, and rural areas, 1960–1990. *Source:* U.S. Bureau of the Census, *U.S. Census of the Population: 1920–1990,* vol. 1; and U.S. Bureau of the Census, *Statistical Abstracts of the United States: 1994* (114th edition). Washington, DC, 1994, p. 8, Table 1.

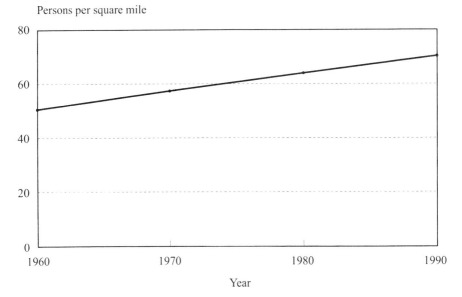

Persons per square mile

Figure 6.12. Population per square mile of land area, 1960–1990. *Source:* U.S. Bureau of the Census, *U.S. Census of the Population: 1920–1990,* vol. 1; and U.S. Bureau of the Census, *Statistical Abstracts of the United States: 1994* (114th edition). Washington, DC, 1994, p. 8, Table 1.

tive growth is uneven over different size urban areas (for example, central city populations have declined in relative terms in recent decades), the overall trends as illustrated is an increase in the urban (or nonrural) population. A related trend is the increase in population density, as measured in persons per square mile (see Figure 6.12). As we discuss later, density may be more proximal correlate of cooperation than urbanicity. Another trend we present here for illustrative purposes is the proportion of single-person households among all occupied housing units. The steady increase in single-person households since 1960 can be seen in Figure 6.13. As we saw in Chapter 5, single-person households are a key correlate of survey cooperation, and have also been identified as a possible cause of the declining census participation rates over the last few decades (see Dillman, 1991; Fay, Bates, and Moore, 1991; Kulka *et al.*, 1991). While the trends for other demographic characteristics (e.g., race/ethnic composition, age distribution, etc.) are less clear, it is nonetheless important to examine response rate trends in the context of possible secular changes in correlates of nonresponse over time.

Another demographic trend hypothesized to have impacted response rates is the decrease in discretionary time availability. While this is likely to affect the likelihood of finding someone at home (see Chapter 4), it may also affect the likelihood of cooperation given contact. Lack of time is a commonly used reason for declining the survey request (see Chapter 8), and to the extent that the perception that the time available for such activities is increasingly constrained, cooperation may be affect-

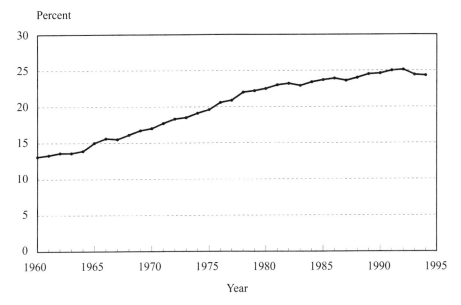

Figure 6.13. Percentage of single-person households among all occupied housing units, 1960–1995. *Source:* U.S. Bureau of the Census, *Historical Abstracts of the United States, Colonial Times to 1970, Bicentennial Edition, Part I.* Washington, DC, 1975, p. 42, Table A335-349; U.S. Bureau of the Census, *Current Population Reports, Series P-20,* No. 467 and earlier reports; U.S. Bureau of the Census, *Statistical Abstracts of the United States:* 1993 (113th edition); Washington, DC, 1993, p. 56, Table 67 and earlier *Abstracts;* and Rawlings, S.W., and Saluter, A.F., *Household and Family Characteristics: March 1994,* U.S. Bureau of the Census, Current Population Reports, P20-483, U.S. Government Printing Office, Washington, DC, 1995, p. x, Table C.

ed. One clear trend that may affect change in contact rates over time is the increased participation of women in the labor force (see Chapter 4), resulting in decreased likelihood of finding an adult member of the household at home. Smith (1995) notes that the increased labor force participation of women might have the added effect of reducing the pool of highly competent field interviewers, thereby lowering response rates.

A related demographic trend likely affecting both contact and cooperation rates is the increase in crime rates and related increase in fear of criminal victimization. House and Wolf (1978) examined trends in response rates across primary sampling units (PSUs) between 1956 and 1972, and concluded that rising crime rates provide a sufficient explanation for increases in nonresponse and refusal in large cities in this period. We believe that increases in crime (and particularly violent crime) not only produce increased fear of victimization, but also lower levels of trust and interpersonal safety. In other words, householders may not only be less likely to open their door to strangers, engage in telephone conversations with strangers, or erect barriers to such contact (leading to increased rates of noncontact, discussed in Chapter 4), but may also be less likely to cooperate with a survey request once contacted.

Another set of hypotheses for declines in survey cooperation relate to increasing alienation, decreasing norms of civic duty, and the like. These are popular hypotheses for declining rates of electoral participation in the United States (Abramson and Aldrich, 1982; Conway, 1991; Rosenstone and Hansen, 1993). To the extent that participation in surveys (particularly those conducted by the federal government or its agents) can be likened to other forms of participation in the polity, similar arguments are used to account for declining levels of cooperation with survey requests (see Mathiowetz, DeMaio, and Martin, 1991; Couper, Singer, and Kulka, 1997). Knack (1990, 1992) characterizes participation in surveys as one form of socially cooperative behavior, along with voting turnout, charitable giving and income tax compliance. He cites declines in these actions over time as evidence of a general erosion of social cooperation in the United States (Knack, 1992; see also Putnam 1995).

Tucker and his colleagues (Tucker *et al.*, 1991; Kojetin, Tucker, and Cashman, 1994) have attempted to relate both these political trends and economic indicators to changes in refusal rates to government surveys. Using the notion of political eras to reflect the fact that response rate declines are not steady over time, they identify three key period for trends in CPS refusal rates. These are the "Sixties or Vietnam" era (1960–1974), the "post-Watergate or Neoconservative" era (1975–1988), and finally the "post-Reagan" era (1989–1991). While the latter is too short to permit meaningful analysis, Kojetin, Tucker, and Cashman (1994, p. 13) find evidence that political and economic conditions were reliably related to CPS refusal rates in the first two eras. However, their work also demonstrates the difficulty of using time trend data to establish causal links between societal conditions and participation in surveys. These kinds of studies remain suggestive at best.

Increasing concerns about privacy and confidentiality have been posited as one cause of declining mail return rates to the U.S. decennial census (Fay, Bates, and Moore, 1991; Kulka *et al.*, 1991; Singer, Mathiowetz, and Couper, 1993). These may be more proximate explanations for changes in survey response rates, as they are more closely linked to the agencies conducting or sponsoring the surveys. Goyder and Leiper (1985) constructed a privacy index, based on a content analysis of privacy-related objections to the decennial census in British, Canadian, and American newspapers over five decades. Not only do they find privacy-related concerns increasing almost monotonically from the 1930s to the 1980s, but that such concerns are negatively related to survey response rates over time (controlling for survey design changes in multivariate analysis). Thus, it appears important not only to track broader societal trends that may shape the broader survey-taking climate, but also to collect trend data on more specific attitudes and reactions to research in general and surveys in particular. The Walker Industry Image Surveys (conducted biannually since 1980) are notable in this regard, tracking both reported levels of exposure to surveys, but also attitudes toward the survey industry and reactions to the interview experience.

A final set of hypotheses for declining cooperation rates can be described as the "over-surveying" effect. The basic logic of the over-surveying hypothesis is that sample persons or households in the United States are a finite resource, and that the increasing number of surveys will lead to multiple approaches to the same persons or households. The problem is best stated by Remington (1992, p. RC-7) who notes that "Americans have a limited reservoir of goodwill to expend on intrusive and unsolicited telephone contacts of any kind—reasonable or unreasonable. Overuse of this reservoir will engender a drought in respondent cooperation with accompanying serious short- and long-term ramifications for the research industry."

Concerns about over-surveying are blamed both on increases in the number and size of surveys conducted in the United States, as well as on the increase of activities such as telemarketing. These two trends, it is argued, have produced increased competition for the scarce resources of householders, leading to increased refusal rates. In a series of telephone surveys on marketing and survey research conducted by Walker Research (1992), respondents were asked whether they had ever participated in a survey and whether they had done so in the past year. Figure 6.14 shows that the number of people who report participating in a survey in the past year has been steadily increasing over the last decade.

Of greater concern appears to be the threat from telemarketing, especially for telephone surveys. This stems from the dual concern of over-use of the telephone as a medium for contacting households (increasing the public's annoyance at such approaches), and concerns about surveys being used for disguised sales pitches (de-

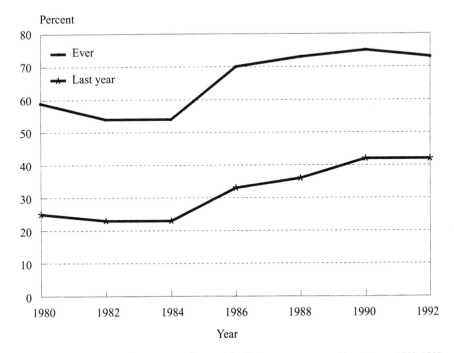

Figure 6.14. Percentage of persons reporting participation in surveys ever and in past year, 1980–1992. *Source:* Walker Research Inc. (1992).

creasing the public's confidence in the legitimacy of surveys). More than 30 years ago, Baxter (1964) reported that 27% of respondents to the California Poll claimed to have experienced selling under the guise of surveying, or SUGGING ("Has anyone ever said they wanted to interview you on a survey—either in person or by telephone—and then tried to sell you something?") Schwartz (1964) reports a similar finding using different methods. About half of those who reported such an incident said that it had occurred in the past year. In a set of four similar surveys conducted in Connecticut in 1966–1967, respondents were asked, "Did you by any chance first think that I was trying to sell you something when I asked to interview you?" Allen and Colfax (1968) report that 25–30% said they definitely suspected an interviewer sales attempt.

Remington (1992) reports on a more recent telephone survey in which respondents reported an average of 26 sales calls in the past year, compared to an average of less than 2 telephone surveys in the same time period. When asked if there was any difference between telephone survey research and telemarketing, 65% of respondents believed they were the same thing or didn't know if they were different. Remington also reports that about one-quarter (24%) of respondents had experienced a disguised sales pitch or SUGGING. Whether or not the incidence of SUGGING or FRUGGING (fundraising under the guise of surveys) has increased in recent years, it is safe to say that the telemarketing industry has experienced tremendous growth over this time period, and the impact of such changing sales approaches on survey cooperation deserves to be watched.

We have seen in this section that disagreement exists over whether response rates have been declining or will continue to do so in the future. There are also a number of plausible explanations offered for the apparent trends observed. However, it is extremely difficult to assemble data to examine long-term trends in survey participation and to evaluate alternative hypotheses for such trends. We are not as convinced as Brown (1994, p. 230), who states that the reasons for the strongly downward trend in response rates in the United Kingdom are ". . . commonplace in market research discussions: declining levels of public co-operation in general; increasing mobility; the fears of admitting a stranger to the home; the impact of doorstep selling, disguised as market research, and the like." Nonetheless, we would argue strongly for continued efforts to collect data on response rates and its components and correlates in ongoing surveys. Despite the difficulty of assembling evidence, we do believe that the factors discussed here contribute to establishing the context in which a survey request is made, and that the survey-taking climate does influence (however indirectly) the decision to participate.

6.3   CROSS-NATIONAL DIFFERENCES IN RESPONSE RATES ON SIMILAR SURVEYS

If assembling time trends of response rates and correlates within a single country is difficult, doing so across different countries is even more so. Nonetheless, such an effort could reveal important differences in response to comparable surveys across countries that could be explained by variation in survey-taking climate. The results of a recent effort by an international group of researchers to assemble such data are reported by de Heer and Israëls (1992; see also de Heer, 1992). This work attempted to obtain response rates for the previous 10 years for labor force surveys in a number of countries (primarily North American and European). In addition, a questionnaire was used to elicit key survey design characteristics of the various surveys. This allowed a comparison of response rates across countries and an evaluation of design differences that could account for varying levels of response across essentially similar surveys. As we noted earlier, this effort again made clear the difficulty of obtaining comparable information on response rate components across surveys and countries. Further, differences in response rates across countries may be explained as much by survey design differences as by differences in the social environment. For example, the Netherlands obtains lower response rates (and higher refusal rates) to its labor force survey than most other European countries. However, organizational policy is not to follow up on reluctant respondents, a practice that is widespread elsewhere. Whether this reflects design differences or underlying societal factors (e.g., different politeness norms on accepting initial refusals) is not clear.

While the effort involved in collecting these comparative data has thus far exceeded the benefit derived from the data in understanding nonresponse and how it may differ across countries, we still believe this type of activity to be useful. However, while the *ex post facto* collection of comparative data on continuous surveys may be difficult, building such an effort into the design phase of cross-national studies may be more productive. If these are accompanied by the assembly of national-level data on social environmental correlates of survey participation (e.g., degree of urbanization, level of political participation, alienation, education, crime rates, etc.), a database could be built to permit comparative analysis of response rates within and across countries. This will permit the identification of environmental factors that may help us understand different levels of survey participation across countries.

6.4 "NATURAL EXPERIMENTS" AT THE SOCIETAL LEVEL

While survey research is rarely at the forefront of public attention, public debate over issues such as privacy can spill over to affect survey response rates, suggesting that the survey-taking climate is susceptible to public opinion on the legitimacy of such enterprises and their perceptions of agencies conducting research. A key example of this is the "Project Metropolit" incident in Sweden. In 1986, it was revealed in the press that a longitudinal study of some 15,000 individuals born in 1953 in the Stockholm metropolitan area had been going on for years without the knowledge or consent of the persons involved, using data from administrative records. The subsequent heated public debate about invasions of privacy had an immediate and lasting effect on willingness to participate in surveys. This affected both government agencies (Statistics Sweden) and private organizations, even though none were connected with the Project Metropolit study (see Dalenius, 1988; Lyberg and Dean, 1992;

Lindström, 1986). While response rates for surveys conducted by Statistics Sweden improved once the debate died down, they never returned to previous levels.

Figures 6.15 and 6.16 show response rate trends for two surveys conducted by Statistics Sweden—the Labor Force Survey and Survey of Living Conditions, respectively (see de Heer and Israëls, 1992). These both clearly show both the immediate effect of the Metropolit debate on refusal rates in 1986, and the long-term effects of this debate on response rates in Sweden.

Another example of the effect of intense public scrutiny on the research endeavor relates to the postponement and eventual cancellation of the German census of 1981 (see Butz, 1985; Dalenius, 1988). Scattered objections to the census for various reasons in the early part of the decade erupted into widespread opposition in 1983, with three of the most influential news publications urging that the census be stopped. A survey conducted at the time revealed that one quarter of the population would break the law and risk substantial fines by boycotting the census (Butz, 1985, p. 90). The root of the opposition centered around concerns that census data would be used against individuals by government agencies. The opposition was so strong that the census was canceled by the Constitutional Court.

While not directly related to surveys, these events remind us that surveys rely on the goodwill of the public, a potentially fragile commodity. In the United States, we have thus far been spared widespread public debate that could erode confidence in

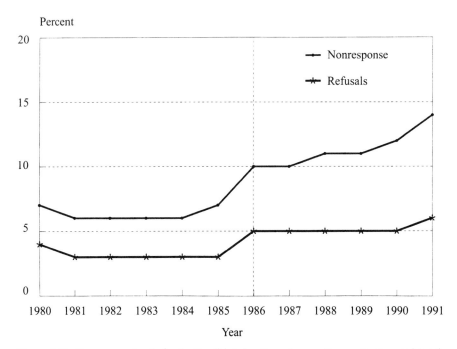

Figure 6.15. Nonresponse trends for the Swedish Labor Force Survey. *Source:* De Heer and Israëls (1992).

Percent

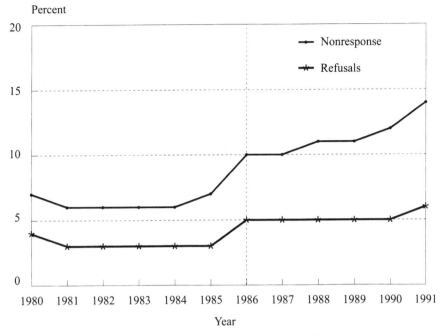

Figure 6.16. Nonresponse trends for the Swedish Survey of Living Conditions. *Source:* De Heer and Israëls (1992).

surveys or in the agencies and organizations that sponsor or conduct surveys. However, the decennial censuses (especially the 1990 census) have been receiving increasing attention from the media, and this may affect survey cooperation in the long run, especially if such debate becomes more heated or negative in the future. While the United States has been fortunate that there have been few such instances as have occurred in Sweden and the Germany, the events nonetheless reveal much about the impact of the survey-taking climate on cooperation rates. Furthermore, the lower than expected mail-return rate to the 1990 U.S. decennial census has sparked renewed interest in possible reasons for this nonresponse. While the census is very different from interviewer-administered surveys, much of this research effort is focusing on macro-level features of U.S. society that may account for this phenomenon. We believe some of this is relevant to our discussion here (see Fay, Bates, and Moore, 1991; Kulka *et al.*, 1991; Singer, Mathiowetz, and Couper, 1993; Couper, Singer, and Kulka, 1997).

6.5 SUBNATIONAL VARIATION IN SURVEY COOPERATION

With cross-sectional studies conducted in a single country (the United States), we cannot examine societal-level impacts on survey cooperation. We need to look at subsocietal level variation (lower levels of aggregation) for evidence of these effects. At the subnational or community level, a variety of contextual factors are hypothesized to influence survey cooperation, including urbanicity, population density, crime rates, and a lack of social cohesion. In the remainder of this chapter, we examine the separate and combined effects of these factors on cooperation.

6.5.1 Urbanicity

One of the most consistently documented ecological correlates of survey cooperation is urbanicity (see Brehm, 1993; House and Wolf, 1978; Goyder, Lock, and Mc-Nair, 1992; Smith, 1983; Steeh, 1981). Residents of small towns are found to cooperate at a higher level than those in large cities, while those in rural areas respond at an even higher rate. Effects have been found for a number of different measures of the urbanicity concept: old city versus new, inner city versus suburb, large city versus small, urban versus rural, and so on. In all of these, the trend is clear: residents of inner-city areas of large metropolitan areas exhibit the lowest levels of cooperation, while those in rural areas have the highest.

The pervasiveness of the urbanicity effect is witnessed in the United States, in Canada (e.g., Gower, 1979; Goyder, 1987), the United Kingdom (e.g., Lievesley, 1988; Market Research Society, 1981), Sweden, Finland (e.g., Ekholm and Laaksonen, 1990), Hungary (Marton, 1995), Poland (Lutyńska, 1987), and Japan (e.g., Jay *et al.*, 1993; Sugiyama, 1991), among others. Consistent effects of urbanicity are also found in the literature on helping or prosocial behavior (see Steblay, 1987).

Although urbanicity is a common correlate of nonresponse, the finding, by itself, does little to explain *why* people in different size communities differ in their likelihood of cooperation with a survey request. We believe that many of the effects of urbanicity found in the literature may be explained in terms of greater population density, higher crime rates, and social disorganization that are often associated with life in large urban areas.

6.5.2 Population Density

A second contextual factor is population density, which is hypothesized to reduce cooperation through the experience of crowding. It is important to distinguish between density, a physical condition, and crowding, an experiential state associated with density. The former is a necessary antecedent rather than a sufficient condition for the experience of crowding (see Stokols, 1976, p. 50). McCarthy and Saegert (1979, p. 55) suggest that the perceptions of crowding are not related to density *per se* but rather to the experience of excessive social encounters in high-density areas that lead to social overload. Social psychological theories of density and crowding (see for example Baum and Valins, 1979; Tittle and Stafford, 1992) suggest that residents of densely populated urban environments face a greater volume and variety of contacts with strangers, most of which are impersonal and fleeting. The stress produced by such overstimulation leads to avoidance of contacts with strangers (Aiello and Baum, 1979). This leads to less helping behavior and greater distrust of

factors singly, without household-level controls, in order to validate the bivariate results found in the survey literature. Next we model the effect of each of these factors, controlling for appropriate household-level variables, to examine the marginal impacts of these contextual factors on cooperation rates. Finally, we combine a number of the key contextual variables in a model, both excluding and including household-level controls.

6.6 ANALYSIS OF ENVIRONMENTAL INFLUENCES ON COOPERATION

Ideally, the modeling of environmental effects together with household-level variables would suggest the use of multilevel models (Bryk and Raudenbush, 1992; Hox, 1995). However, the models we use here are fixed-coefficient models, not random-coefficient models. The variance estimates of the coefficients for the urbanicity variables reflect the clustering of the samples into geographical areas (many of which are completely homogeneous on the urbanicity measures). The statistical tests thus reflect the inter-area component of variance (see Section 3.4.6).

The household-level variables included as controls here are those discussed in Chapter 5. To these variables are added the social environment indicators that the reader saw in Chapters 4 and 5. The first is an indicator reflecting three categories of urbanicity: central city of a consolidated metropolitan statistical area (CMSA) (the largest 18 metropolitan areas of the United States), balance of the CMSA, and other areas (including both metropolitan statistical areas (MSAs) containing cities of 50,000 or more and non-MSAs (less urban areas)). Earlier analyses treated these last two (other urban and rural) as separate categories, but they were combined after no significant differences were found between them.

We use a county-level measure of serious crimes as an indicator of fear of crime. The measure used in the analysis is the crime rate of the household's county of residence, as measured in 1988 by the FBI Uniform Crime Reports of incidents reported to the police. It would be preferable to have a measure of crime rates at lower spatial aggregation but none was available nationally. Crime rates are scaled to measure serious crimes per 1,000 population.

As an indicator of crowding, we use population density (1,000 persons per square mile) measured at the city level. For sample cases in unincorporated areas, we use county-level density measures. The density measure is scaled to thousands of people per square mile. As with crime rates, the density measure fails to reflect intracity variation in population density.

Without direct measures of social cohesion at the neighborhood or community level, we must again make use of indirect indicators. We argue that areas with a large percentage of persons residing in group quarters exhibit a greater transience of the population, and hence reduced attachment to the community. Similarly, the larger the proportion of homes that are owner-occupied, the greater the permanence and commitment of residents to the neighborhood (see Sampson, 1988). Single-unit dwellings may also be associated with greater cohesion, relative to multiunit structures (and particularly high-rise apartment complexes) (see Smith and Jarjoura,

1989). Race diversity is another indicator of a lack of social cohesion. It is argued that greater racial diversity in a community reduces the cooperation among groups, and hence reduces social cohesion (see Miethe, Hughes, and McDowall, 1991; Sampson and Groves, 1989). We use the percentage of persons of minority race as an indicator of race diversity. Finally, we argue that children act as catalysts for the involvement of parents in community-based activities (with schools often being the center of those activities). We thus use the proportion of children (and particularly those of school-going age) as a further indicator of social cohesion (see Lievesley, 1988). Wilson (1985) found that the presence of children in urban and suburban neighborhoods had a positive effect on perceptions of trust and helpfulness. The geographical aggregate census data are limited to broad age categories, forcing us to use the percentage of persons under 20 as an indicator of the presence of children. All of these social cohesion indicators are measured at the level of the census block. This is an area comprising approximately 100 housing units, roughly equivalent to a city block in urban areas.

As noted earlier, our analytic plan for examining these data has the following three goals:

1. to examine whether the data exhibit the same bivariate correlates of nonresponse found in the past literature
2. for each of the single social environment correlates, to examine whether their effects can be explained by household-level attributes
3. to estimate a model including all social environmental and household-level influences.

We tackle each of these in turn.

6.7 BIVARIATE RELATIONSHIPS OF SURVEY COOPERATION AND ENVIRONMENTAL FACTORS

The first step of the analysis tests whether findings of social environmental differences in nonresponse common to the past literature apply to these data as well. Table 6.1 contains the results of the logistic regressions of cooperation (response given contact) on each of the social environment factors singly. Dummy-variable indicators for each of the surveys were included in the models to reflect differences in cooperation rates across surveys attributable to a variety of design differences; however, these are not presented here. The logistic coefficients for the social environment variables are presented, with the standard errors in parentheses. The likelihood ratio test (see Hosmer and Lemeshow, 1989) is also presented, for the significance of the social environment variables, relative to the base model with only dummy indicators for the different surveys.

The findings in Table 6.1 support those of House and Wolf (1978) and Goyder, Lock, and McNair (1992) with respect to urbanicity, population density and crime rates, respectively. Residents of large metropolitan areas (and particularly the core

Table 6.1. Coefficients of logistic models predicting cooperation versus refusal by various sets of environmental attributes (standard errors in parentheses)

Independent variables	Urbanicity	Population density	Crime rate	Social isolation
Constant	2.68**	2.66**	2.76**	2.21**
	(0.14)	(0.14)	(0.16)	(0.21)
Urbanicity				
Central City	–0.56**			
	(0.12)			
Balance of CMSA	–0.35**			
	(0.11)			
Other	—			
Population density[a]		–0.041**		
		(0.0083)		
Crime rate[b]			–0.041**	
			(0.015)	
Percent persons in group quarters				0.010
				(0.011)
Percent homes owner occupied				0.00037
				(0.0019)
Percent persons of minority race				–0.00090
				(0.0015)
Percent single-unit detached dwellings				–0.00049
				(0.0016)
Percent persons under 20 years old				0.016**
				(0.0044)

Note: Dependent variable coded 1 = interview, 0 = refusal. Coefficients for dummy variables for individual surveys omitted from table.
[a]Measured in thousands of people per square mile.
[b]Measured in serious crimes per 1,000 population.
*$p < 0.05$.
**$p < 0.01$.

cities of such areas), those in densely populated areas, and those in areas with high crime rates are less likely to accede to an interviewer's request to participate in a survey.

This is consistent with the literature that crowding, fear of crime, and high levels of stimulus input associated with urban areas lead to avoidance of contact with strangers. This may be especially true of visits by interviewers to sample persons' homes or primary environments. The home as a primary environment gives individuals a measure of control over the number and type of people with whom they must interact and the number of social interruptions they are likely to experience (see Schiffenbauer, 1979). Intrusions into such primary environments may be particularly threatening for those in high-crime, high-density urban areas that may provide little opportunity to retreat from such intrusions. In addition, there is anecdotal evidence that concerns about personal safety on the part of interviewers may alter their persuasion behavior in such neighborhoods, leading to greater acceptance of reluctance to participate.

It is noteworthy that population density and crime rates appear to have greater predictive power than the urbanicity indicators themselves (as judged by the log likelihood statistics and their degrees of freedom). Most surveys can provide comparisons of response rates in urban versus rural areas, and, almost uniformly, it is found that lower rates apply in urban areas. The more powerful predictive value of population density and crime rates may suggest that those attributes of urban areas underlie the lower urban participation rates.

Table 6.1 also presents coefficients for measures that might be considered indicators of social cohesion: the percentage of persons in group quarters, the percentage of homes that are owner occupied, the percentage of persons of minority race, the percentage of single-family housing units, and the percentage of persons under 20 years old. These variables were entered together in the final model in Table 6.1, in order to measure the net effect of social cohesion measures on cooperation. As measured by the log likelihood test, the variables as a group explain measurable amounts of variation in rates of cooperation ($G = 24.91$, $df = 5$, $p < 0.01$). However, few of these predictors individually reach traditional levels of statistical significance. The single exception is the percentage of persons under 20, which has a positive effect on survey cooperation. This supports the findings of Lievesley (1988) in the survey context and Wilson (1985) in terms of trust and helping behavior among younger households.

At this point in the analysis, most of the speculations and findings from the past literature on environmental influences on survey cooperation have been supported with these data. Despite the fact that the six surveys have higher overall cooperation rates than do most of the surveys studied in the past literature, the sets of social environmental correlates are largely similar.

6.8   MARGINAL EFFECTS OF INDIVIDUAL ENVIRONMENTAL FACTORS

The final step in the analysis examines whether the combined effects of social environmental and household-level variables (discussed in Chapter 5) are consistent with the theoretical propositions above. Two multivariate combinations of predictors are explored. The first addresses the question of whether the inference using single-predictor models of social environmental effects would change when multiple predictors are examined jointly. The second examines whether individual environmental predictors maintain their influence on cooperation in the presence of controls for household-level predictors (those discussed in detail in Chapter 5).

6.8.1   Combined Effects of Environmental Influences

We know that the central cities have higher crime rates ($r = 0.56$). Does the lower cooperation rates of the central cities merely reflect the higher crime rates of those areas? In the first model ("environment model") in Table 6.2, all the social environmental predictors found useful individually are included. The size of each coefficient declines relative to its size in the models in Table 6.1.

The largest reductions in size of effect appear for the crime rate variable (a reduction from –0.041 to –0.011). From other analyses (not presented here), we know that this is largely the effect of controls on population density and urbanicity. This is consistent with the notion that the anonymity and reluctance to interact with strangers, common to densely populated urban areas, is sufficient for reduced survey cooperation, and that the influence of crime rates on reluctance is marginally insignificant. (We remind ourselves that another interpretation is the fact that the crime rate indicator may be measured at the wrong level to observe effects.)

The model that combines all social environmental effects also shows reduced effects of the urbanicity indicators. This shows that in locales that have similar crime

Table 6.2. **Coefficients of logistic models predicting cooperation versus refusal, excluding and including household-level controls**

Independent variables	Environment model		Full model	
	Coefficient	(Standard error)	Coefficient	(Standard error)
Constant	2.44**	(0.20)	2.72**	(0.29)
Social environment				
Urbanicity				
Central city	–0.31*	(0.15)	–0.27	(0.17)
Balance of CMSA	–0.23*	(0.11)	–0.14	(0.13)
Other	—		—	
Population density[a]	–0.021*	(0.010)	–0.022*	(0.011)
Crime rate[b]	–0.011	(0.017)	–0.0053	(0.019)
Percent under 20 years old	0.012**	(0.0041)	0.0096*	(0.0047)
Household				
Owner occupied			–0.20	(0.18)
Monthly rent for renters[c]			–0.044	(0.033)
House value for owners[d]			–0.011	(0.0060)
Household age:				
All persons <30 years			0.67**	(0.14)
Mixed householder ages				
All persons >69 years			0.42*	(0.19)
Single person household			–0.36**	(0.13)
Children <5 in household			0.63**	(0.15)

Note: Dependent variable coded 1 = interview, 0 = refusal. Coefficients for dummy variables for individual surveys omitted from table.
[a] Measured in thousands of people per square mile.
[b] Measured in serious crimes per 1,000 population.
[c] Measured in units of $100.
[d] Measured in units of $10,000.
*$p < 0.05$.
**$p < 0.01$.

rates and population density, the contrasts of central city, suburb, and other locales are not as important. The direction of the urbanicity effects are as found in the simpler model, but somewhat smaller, achieving levels of statistical significance at the 0.05, rather than 0.01 level.

Finally, the variable least affected by the presence of other social environmental variables is "percent of persons under 20 years of age," a proxy for the density of families with children in the immediate area of the sample household. Such neighborhoods show strong tendencies toward cooperation with the survey requests, wherever they are located, whatever the density and crime rates.

6.8.2   Environmental Effects Controlling for Household-Level Predictors

The final step in the analysis is to examine all the social environmental predictors in the presence of the household-level control variables. With that analysis, we again compare the values of the coefficients for each of the social environmental variables and their marginal effect on the overall predictive value of the model. (The reader has already seen this analysis earlier, as the last model in Table 5.21.)

The second model in Table 6.2 shows that the effects of the urbanicity variables decline to nonsignificant levels when controlling for household-level variables. The indicator for suburbs of CMSAs (from –0.23 to –0.14) is greatly diminished. In contrast, population density remains a negative influence on cooperation. These results, taken in combination, may provide insight into the common finding across the world that urban areas achieve lower cooperation rates than do nonurban areas. Some of these effects reflect that some parts of urban areas are very densely populated. The high volume of impersonal contacts with strangers may lead to lower cooperation with survey requests. Second, urban areas are also populated by households that have characteristics related to low cooperation (e.g., single-person households), and that also contributes to lower urban cooperation rates.

The impact of the percentage of persons under 20 years old in the neighborhood of the household remains positive, controlling on household attributes. The controls do radically decrease the value of the coefficient (from 0.012 to 0.0096), but it remains significant at the 0.05 level. This is not unexpected either statistically or theoretically. The Lievesley (1988) hypothesis on this attribute is that families with young children tend to grant interviews, and the household variables contain two measures—whether the household has children and whether all persons in the household are less than 30 years old. In the presence of these household attributes, the social environmental measure of whether there are young persons in the neighborhood is still a significant predictor of cooperation. The data support the original theoretical proposition; the locus of its effects are partially at the household, and partially at the neighborhood level.

Finally, we note that attention to individual coefficients in the full model in Table 6.2 may mislead. It is true that no single coefficient for the social environmental variables achieves traditional levels of statistical significance. It does not imply, however, that different social environments are similar on their rates of cooperation, controlling on household attributes. The social environmental variables as a group *do* significantly improve the fit of the overall model (likelihood ratio $\chi^2 = 17.04$, *df* $= 4$, $p < 0.01$) after household-level effects are accounted for. The importance of individual environmental variables, however, is diminished. One way to quantify this is to compare the change in the two likelihood ratio statistics associated with the social environmental variables, one without the household-level controls, and one with the household-level controls. (We use as the base model one with a constant and the survey dummy-variable predictors.) The change due to environmental predictors is 39.84 without the household controls, and 17.04 with the controls. In short, about $(39.84 - 17.04)/39.84$ or 57% of the measure of fit of the model associated with the environmental variables seems to overlap with the household predictors. We might roughly characterize the household-level influences as explaining over half of the effect of the environmental predictors.

The reader will recall that the environmental variables appear to have different effects on contact (see Table 4.13) than on cooperation. In predicting contact, the urbanicity indicators remained influential, controlling on population density and household-level variables. In predicting cooperation, those variables appear to explain the urbanicity effect. We suspect that population density and the household-level indicators are useful indicators of the lack of social connectedness, which seems to affect cooperation. The contact model, on the other hand, probably misses measures of at-homeness associated with urbanicity (e.g., commuting time, time away for errands).

In one final model, we examined whether the conclusions about the effects of social environment are changed when attributes of interviewers used in the different areas are brought into the analysis. In a model not presented here, we found no important changes in interpretation of environmental effects. Instead, it appears that effects of interviewers are relatively independent of those of the social environment.

6.9 SUMMARY

We believe that surveys are not immune from the influences of societal change, whether these are changes in the demographic composition of the society, or of shifts in attitudes, norms, or values among members of the society. Even though surveys are not deemed a central feature of society and its institutions, they may be indirectly affected by changing relations with such institutions, manifested in increased alienation or cynicism, reduced political efficacy or trust in government, loosened obligations or sense of civic duty, and so on. To the extent that such trends occur in society, it is important to track them and their impact on survey participation over time. While we agree with Lyberg and Dean (1992) that the survey-taking climate is an important element in understanding survey participation rates both cross-nationally and over time, we have noted the difficulty of assembling suitable data to tests these hypotheses directly.

At a subnational level, we are better able to examine the effect of the social environment on survey participation. While we still do not have measures of social psychological attributes (e.g., fear of crime, social isolation, trust, civic duty, etc.), we make use of key ecological correlates. Large urban areas in the United States are often densely populated, often with high crime rates, and sometimes with weakened community ties among their residents. These areas tend also to be populated by single-adult households and households without children. These attributes, singly, are related to lower cooperation with surveys but also reduced helpfulness and prosocial behavior in general. Thus, what to survey researchers forms a limitation on statistical inference is a powerful force affecting everyday behaviors of urban dwellers.

With the data from the decennial match study, we find support for prior findings that large urban areas, dense populations, and areas with high crime rates exhibited lower cooperation on the surveys. We find only modest impact of the indicators of social cohesion on cooperation. Independent effects of the urbanicity, density, and crime rates are measurable. The fact that these three variables are so highly correlated underscores the large net differences between urban and rural areas in survey response rates.

Adding a set of household-level controls identified in earlier analyses as key correlates of survey cooperation, however, substantially reduces the strength of the social environment effects, but does not eliminate them entirely. Roughly speaking, about half of the effect of the social environmental variables arises because of differences in household structure, race, age of household members, presence of children, and socioeconomic attributes of households. That a large portion of the environmental differences in response rates are explained by person-level attributes is important in locating the causes of survey participation. In a practical sense, this means that survey administrators should go below the level of sample areas in assessing the likely cooperation rates in a survey, to subclasses within urban areas, for example, that will have behaviors not dissimilar to members of the same subclass in other areas. From a theoretical perspective, the finding reminds us that a theory of survey participation will be likely to have its powerful insights at the person or household level, less so at higher ecological levels.

On the other hand, the result that some of the lower cooperation among residents of high-density areas remains unexplained by household-level attributes underscores the fact that the data do not identify all the household-level attributes leading to the decision to accept or refuse a survey request. These areas do appear to foster behaviors unlike those in other areas. We believe, along with much of the past literature on crowding effects on social interaction, that the missing variables are a set of social psychological attributes. For example, we would hypothesize that measures of fear of crime at the household level would explain away much of the effect of crime rates. In the absence of these and other such variables, however, we must speculate on the social psychological influences on survey participation.

It is important to remember that the social cohesion indicators (with one exception) do not perform as expected. One possibility is that they are weaker indicators of the underlying concepts than the other social environment variables. The links between crime rates and fear of victimization, and between population density and avoidance of contact with strangers, may be stronger than those for the indicators of social cohesion available to us. The impact of community cohesion in shaping the context in which the request for survey cooperation takes place merits further exploration.

In short, we have learned the following through the analyses presented in this chapter:

- Densely populated, high-crime urban areas tend to have lower cooperation rates. Much of the urban–rural contrast appears to be explained by other ecological variables.

- The remaining tendencies for lower urban cooperation rates appear to be explained by different household compositions in urban than rural areas. Urban areas are populated by greater relative numbers of single person and childless households, who tend not to cooperate.

- Ecological effects on cooperation also remain for areas with children and young adolescents. We view this as added evidence for the effect of social cohesion arising from households with children being spatially clustered (see Chapter 5).

6.10   PRACTICAL IMPLICATIONS FOR SURVEY IMPLEMENTATION

This chapter opened with the display of several time series of response rates, separated into noncontact and refusal components. It would be a fair inference from those displays to conclude that over the past few decades, surveys with high response rate goals have maintained their response rates only with greater efforts in the field. These appear to have reduced the noncontact rates, while the refusal rates have increased. These time series alone give guidance to survey designers. Given our knowledge based on the empirical analysis of Chapters 4 and 5, we know that the decision to maintain overall response rates by reducing noncontacts will change the composition of the nonrespondent pool. For example, nonrespondents will be less likely to reside in multiunit structures or have other impediments to access. The time trends suggest that more attention to calibrating efforts between the reduction of noncontact and of refusals would be wise. Simply focussing on one source of nonresponse (e.g., noncontacts) may change the composition of the nonresponse pool and hence, potentially, change nonresponse error.

On first consideration, it would seem that the social environmental influences on survey response are fixed attributes that the survey designer is powerless to control. Some might react that nothing could possibly be done about these exogenous factors and thus they should be ignored. We believe this to be unwise for two different reasons. First, even though the factors are not under designer control, their salience to the decision of the householder can be altered by how the survey request is framed. For example, if a scandal regarding breaches of confidentiality of medical records is currently affecting response to surveys, then interviewer protocols must distinguish the protection of the survey data from that violated in the scandal. One strategy would be to emphasize aspects of the survey not involving confidentiality pledges but arm the interviewers with detailed explanations to address explicit householder concerns, when they are raised.

A second reason justifying attention to the social environmental factors is that their importance changes over time. As we saw in the CPS trends in Figure 6.2, there appears to be a ten-year period in the response rates of several surveys conducted by the Census Bureau, such that survey response rates are higher during the advertising campaign for the decennial census. This is a time when widely visible efforts are made to place the survey sponsor in the most favorable light possible. Although the advertising concentrates on the decennial census, its benefits extend to other activities of the same organization. Similarly, in longitudinal surveys, events can occur between waves that affect how respondents view their participation in the survey. In government surveys, these might involve notions of political efficacy or trust in government. In commercial customer-satisfaction surveys, legal actions taken against the service providing firm can negatively affect the reactions of sample customers to survey requests. In this case, the value of using a survey organization distinct from the service firm is higher.

Because the survey climate can change over time, it should be monitored. In the United States, we expect that the growing use of computer files of administrative data for statistical purposes, sometimes matched to survey interview records, may lead to greater concerns in the household population to survey data confidentiality. It would be useful to have a social indicator of such concerns in order to be forewarned about such changes.

The third reason for survey designers to consider social environmental factors, as we have shown in the empirical analysis in this chapter, is that there is variation in these factors within a population. Densely populated neighborhoods appear more hostile to survey participation than do others. Neighborhoods with younger children have higher participation rates, other things being equal.

The wonderful attribute of social environmental effects, relative to the household-level influences we observed in the last chapter, is that some of them are easily observable before contact attempts are mounted. In area frame surveys, ecological socio-demographic data can be appended to address listings, providing the survey managers with forewarning of some environmental attributes. In telephone surveys, census data can be appended (albeit at a less informative level of aggregation) to telephone numbers and used to guide assignment of interviewers, scheduling of first calls, etc.

In both modes, interviewers with more experience with the pressures on residents of densely populated areas would be useful. Introductory protocols should give interviewers flexibility in methods to distinguish their approach from those of the hundreds of strangers encountered each day by urban dwellers, to communicate the absence of threat of harm or fraudulent intent, and to adapt the request to the limited time availability of the urban dweller. We would speculate that a key to success in urban areas is destroying the impression that the survey request is another fleeting encounter that serves only the narrow purposes of the requester. This implies that repeated mailings (personalized whenever possible), courtesy visits that do not request anything, but inform the household of the socially useful purposes of the survey organization, as well as other strategies may be disproportionately beneficial in urban areas. These acts might communicate that the organization is willing to invest several visits into addressing the interests, concerns, and lifestyle constraints of the sample household.

Interviewers working in difficult environments encounter different types of persons and different reactions to the survey request. Some strategies of recruiting respondents may be more effective in urban areas, for example, than rural areas. Recognizing these differences during interviewer training and teaching interviewers to be keen observers of both neighborhood and household characteristics may improve rates of survey participation they produce.

CHAPTER SEVEN

Influences of the Interviewers

7.1 INTRODUCTION

In this chapter, we turn our attention to the survey interviewer, examining data from a series of interviewer questionnaires administered to those working on the match studies described in Chapter 3. Our initial goal with these questionnaires was to identify relatively fixed interviewer attributes associated with success in gaining cooperation. However, over the course of our work, we began to place greater emphasis on the interaction between interviewer and householder at the time of the survey request (see Chapters 8 and 9). In other words, we have increasingly come to believe that main effects of interviewers on survey cooperation are likely to be rare, and it is the interaction (in both social and statistical senses) that is critical.

This is by no means meant to imply that interviewers do not play a vital role in surveys. Indeed, it is the behavior of interviewers at individual households and during interactions with householders that we believe to be critical. Despite this view, we present the interviewer-level results here in order to provide a context for the later discussion of the householder–interviewer interaction, and to show how and why we have come to modify our views on the importance of interviewer-level effects.

We first present a more detailed overview of the role and tasks of the survey interviewer in gaining cooperation, identifying the likely sets of factors associated with the interviewer. We then explore each of these sets in turn, discussing demographic characteristics of interviewers, their personality, experience, and expectations. We then turn to a discussion of interviewer behavior across households. Finally, we fit a set of multivariate models to examine the combined effects of these sets of factors on interviewer-level response rates and on household-level cooperation rates. We attempt to translate these results to the household level, using data from the match study, to explore interviewer-level correlates of survey cooperation at the household level.

7.2   INTERVIEWER EFFECTS ON COOPERATION

Survey interviewers plays a critical role in the process of obtaining cooperation to survey requests. Wood (in Hoinville, Jowell, and Associates, 1977) notes that:

> Interviewers spend their days approaching strangers, asking them for answers to questions in which they may have no special interest, for a purpose they may find obscure on behalf of an organization they may not have heard of. How the interviewer does this is critical to obtaining respondents' cooperation in particular projects and to public acceptance of survey research in general.

In addition to the direct influences that interviewers have on the process of gaining cooperation (see Figure 7.1), they are also the medium through which many aspects of the survey design are communicated to the householder. Interviewers serve as the agents for the data collection organization in its representations to the householder. As the predominant (or often sole) link between researcher and respondent, the interviewer also conveys to the survey organization information about the household-

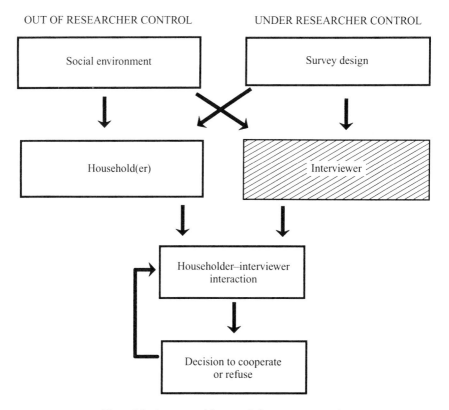

Figure 7.1. A conceptual framework for survey cooperation.

er's reaction to the survey request and, in this way, influences further actions that the organization may take to gain cooperation from the householder. Finally, as discussed in Chapters 8 and 9, the interviewer is a key participant in the interaction with the householder, often controlling those brief moments during which the decision to cooperate with or refuse the survey request is made.

Despite the importance of interviewers in survey participation, relatively little research has studied their role in gaining cooperation. One reason may be the difficulty of conducting such research. There is generally restricted variation in the characteristics of interviewers hired, making studies of the effects of such characteristics difficult. Furthermore, the small number of interviewers (relative to sample persons) used in most studies poses statistical difficulties for investigations of all but the largest interviewer effects. In addition, one needs random assignment of interviewers to sample persons to avoid confounding of interviewer effects with other attributes of the sample.

Cross-sectional designs to examine the effects of stable interviewer characteristics or attributes are also hampered by the problem of self-selection. Are more experienced interviewers better because they have been interviewing longer, or is it because better interviewers tend to remain on the job longer? Finally, interviewer behavior has proven to be notoriously difficult to control in an experimental setting, making exploration of this aspect of their work less tractable. Thus, much of what we believe about the impact of the interviewer on survey participation remains untested.

It is indeed difficult to separate a discussion of interviewers from other influences on survey cooperation addressed in this book. As agents of the survey organization, interviewers implement aspects of the survey design covered in Chapter 10. Similarly, almost all of what we know about householders and their reaction to the survey request comes by way of the interviewer. Nonetheless, this chapter is focused on the role-restricted (affected by training) and role-independent (not restricted by training) aspects of interviewer influences on cooperation (see Brenner, 1982; Hagenaars and Heinen, 1982).

We present in Figure 7.2 a more detailed model of the interviewer's role in survey cooperation. The highlighted relationships are those that we can empirically examine with our data. As is noted in the figure, interviewers can have both direct and indirect effects on cooperation. For example, householders may react directly to various fixed attributes of interviewers such as age, gender, or race. Short of switching interviewers, these will be constant across all households that an interviewer visits, although the reactions to these attributes may differ across households. Similarly, survey designs can serve both as constraining and facilitating factors on interviewer behavior. However, some aspects of the design are communicated directly to sample householders (through an advance letter, for instance) without any filtering on the part of the interviewer.

We believe that interviewer behavior is a key component in gaining cooperation. Interviewers vary in their effectiveness in gaining cooperation from different types of households, different householders, in different types of areas, and for different kinds of surveys. The behavior of interviewers is determined in part by their atti-

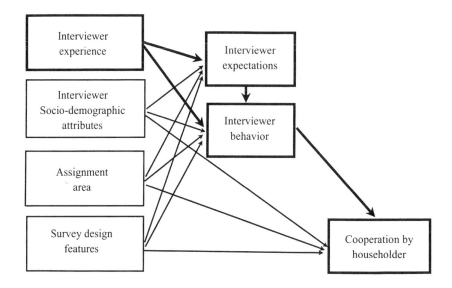

Figure 7.2. Interviewer influences on survey cooperation.

tudes and expectations, motivation, and experience. Thus, it is likely that the interviewer influences survey nonresponse in many ways and is a crucial element in understanding why some people agree to cooperate with survey requests while others decline.

A brief word about the data sources we use here before we turn to the analyses. The data used in this chapter are primarily from questionnaires administered to interviewers working on each of the surveys in the decennial nonresponse match study, with the exception of CPS (see Section 3.6). These data are supplemented with similar questionnaires administered to interviewers working on the National Survey of Health and Stress (NSHS), also described in Chapter 3. While this combined set of interviewers cannot be regarded as representative of all survey research interviewers in the United States (or even those working for academic survey units or the Census Bureau), these data nonetheless permit comparisons across organizations on interviewer attributes, experience, attitudes, and expectations relating to survey nonresponse.

The data are further restricted in two key ways. First, we were unable to obtain demographic information for interviewers working on the three Census Bureau surveys. In order to permit combined analyses of interviewers working on these six surveys, we exclude demographic variables from all but the initial descriptive analyses. Second, we were unable to obtain response rate details for these three surveys, so we could not calculate interviewer-level cooperation rates. The dependent variable we focus on for most of this chapter is the interviewer-level response rate, which includes both noncontacts and refusals, as well as other types of nonresponse. For the three ongoing Census Bureau surveys, the response rates are a six-month aggregate of rates at the time of the survey-census match project, whereas for the remaining three surveys, these are the overall response rates for the surveys themselves.

7.3   THE ROLE AND TASK OF INTERVIEWERS

The tasks that an interviewer is expected to perform are many and varied. These range from listing area segments to locating and contacting sample units, persuading householders to cooperate, asking questions and recording answers, and performing a variety of administrative functions.

There appear to be some contradictions between the prescribed behavior of interviewers in gaining cooperation and the prescribed behavior in administering structured interviews. The interview itself is usually highly structured, with little deviation from the questionnaire script permitted on the part of the interviewer (see, e.g., Fowler and Mangione, 1990). In contrast, the behavior of interviewers in survey introductions is largely unscripted and adaptive to householder objections or concerns. Within general guidelines on what is permissible or not in this stage of the interaction, interviewers are generally left to their own devices in their efforts to gain cooperation from sample householders.

These different roles have implications for the kinds of interviewers hired and the characteristics considered desirable. They also have implications for the way one evaluates and rewards interviewers. It is important to acknowledge the multiple roles of interviewers, even when focusing on only one aspect of their jobs—gaining cooperation. Which skills are more important: obtaining high response rates or collecting high quality data from respondents? Are these mutually compatible objectives? What proportion of training time and effort should be spent on these various activities? Which skills are more easily trainable, which should be present already among desirable recruits? These questions have received too little research attention.

Aside from the self-selection of people attracted to the work of interviewing, we believe that most of the acculturation process of producing effective interviewers occurs during training and on the job. Thus, it is important to examine how survey organizations typically approach the training of interviewers for the task of gaining cooperation.

While it is difficult to assemble empirical evidence of the relative attention paid to various topics in interviewer training, our review of training manuals from the organizations involved in the decennial match study suggests that more emphasis is placed on the questionnaire or instrument than on gaining cooperation. In the words of one interviewer: "A lot of time is spent teaching intelligent people to read questions, when the interaction, which is so much more difficult and important, is left to chance." For example, in the week-long training for the CPS (40 hours of classroom training), less than one hour is devoted to the topic of getting in the door. In the U.K. Office of Population Censuses and Surveys' handbook for interviewers (Mc-Crossan, 1991), one chapter (out of nine) is devoted to gaining cooperation. Similarly, the Survey Research Center's General Interviewer Training manual in 1992 had only one of 11 chapters focussed on nonresponse issues.

Luiten and de Heer (1994) attempted to obtain systematic data from government statistical agencies on their practices with regard to interviewer hiring, screening, training, remuneration, supervision, and so on. What is striking from this study is the wide range of organizational practices across the twelve countries. For example, the length of interviewer training ranges from 1 day (for the Slovenian labor force survey) to 8 days (Statistics Sweden), the latter including three days of nonresponse instruction. Similarly, half of the countries pay interviewers an hourly rate, while the other half pay per (completed) interview. Organizations also vary in whether interviewers are treated as permanent or temporary staff, in the level of pay, the benefits offered to interviewers, career structures, degree of control, supervisory ratios, and so on. Similar results were found for U.S. organizations by Miller-Steiger and Groves (1997).

In summary, organizations differ in the expectations they have of their interviewers, the training and supervision they provide, the reward structure for interviewer work, and the emphasis on components of the survey process (e.g., nonresponse versus data quality). All of these factors form the organizational milieu within which interviewers conduct their work. However, little is known about the climate within which interviewers carry out their tasks and the effects of such factors on the motivation and performance of interviewers. Almost all previous studies of interviewer effects on survey cooperation have been conducted within a single organization (and even a single survey, in most cases). A comparative analysis across organizations may reveal differences in how the interviewers approach their jobs that reflect such organizational differences.

In the next few sections we present descriptive data on the characteristics of interviewers for the six interviewer surveys. Given the similarities we found among interviewers working on the three Census Bureau surveys on the one hand, and among the three non-Census surveys, on the other (see Couper and Tremblay, 1991), we collapsed the analyses into these two groups. We examined variation in interviewer background characteristics, personality, experience, and expectations, and the effect of each of these sets of factors on interviewer-level response rates.

7.4 SOCIO-DEMOGRAPHIC CHARACTERISTICS OF INTERVIEWERS

What is the demographic profile of the interviewer labor force? Unfortunately, we were precluded from asking demographic questions (age, gender, race) directly of interviewers on the Census Bureau surveys, but we have data from a variety of other sources. First, it is clear that the interviewer workforce is predominantly comprised of women. This ranges from 97% women on the NHSDA and 93% on SCP to 79% of NSHS interviewers. In Luiten and de Heer's 1994 survey, they found that 98% of Statistics Netherlands interviewers were female, 99% of those at Statistics Canada, 93% at Statistics Sweden, and 87% of interviewers at the Australian Bureau of Statistics. In a survey of 380 interviewers working on the CATI-CAPI Overlap Study (the precursor to the automated CPS in 1993), Couper and Burt (1994) similarly found that 82% of these Census Bureau interviewers were female. Barker (1987) reports that 95% of market research interviewers in Ohio and New Zealand are women. By way of contrast, in 1990, women comprised only 45% of the U.S. civilian labor force (U.S. Bureau of the Census, 1995). We expect the female–male ratio to be less skewed in centralized telephone facilities. However, in 1997 only 25% of telephone interviewers in the University of Michigan's Survey Research Center were men.

There are some notable exceptions to this female-dominated workforce. For example, Peneff (1988) reports that only a third of the interviewers working for the French national statistical office (Institut National de le Statistique et des Etudes Economiques, or INSEE) were female. Similarly, 35% of interviewers on the German ALLBUS (Allgemeine Bevölkerungsumfrage der Sozialwissenschaften) survey were female (Koch, 1991). Finally McCrossan (1992) noted that at one time interviewers working for the U.K. Office of National Statistics were almost exclusively women, but in recent years many more men have joined the workforce, with men comprising 27% of the interviewer workforce in 1992. A major reason for this trend appeared to be a downturn in the U.K. economy at the time.

The mean age of interviewers working on the SCP, NHSDA, and NSHS were 52, 49, and 52 years respectively. On each of these three surveys, the median age was just over 50 years (53, 51, and 53, respectively). The same was found for CATI-CAPI Overlap Study interviewers (Couper and Burt, 1994). With few exceptions, this parallels the findings of Luiten and de Heer for government statistical agencies around the world. Anecdotal evidence suggests that the average age of interviewers in centralized telephone facilities (especially university-based organizations) may be much lower, given the more intermittent nature of the work. For example, the average age of University of Michigan Survey Research Center telephone interviewers in 1997 was 29.

We have data from all six of the interviewer surveys regarding their educational background (see Table 7.1). We note that Census Bureau interviewers appear to have lower levels of education than their non-Census counterparts. However, both groups have higher levels of education than the general U.S. population in 1990. Among those 18 years old and older, 57% have a level of high school or less, 25% have some college, and 18% have four or more years of college (U.S. Bureau of the Census, 1993).

Table 7.1. Percentage of interviewers in education categories by survey type

	Census surveys	Non-Census surveys
High school or less	26.5%	12.4%
Some college	45.7	43.4
Four or more years of college	27.8	44.2
Total	100.0%	100.0%
(*n*)	(729)	(516)

Table 7.2. Percentage of interviewers reporting various reasons for job and existence of another job by survey type

	Census surveys	Non-Census surveys
Primary reason for interviewing job		
For basic living expenses	42.5%	34.6%
As supplemental income	44.1	46.1
For reasons other than money	13.4	19.3
Total	100.0%	100.0%
Have another paid job?		
Percent yes	23.5%	37.3%

Given the demographic distribution of interviewers and the often intermittent nature of their work, we were interested in the extent to which they regard their work as a primary source of income or career. Interviewers were asked which of three statements came closest to the reason that they have their interviewing jobs (see Table 7.2). It can be seen that the majority of interviewers consider their interviewing work as something other than a primary source of income. A substantial minority of interviewers (24% of Census Bureau interviewers and 37% of non-Census interviewers) have another job in addition to interviewing.

In combination, these results suggest that interviewing is not regarded as a career by most interviewers. Their expressed views on the interviewing jobs suggest that many are not primarily motivated by money. While their pay may be compared with many hourly-paid, low-skill jobs, the education of the interviewer workforce is relatively high. Survey organizations are increasingly concerned about how to attract and maintain such a high-quality workforce during a time of increasing flexibility in the job market regarding work scheduling and part-time appointments.

We do not examine the relationship between stable interviewer characteristics such as race, gender, and age and survey participation. We believe that these are largely out of the control of the researcher, as are most other interviewer attributes. Furthermore, these are likely to produce few interpretable main effects on cooperation, but are more likely to interact with householder characteristics and survey topic effects (the argument behind race and gender matching, for example) that we are unable to explore here. We further expect the effects of such fixed interviewer attributes on survey participation to be largely mediated by interviewer experience, attitudes, and behavior, as noted in Figure 7.2. We return to this issue later in the chapter.

7.5 INTERVIEWER PERSONALITY

Another stable set of interviewer characteristics hypothesized to impact on performance is personality. This has long been a topic of interest (see, e.g., Hyman, 1954;

Axelrod and Cannell, 1959). A number of efforts have been undertaken to identify interviewer personality types associated with success in gaining cooperation and other interviewer activities, with the goal of using these measures in the recruitment and selection of interviewers.

For example, McFarlane Smith (1972) administered a series of tests to 123 interviewers selected randomly from the existing field force. In addition to the initial rating of the interviewer obtained from a standardized personal interview, two measures (extroversion and neuroticism) from a personality inventory and other standardized measures (clerical, arithmetic, and vocabulary tests) were administered. These were then compared to supervisor ratings of the quality of interviewer work based on observations in the field. These assessments did not distinguish between gaining cooperation and other aspects of interviewer performance (reading questions as worded, neutral tone, etc.). McFarlane Smith found that the initial rating from the recruiting interview by itself was a good predictor of performance. She also found that emotional stability (the absence of neuroticism) and a tendency toward introversion was associated with interviewer ability. The latter echoes the findings of Hyman (1954) and Axelrod and Cannell (1959), although none of these studies focused on gaining cooperation.

In an ambitious project at the U.S. Census Bureau, Johnson and Price (1988) report on a study in which a variety of personality assessment instruments was administered to 623 incumbent interviewers in 1985. These included the Jackson Personality Inventory, containing 15 subscales or traits (Jackson, 1976). Responses were obtained from 466 interviewers (for a 75% response rate). Performance data and subjective evaluations from supervisors were also obtained for these interviewers. Johnson and Price (1988) found two personality factors positively correlated with interviewer-level response rates: social adroitness (described as skillful in persuading others), and organization (described as making effective use of time and not being easily distracted). However, the correlations for all traits with response rates were modest, the highest being 0.26 and the average being around 0.08. The project was designed to be the start of an ongoing assessment of interviewer performance and retention, but no follow-up data were collected on these interviewers.

Turning to our own work, in developing the concept of tailoring, we wondered whether tailoring (see Chapters 2 and 9) was a learned behavior or a personality characteristic. In other words, are some people innately better at tailoring than others? This is an important question, as it addresses the issue of whether we should be recruiting interviewers with particular characteristics, or whether we can develop training procedures to develop (or enhance) the tailoring skills of interviewers. Similar questions were raised by Morton-Williams (1993) about the notion of social skills, a concept similar to tailoring.

At one point in our theoretical work, it appeared that the notion of tailoring was similar to the social psychological concept of "self-monitoring" (Snyder, 1974). Self-monitoring refers to the extent that persons "regulate their self-presentation by tailoring their actions in accordance with immediate situational cues" (Lennox and Wolfe, 1984, p. 1349). Similarly, Snyder (1980, p. 36) notes, "High self-monitoring individuals regard themselves as rather flexible and adaptive individuals who

shrewdly and pragmatically tailor their social behavior to fit situational and interpersonal specifications for appropriateness." Persons high on self-monitoring will
be alert to the actions and words of others in an interaction and adjust their behavior
based on those actions and words. Self-monitoring is conceptualized as a stable personality trait rather than learned behavior. These traits seemed to us to be similar to
that we have observed in good interviewers.

In order to test whether self-monitoring was related to tailoring behavior and
success in gaining cooperation, we administered the self-monitoring scale to 138 interviewers during their training for the survey on Asset and Health Dynamics of the
Oldest Old (AHEAD) (see Chapter 3, Section 3.5.4). Our first surprise was that interviewers scored low on self-monitoring, lower than two students samples and con-
siderably lower than a group of actors. But this may mirror the earlier findings of a
tendency toward introversion among survey interviewers (Hyman, 1954; Axelrod
and Cannell, 1959).

As we might have expected, we found self-monitoring to be positively related to
survey experience. We also found that self-monitoring was related to self-reports of
tailoring behavior. For example, 82.8% of high self-monitors reported varying their
introductions at different households, compared to 78.4% of low self-monitors.
Similarly, 70.3% of high self-monitors felt it was better to persuade reluctant respondents to participate than accept a refusal, compared to 55.4% of low self-monitors. Similar effects were found for a variety of other self-reported measures. However, when we modeled the effect of self-monitoring on cooperation rates in
AHEAD, both singly and controlling for other variables, we found little effect on interviewer performance. The self-monitoring scale consists of three subscales (other-
directedness, extroversion, and acting ability). Our findings suggest that of these
three, only other-directedness appears related to interviewer success in gaining cooperation.

To summarize, no research, including our own, has yet found strong links between stable interviewer-personality characteristics and success in gaining cooperation. One reason for this may be that interviewers are a relatively homogenous
group, and there is little variation in key personality attributes to find such effects. It
could also be that tailoring, social skills, and other adaptive behaviors can indeed be
learned on the job. Morton-Williams (1993) argues that social skills can be taught,
and offers an outline of such a training program in her book. This is clearly still an
area of research need, and has implications for interviewer recruitment, selection,
training, and evaluation. But we do not have the data to explore this issue here, and
the role of interviewer personality in success in gaining cooperation must remain
unresolved for now.

7.6   INTERVIEWER EXPERIENCE

The prevailing belief in the survey industry is that interviewer experience is a critical factor in gaining cooperation from sample persons. In a highly cited study,
Durbin and Stuart (1951) found experienced interviewers to be "decidedly superior"

to untrained student volunteers in terms of the response rates they obtained. They compared 55 unpaid student volunteers (who were given travel expenses and subsistence allowances) to 46 professional interviewers (27 from a market research company and 19 from government). However, the two groups differed in terms of other characteristics that weren't controlled in the study (for example, 56% of students versus 30% of professionals were male; only 7% of students were older than 25, while 33% of professionals were under 30 years old). Finally, the student volunteers were given "no training or instruction," other than 3 briefing meetings of 3 hours each (Durbin and Stuart, 1951).

Despite these flaws, the findings of Durbin and Stuart have been echoed in a number of other studies that have found a positive relationship between interviewer experience and response rates. Groves and Fultz (1985) found that novice telephone interviewers (those with less than 6 months' experience) obtained the lowest cooperation rates. In a study of U.S. Census Bureau interviewers trained in 1962 and 1963, response rates increased steadily over the first few months of service, reaching the level of experienced interviewers after 22 months (see Inderfurth, 1972). Bank and Landsmeer (1990), studying market research CATI interviewers in the Netherlands, found a positive initial effect of experience on response rates in the first few months, but that this tapered off over time. Lievesley (1988) examined average response rates across a number of studies for over 960 interviewers working for Social and Community Planning Research (SCPR) in the United Kingdom. She found that interviewers who had worked for SCPR for more than 5 years had lower refusal rates than more recent recruits, but found no differences with respect to rates of noncontact, a finding echoing our conclusions in Chapter 4.

Two pieces of contrary evidence regarding interviewer experience and response rate can be mentioned. In a telephone survey conducted by 35 interviewers, Singer, Frankel, and Glassman (1983) compared the response rates of three groups: no experience, less than 1 year, and 1 or more years of experience. They found the latter group to have significantly lower response rates. However, this is based on only six interviewers in this group. In a study of newspaper readership in Sweden, Schyberger (1967) compared the performance of 16 experienced interviewers (at least 3 months of fieldwork) with 16 new hires. The experienced interviewers had higher nonresponse rates (54% versus 46%) in the study. After additional nonresponse follow-up, this difference largely disappeared.

Why do we expect experienced interviewers to be better at gaining cooperation than inexperienced interviewers? First, we assert that it is the knowledge of the interviewing task gained from experience that is the likely cause of tenure effects. What are the indicators of knowledge-based differences across interviewers? We believe there are both mode and design differences to describe before turning to the empirical effects of experience.

In some survey designs, especially among establishment surveys and surveys of elites, it is common to recruit interviewers with knowledge of the substantive area. For example, recruits with farm backgrounds are often used in agricultural surveys; those with business degrees, in surveys of businesses. This practice is consistent with valuing the ability of interviewers to understand the concerns of sample persons, to speak their language, and thus to be more credible agents for the survey request.

Related to professional or technical knowledge is knowledge of the culture or lifestyle of the population studied. Thus, studies of ethnic or racial minorities often employ indigenous interviewers in order to attain greater trust between sample persons and interviewers.

Face-to-face interviewers distinguish themselves from telephone interviewers on yet another type of knowledge. Since they are typically hired to interview persons in the same geographical area in which they live, they often possess "local knowledge" of the ethnic, socioeconomic, lifestyle, and demographic character of the sample neighborhoods. This local knowledge is most often stripped away from telephone interviewers, who are typically assigned cases without concern about their personal backgrounds.

Professional or technical knowledge, cultural knowledge, and "local" knowledge are three types of knowledge that can aid an interviewer in obtaining the cooperation of sample households. To the extent that all of these are used as hiring criteria, we would expect the effects of tenure on participation rates to be smaller. In essence, these types of knowledge provide interviewers with advantages that other interviewers must obtain only through experience.

If our notion of tailoring is correct, interviewers learn these skills through exposure to a wide variety of circumstances and through trial and error of different approaches. Similarly, confidence is gained through experience in successfully dealing with resistance in the field. Exposure to alternative training guidelines and practices on different surveys and for different organizations may also serve to increase the repertoire of techniques available to an interviewer. Experience can thus have two components: length of experience (or number of years employed as an interviewer) and breadth (number of different surveys or organizations). Most of the studies reviewed above focus only on the former. We argue that both length and breadth of interviewing experience serve to increase the variety of different interviewing situations to which an interviewer is exposed. However, working for a large number of different organizations may be an indicator of other interviewer problems.

Furthermore, we expect the relationship between tenure (years of experience) and survey participation (particularly cooperation) to be nonmonotonic. Specifically, we expect the greatest gains of experience to be found in the first few months of interviewing. It is during this critical period that interviewers gain practical experience and build their confidence in gaining cooperation. Many interviewers learn their craft on the job, rather than exiting training fully prepared for the rigors of interviewing. Every new situation interviewers encounter may add to their repertoire of tailoring skills. There are a finite number of unique situations an interviewer is likely to encounter, and thus the increase in performance should not be monotonic over time, and should flatten out as certain experience levels are reached.

Using the same logic, since the cues provided over the audio medium alone are restricted, we'd expect longer learning curves for telephone interviewers than face-to-face interviewers. We might also expect smaller overall tenure effects in telephone surveys than in face-to-face surveys. Further, given the higher turnover rates common in centralized telephone facilities, we expect any such tenure effects to be harder to detect.

Some have speculated that the relationship between experience and performance is curvilinear, with response rates declining as interviewers settle into the routine of their job and perhaps reduce the amount of effort and energy they put into the task of persuading sample persons to participate. However, we have found little evidence of this, either in our discussions with interviewers, or in our analyses of the interviewer questionnaire data. Highly experienced interviewers appear to take great pride in their skills and ability to convince even the most hardened refusers to participate, and as a consequence are often given the harder assignments.

We should again caution against drawing conclusions about interviewer experience in cross-sectional studies such as this one. More experienced interviewers may be better simply because those who are better at gaining cooperation tend to stay with the survey organization, while those who have less success at persuading householders to participate choose other careers.

As noted above, interviewer experience can be conceptualized in various ways. Two of these are length and breadth of experience. Length of experience can be measured in the time an interviewer has worked at a particular organization or on a particular survey, or the total number of years of interviewing experience. Breadth of experience can be measured in terms of the variety of different organizations for which they have worked, or the variety of surveys within an organization. The levels of experience on these two dimensions of the interviewers in our dataset are presented in Table 7.3.

It is clear from Table 7.3 that the Census Bureau interviewers in our survey are a more experienced workforce than those at the other organizations. By way of con-

Table 7.3. Percentage of interviewers reporting various experience levels by survey type

	Census surveys	Non-Census surveys
Years worked for survey organization		
Mean	7.0 years	3.3 years
(Standard error)	(0.21)	(0.14)
Total interviewing experience		
Less than 1 year	11.5%	22.5%
1–2 years	15.3	10.6
3–5 years	19.2	15.7
More than 5 years	54.1	51.2
Total	100.0%	100.0%
Number of organizations worked for in last 5 years		
One	76.5%	31.4%
Two	15.3	24.2
Three or more	8.2	44.4
Total	100.0%	100.0%

trast, 70% of telephone interviewers at the University of Michigan's Survey Research Center have less than one year of interviewing experience. It is interesting to note from Table 7.3 that there is considerable movement of interviewers among survey organizations. Almost 24% of Census Bureau interviewers and 69% of non-Census interviewers have worked for other survey organizations in the past 5 years. In fact, 11% of Census and 49% of non-Census interviewers reporting working for some other organization in the 6 months prior to the survey, suggesting that there are a fair number of interviewers who work simultaneously for multiple organizations.

How does interviewer experience relate to the response rates interviewers obtain? Before we examine this, we should note that the size of interviewer assignments vary, and hence affect the variance of the interviewer-level response rates. For this reason, we employ weighted analyses, where the weight is the assignment size (the denominator or the response rate). All subsequent analyses examining response rates are based on weighted estimates. As expected, we find a significant positive correlation between experience within the organization and response rates ($r = 0.255$). However, this appears to be dominated by the three Census Bureau surveys; the correlation between experience and response rates for the three non-Census surveys combined is not significantly different from zero.

Using a transformation of the experience measure [log (years worked + 1)] to reflect the hypothesized curvilinear relationship, we again find a positive effect overall ($r = 0.273$), but again this is dominated by Census Bureau interviewers ($r = 0.189$), while the effect for non-Census interviewers is modest at best ($r = 0.0025$). A possible explanation for this difference may be that the Census Bureau does not devote much time to persuasion and gaining cooperation in its initial training, but interviewers are accompanied by supervisors on their first few days in the field, so these issues may be covered in on-the-job training. In contrast, the other organizations tend to devote more training resources to the issue of gaining cooperation. This may result in a steeper post-training learning curve for Census interviewers. This is partially borne out by interviewer responses to a question on the adequacy of training for gaining cooperation. Almost twice as many Census Bureau interviewers reported receiving too little training in this area than non-Census interviewers (28% versus 15%, respectively).

Table 7.4 shows the relationships between the remaining experience measures in Table 7.3 and response rate. In these bivariate analyses we find support for the effect of interviewer experience (length of experience) on response rates. However, in terms of breadth of experience, we find a negative relationship between number of organizations and interviewer-level response rate. This suggests that working for many different organizations may not just entail a greater variety of interviewing experiences, but may also reflect underlying interviewer problems associated with turnover and low retention.

This finding parallels that from an early study of interviewer performance. Hyman (1954) reported that NORC interviewers who had more than 5 years' experience with other organizations showed poorer than average ratings, while Sheatsley (1951) reported that the performance of the same interviewers improved with expe-

Table 7.4. Mean interviewer-level response rates by two experience measures

	Interviewer-level response rates	
	Mean	(Standard error)
Total interviewing experience		
Less than 1 year	89.7%	(0.74)
1–2 years	93.8%	(0.45)
3–5 years	93.9%	(0.45)
More than 5 years	94.5%	(0.24)
$F = 10.85, df = 3, 1213, p < 0.01$		
Number of organizations worked for in last 5 years		
One	94.8%	(0.20)
Two	92.8%	(0.52)
Three or more	90.6%	(0.62)
$F = 20.10, df = 2, 1239, p < 0.01$		

rience within NORC. However, we should note that the analyses in Table 7.4 do not have appropriate controls for other factors that may lead to a greater breadth of experience.

We explore these relationships in greater detail later in multivariate models predicting interviewer-level response rates. But first we examine marginal distributions of interviewer attitudes and expectations regarding nonresponse.

7.7 INTERVIEWER ATTITUDES AND EXPECTATIONS REGARDING NONRESPONSE

What effects do interviewer attitudes and expectation have on the response rates they obtain? There is some evidence that interviewer expectations affect the quality of data they obtain in survey interviews (see Sudman *et al.*, 1977; Singer and Kohnke-Aguirre, 1979), but few studies have explored the issue with regard to nonresponse. A notable exception is the study by Singer, Frankel, and Glassman (1983), in which they find that interviewer expectations of difficulty in gaining cooperation were significantly related to response rates in an RDD survey. In a study in Finland, Lehtonen (1995) compared a group of 120 professional interviewers to a specially recruited group of 93 public health nurses, both working on the Finnish Health Security Survey. Table 7.5 shows the proportion of each group agreeing or strongly agreeing with each of statements presented. These suggest that the attitudes of professional interviewers may be different from those of other professions (at least nurses).

While he does not report on the direct relationships between the responses to these items and interviewer-level cooperation rates, Lehtonen notes that the response rate obtained by the professional interviewers was some 14 percentage

Table 7.5. Percent of interviewers agreeing with statements, by type of interviewers

Statement	Professional interviewers	Public health nurses
1. Reluctant respondents should always be persuaded to participate	60%	25%
2. With enough effort, even the most reluctant respondent can be persuaded to participate	29%	15%
3. An interviewer should respect the privacy of respondents	96%	99%
4. If respondent is reluctant, refusal should be accepted	27%	82%
5. Voluntariness of participation should always be emphasized	35%	87%

Source: Lehtonen (1995).

points higher than that obtained by the public health nurses on a matched sample (88% versus 74%). These findings suggest that the effect of interviewer experience on response rates may be moderated by their attitudes and behaviors, as specified in Figure 7.2.

Interviewers appear to be fully aware of the role of positive expectations in gaining cooperation. In numerous discussions with interviewers, the importance of a positive attitude is stressed. For example, one interviewer stated: "The way to succeed in this is to expect success. A positive attitude is worth gold." Another expressed the following: "I do not have much trouble talking people into cooperating. I love this work and I believe this helps 'sell' the survey. When I knock on a door, I feel I'm 'gonna' get that interview!" This sentiment is echoed by another: "90% is attitude. I always believe I will get the interview and usually do. Approach the door with a smile, confident—be upbeat and pleasant. Who wouldn't want to let me inside?" Interviewers repeatedly tell us that if they don't believe in the general importance of the work they do and in the particular study they are conducting, they cannot convey this to the sample householder.

Before we examine the effect of interviewer expectations on response in our data, it is useful to examine descriptive data on interviewers' attitudes and expectations surrounding the issue of survey cooperation. A number of measures were included in the surveys of interviewers, and we present a few selected items here (see also Couper and Tremblay, 1991).

Given an assumption of greater perceived authority and legitimacy on the part of the federal government relative to academic survey organizations (see Chapter 10), and the typically higher response rates achieved for government surveys (see Chapter 6), we expect Census Bureau interviewers to have greater expectations of success and thus greater confidence in their own abilities than their counterparts in academic organizations. Census Bureau interviewers are expected to have stronger beliefs in the legitimacy of their work, and this may affect their behavior when encountering obstacles to completing their assignments.

In our focus groups with Census Bureau interviewers, this was borne out on a number of occasions. Census Bureau interviews tended to express the belief that they had the right to enter any property in pursuit of the interview. They perceived themselves as agents of the federal government, legally entitled to make contact with the sampled household and to describe the survey to householders. Such views were not as frequently expressed by interviewers at other organizations.

The confidentiality of the data collected by the Census Bureau is explicitly protected by Title 13 of the United States Code (Title 15 in the case of the NHIS). In other organizations, while certainly taken no less seriously, confidentiality is protected by pledges based more on ethical than legal grounds. We expect this difference in the degree of protection of respondent confidentiality to be reflected in interviewer beliefs, and to be a factor in interviewer efforts to reassure and persuade sample persons.

In summary, we expect Census Bureau interviewers (and indeed, government interviewers in general) to have a stronger sense of self-confidence and belief in the legitimacy of their work than those working for academic or private research organizations. Furthermore, we also expect these attitudes to be reflected in differences in reports of behavior.

We asked interviewers, based on their experiences, how many respondents believe their answers are truly confidential. Their responses to this question are presented in Table 7.7. In another question on confidentiality, we asked interviewers the following: "Do you think there are any situations under which [your agency] would give individual survey responses to the following agencies: FBI, CIA, INS, IRS, State or local government agencies?" The percentage of interviewers saying "yes" to any of these is also presented in Table 7.6.

Although not evident in Table 7.6, the responses from SCP interviewers appear closer to those of interviewers for the Census Bureau. This may be explained in part by the fact that the Survey of Census Participation was conducted on behalf of the Census Bureau, and the data were collected under Title 13, USC. While interviewers on all surveys share generally the same views of respondent perceptions of confidentiality (that a substantial proportion don't believe the confidentiality pledges),

Table 7.6. Percentage of interviewers reporting different beliefs regarding confidentiality pledges, by survey type

	Census surveys	Non-Census surveys
Respondent belief in confidentiality		
All, or almost all	15.2%	13.0%
A majority	46.6	46.7
Other	38.1	40.2
Total	100.0%	100.0%
Give data to any agency		
Percent yes	16.7%	32.2%

interviewers at non-Census organizations appear more likely to believe the data can be given to certain government agencies.

The Census Bureau results replicate the findings from a similar survey of over 800 Census Bureau interviewers conducted in 1987 (Lavin, 1989). Results from the National Academy of Sciences panel on privacy and confidentiality (1979, pp. 34–35) suggest that interviewer beliefs regarding the confidentiality of the data are conveyed to householders they contact.

In an explicit attempt to get interviewers to think about the trade-off between response rate and data quality, we asked them to select which of the following two statements came closest to their own views:

(a) "It's better to persuade a reluctant respondent to participate than to accept a refusal, even when you think they won't give very accurate answers."

(b) "It's better to accept a refusal from a reluctant respondent than to persuade them to participate when you feel they won't give very accurate answers."

The percent of interviewers agreeing with the first statement (better to persuade) is presented in Table 7.7. As a measure of self-confidence, interviewers were also asked the extent to which they agreed with the following statement: "With enough effort, I can convince even the most reluctant respondent to participate." Table 7.7 has the percentage agreeing with this statement. Finally, to measure the effort interviewers may go to in order to obtain a completed interview, we asked how likely they were to do various things if some of their sample were in an unsafe area. The percentage of those saying they were likely to "proceed as usual" is also presented in Table 7.7.

Table 7.8 presents the bivariate relationships between these expectation variables and interviewer-level response rates. These results support the notion that more positive interviewer expectations (greater belief in confidentiality of the data, belief in the importance of converting refusers, and willingness to proceed as usual in the face of obstacles) are associated with higher response rates, at least at the interviewer level. Separate analyses for Census and non-Census surveys produce similar patterns. We should again caution that these analyses require careful interpretation. We

Table 7.7. Percentage of interviewers reporting different attitudes and behavior regarding reluctant householders, by survey type

	Census surveys	Non-Census surveys
Rate/quality		
Percent better to persuade	87.3%	62.9%
Confidence: can convince almost anyone		
Percent agree	56.6%	62.6%
Effort: proceed as usual		
Percent likely	58.6%	42.8%

Table 7.8. Mean interviewer-level response rates by interviewer attitudes and behavior

	Interviewer-level response rates	
	Mean	(Standard error)
Respondent belief in confidentiality		
All, or almost all	94.6%	(0.44)
A majority	94.1%	(0.27)
Other	93.5%	(0.34)
$F = 3.76$, $df = 2$, 1205, $p < 0.05$		
Give data to any agency		
Yes	92.8%	(0.49)
No	94.1%	(0.21)
$t = 2.39$, $df = 1$, $p < 0.05$		
Rate/quality: better to persuade than accept refusal		
Agree	94.3%	(0.20)
Disagree	91.8%	(0.59)
$t = 4.07$, $df = 1$, $p < 0.01$		
Confidence: can convince almost anyone		
Agree	94.1%	(0.25)
Disagree	93.8%	(0.29)
$t = 0.97$, $df = 1$, n.s.		
Effort: proceed as usual		
Likely	94.5%	(0.24)
Unlikely	93.1%	(0.31)
$t = 3.59$, $df = 1$, $p < 0.01$		

need to include appropriate controls for differences in assignment area. It could be, for example, that higher response rates associated with certain areas lead to more positive interviewer expectations.

7.8 INTERVIEWER BEHAVIORS

Figure 7.2 hypothesizes that a key interviewer-level determinant of likelihood of success in gaining cooperation is the behavior of interviewers. As we note in Chapters 2, 8, and 9, we believe that it is interviewer behavior at the household level, and particularly interviewer behavior tailored to individual household situations, that is critical. However, in this chapter we are interested in explaining whether there are behavioral tendencies (habits, standard procedures, working rules) at the interviewer level that affect cooperation. In other words, are there things that some interviewers typically or routinely do and others do not that are useful in gaining cooperation from sample households? These would typically be the behaviors or actions of interviewers that would be covered in interviewer training.

In Chapter 3 we compared NSHS interviewer reports of what they usually or typically do (as measured in the interviewer questionnaire) with what they reported doing on a particular contact with a householder. We found that there appears to be overreporting of behaviors on the interviewer questionnaire, but that the two sets of reports are well matched ordinally. In other words, those who say they usually do things more often tend to report these behaviors more frequently at the contact level. Given that we have contact-level reports of interviewer behavior only for the NSHS, we examine the interviewer-level reports of behavior here. We thus use general interviewer-level reports of behaviors as weak proxies for what interviewers actually do in the field.

We attempted in the interviewer questionnaires to measure the extent to which interviewers used various compliance-enhancing techniques (see Cialdini, 1984) in gaining cooperation. These techniques are discussed in greater detail in Chapter 2, and their operationalization is detailed in Couper and Groves (1992a). The compliance principles we examined are authority, reciprocation, social validation or social proof, scarcity, consistency, and saliency.

As our research on survey cooperation has developed, we have become increasingly convinced that we are unlikely to find main effects of these interviewer behaviors on survey cooperation. In other words, it is not whether an interviewer uses each of these techniques or not, but rather when and how they are used that is important. We believe that these approaches, along with other techniques, comprise tools in an interviewer's toolbox, and that the appropriate application of a particular tool at the appropriate time will increase the likelihood of cooperation. We have termed this selection and application of appropriate tools "tailoring," and we believe this is primarily an interaction-level phenomenon. However, we also constructed a crude indicator of tailoring and the repertoire of techniques at an interviewer's disposal, to test along with interviewer reports of compliance-enhancing behaviors. We do not present the details of these analyses here (see Couper and Groves, 1992a), but briefly summarize the findings. In summary, we examined variation in interviewer self-reports of the use of these techniques and examined the bivariate relationships of these indicators with interviewer-level response rates.

When turning to the bivariate relationships between the reports of various behaviors and interviewer-level response rates, we find no interpretable effects. This holds both for the full group of interviewers and for separate analyses of Census and non-Census interviewers. That is, there is no observable relationship between reported use of the compliance enhancing techniques and interviewer-level response rates.

While a number of explanations can be offered for the lack of effects (e.g., poor measurement, inappropriate specification of the dependent variable, or incorrect level of analysis), we are increasingly convinced that we should not expect to find effects at the interviewer level if our theory is correct. In other words, main effects of interviewer-level behavior is inconsistent with the notion of tailoring, which is inherently interactive (in both the social and statistical sense).

In examining variation in reported use between Census and non-Census interviewers, we surmised that interviewers representing a government organization may have an easier time persuading householders to participate, other things being equal. The legitimacy of the organization they represent, the authority with which they are invested to carry out their work, the legal protections of confidentiality, and the perceived importance to society of the data they collect may have all been already established, at least in the mind of the interviewer. In contrast, interviewers representing an academic organization may need to invest greater effort in persuading householders to participate. Such interviewers may have to rely more on their skills and abilities in obtaining cooperation. We find some support for this in that non-Census interviewers appear to make more use of compliance-enhancing techniques than Census interviewers (e.g., 68% of non-Census and 58% of Census interviewers report using reciprocation, 73% of non-Census and 31% of Census interviewers report using social validation).

7.9 MULTIVARIATE MODELS OF INTERVIEWER-LEVEL EFFECTS

We have seen earlier that interviewer experience and expectations appear to be related to the response rates they obtain. In this section we examine the combined effects of these and other variables on interviewer-level response rates. We do this for the important reason that the bivariate analyses described above may confound the effects of several variables. We do not employ an interpenetrated design to measure interviewer effects, and thus control for assignment-area differences across interviewers. We also need to control for response rate differences across the surveys. Even then, we must acknowledge that some differences in likelihood of participation will remain across interviewer assignments.

We present two sets of models in this section. The first are multivariate models predicting interviewer-level response rates, using the same data presented in bivariate analyses above. This is to examine whether these effects hold in the presence of appropriate multivariate controls.

Next we examine the same set of interviewer variables in the decennial nonresponse match dataset. This is done for two reasons. First, we want to see if the interviewer variables we identified as affecting interviewer-level response rates are also important when examining household-level cooperation rates (our primary analytic focus of interest in this book). Second, we include a set of household-level controls (e.g., household size, age of household members) to replace the weak assignment-area controls of the interviewer-level models. If in the presence of these controls we still find effects of interviewer-level variables, we will be more convinced of the role of these factors in survey participation.

7.9.1 Interviewer-Level Models

To turn to the first set of analyses, Table 7.9 presents the multivariate models predicting interviewer-level response rates. The assignment-area controls are measured at the county level. This is the lowest level of aggregation we could obtain for these data. Typically, the primary sampling unit (PSU) in which an interviewer works consists of one or more coterminous counties. Obviously these measures can only ac-

Table 7.9. Coefficients of regression models predicting interviewer-level response rates (standard errors in parentheses)

Independent variables	Model 1		Model 2	
Constant	92.69**	(3.60)	92.18**	(3.69)
Assignment area				
Population density[a]	−0.17**	(0.030)	−0.18**	(0.030)
Crime rate[b]	−0.030**	(0.0067)	−0.031**	(0.0068)
Percent 65 or older	0.057	(0.059)	0.029	(0.061)
Percent under 5	0.58**	(0.19)	0.45*	(0.19)
Household size	−2.52	(1.42)	−2.48	(1.43)
Interviewer experience				
Log(tenure)	0.68**	(0.23)	0.65**	(0.22)
Interviewer expectations				
Confidentiality			0.27	(0.41)
Rate/quality			−0.042	(0.44)
Confidence			0.65**	(0.18)
(*n*)	(1,155)		(1,080)	

Note: Coefficients for dummy variables representing different surveys are omitted from table.
[a]Measured in thousands of people per square mile.
[b]Measured in serious crimes per 1,000 population.
*$p < 0.05$.
**$p < 0.01$.

count for gross differences among interviewer assignments. We include as controls key environmental correlates of cooperation (see Chapter 6) that are measured at the county level and available for all surveys in our dataset. In addition, we include a set of dummy variables to account for design and response rate differences across the six surveys.

We first model the effect of interviewer experience, controlling for assignment-area characteristics and dummy variables for the surveys (CEQ being the omitted survey). We also included the breadth of experience measure but found that, in the presence of the multivariate controls, this variable had no significant effect on response rate, so we omit this variable from further analyses. We can see from Model 1 in Table 7.9 that interviewer experience (measured as the natural log of the number of years worked for the organization) has a significant positive effect on interviewer-level response rates, controlling for differences in assignment area. This is consistent with the bivariate results described above, and with previous findings that show a positive effect of experience on response rates. Interviewer-level differences in response rates appear to be more than simply artifacts of differences in the area to which they are assigned, and experience appears to play a key role in such differences.

In the second model in Table 7.9 we add a subset of the interviewer expectation variables discussed in Section 7.6 above, following a series of preliminary model-fitting activities not shown here. In the multivariate model, only interviewer confidence ("I can convince almost anyone to respond") reaches significance. However, we have now found across several surveys and in a variety of models, that interviewer confidence is positively associated with response rates, even given the weak measure we employ here.

Finally, we added to Model 2 in Table 7.9 the set of interviewer behaviors discussed earlier. These results are not shown, but as expected, none of these variables are significantly related to interviewer-level response rates, after controlling for assignment area, experience, and expectations. Our crude operationalization of tailoring at the interviewer level also fails in these models, again suggesting that it may be household- or contact-level behavior that is important.

7.9.2   Case-Level Cooperation Models Measuring Interviewer-Level Influences

As a final step, in order to link to the analyses in other chapters of this book and to explore the impact of interviewer-level measures on household-level cooperation rates, we fit a similar set of models using the decennial census match data (restricted to the five surveys with interviewer questionnaire responses). The dependent variable in these logistic regression models is the likelihood of cooperating with the survey request, given contact.

Table 7.10 contains two of the models. The first model fits the full set of interviewer variables tested at the interviewer level, with one important difference. The interviewer of record for many of the decennial census match dataset cases is defined as the last person to work the case. More experienced interviewers tend to be assigned more difficult cases to convert; we identified those interviewers likely to be reassigned cases (designated refusal converters, or, in the case of the Census Bureau, supervisory field representatives who tend not to be given their own regular assignments). We then created an interaction term for this reassignment measure to reflect the increased difficulty of cases these interviewers are given. The model controls only for the survey indicators to reflect gross response rate differences across surveys.

We can see from the first model in Table 7.10 that interviewer experience has a significant ($p < 0.05$) positive effect on the likelihood of cooperation. The significance of the reassignment by tenure interaction term confirms the need to take such reassignment into account. The confidence measure just fails to reach traditional significance levels ($p = 0.059$).

In the second model in Table 7.10 we add in as controls the set of household- and environment-level variables found to be important in those analyses. We again see that in the presence of this set of variables interviewer tenure still has a significant effect on cooperation. The effect of tenure still appears to be modified by the reassignment of cases to more experienced interviewers. The expectation measures are in the expected direction, but fail to reach significance (coming close in the case of the confidence measure).

(A careful reader will note that the second model in Table 7.10 has a similar form to that of the final models in Tables 5.20 and 6.2. There are, however, important differences. The case base of Table 7.10 omits the CPS sample. This produces some minor changes in the values of coefficients, but also increased standard errors of the

Table 7.10. Coefficients of logistic models predicting household-level cooperation versus refusal (standard errors in parentheses)

Independent variables	Model 1		Model 2	
Constant	1.95**	(0.31)	2.03**	(0.42)
Social environment				
Urbanicity				
Central city			−0.14	(0.18)
Balance of CMSA			−0.11	(0.15)
Other			—	
Population density[a]			−0.032*	(0.012)
Crime rate[b]			−0.057	(0.21)
Percent persons under 20			0.0084	(0.0053)
Household				
Owner occupied			−0.19	(0.21)
Monthly rent for renters			−0.035	(0.040)
House value for owners			−0.013	(0.0066)
Household age				
All persons under 30			0.76**	(0.16)
Mixed householder ages			—	
All persons over 69			0.31	(0.20)
Single-person household			−0.31*	(0.14)
Children <5 years in household			0.70**	(0.17)
Interviewer experience				
Log(tenure)	0.14*	(0.066)	0.17*	(0.073)
Reassignment indicator	−0.27	(0.15)	−0.25	(0.15)
Tenure × reassignment	−0.17**	(0.057)	−0.16**	(0.056)
Interviewer expectations				
Confidentiality	0.22	(0.14)	0.15	(0.15)
Rate/quality	−0.013	(0.11)	0.057	(0.12)
Confidence	0.13	(0.070)	0.15	(0.079)

Note: Dependent variable coded 1 = interview, 0 = refusal. Coefficients for dummy variables representing different surveys are omitted from table.
[a]Measured in thousands of people per square mile.
[b]Measured in serious crimes per 1,000 population.
[c]Measured in units of $100.
[d]Measured in units of $10,000.
*$p < 0.05$.
**$p < 0.01$.

coefficients. We judge that these differences have no substantive importance to our analyses.)

These analyses provide support for the contention that interviewer experience does matter in an existing pool of interviewers, even in the presence of relatively powerful controls. We suspect that interviewer expectations and behavior are also important in affecting household-level cooperation, but do not see these effects in the data analyzed here.

7.10 SUMMARY

We have seen that survey interviewers can play a crucial role in reducing nonresponse. Especially in face-to-face surveys, interviewers can also be of great help in collecting information to facilitate postsurvey adjustment for nonresponse. Despite this, the role of the interviewer in nonresponse has not yet received the research attention it deserves. There is much we don't know about how to recruit successful interviewers, how to train interviewers to persuade sample householders to participate, how to convert reluctant respondents, and so on.

In short, from the empirical analyses in this chapter we have learned the following about the interviewer's role in survey cooperation:

- Those with greater interviewing experience tend to achieve higher rates of cooperation than those with less experience. We do not know from our analysis whether this is the result of less productive interviewers terminating their employment earlier or evidence of the benefits of coping over time with diverse situations in recruiting respondents.
- There is some suggestion in the empirical data and strong support from our qualitative investigations that interviewers who, prior to the survey, are confident about their ability to elicit cooperation tend to achieve higher cooperation rates.

We have speculated in this chapter and elsewhere (see Chapter 9) on the possible mechanisms through which greater interviewer experience and a heightened sense of confidence may translate into success in the field, but these remain for now largely untested assumptions. We are convinced that it is less the stable characteristics of interviewers that determine the likelihood of success than the interaction of these characteristics and interviewer behavior with householders that are important in gaining cooperation. We explore this issue in greater detail in the following two chapters.

7.11 PRACTICAL IMPLICATIONS FOR SURVEY IMPLEMENTATION

By way of summarizing the last four chapters, the research has implied that influences on contactability of sample households are distinct from those on cooperation. Plans should be made to collect information on those attributes, and to calibrate efforts at cooperation and contact to different subgroups. At the household level, we examined socio-demographic attributes that reflect social psychological predispositions to cooperate with the survey request. Most of these attributes can be observed only through repeated calls on the sample household. Once known, however, they offer indirect information on the possible concerns of the sample household regarding survey participation. Introductory approaches should be altered to address those concerns. At the social environmental level, we saw that issues regarding cooperation are distinct in densely populated urban areas. Approaches need to dispel the default cognitive script of householders that the survey request visit is another fleeting call on their limited resources that benefits only the requestor. Every one of these prescriptions has implications for the behavior of survey interviewers.

Further, in this chapter we learned that interviewer tenure remains a positive predictor of cooperation in the context of controls on key household attributes. Our theories lead us to interpret this result as reflecting the higher skills at tailoring the introductory conversation with householders to the concerns they have about cooperation. Other things being equal, therefore, surveys that use more experienced interviewers will achieve higher cooperation rates. This finding, however, may be of little practical guidance to a survey researcher, because tenure is not usually an attribute under the direct control of the researcher. Instead, the practical implications of these findings may be that training regimens for interviewers should attempt to provide some of the tailoring lessons that interviewers learn through trial and error over time.

The finding on interview experience leads to explicit tradeoffs of cost and errors. This is especially true in centralized telephone facilities, where interviewer turnover is more of a problem, often associated with fewer benefits and lower pay than field interviewers. Experienced interviewers tend to receive higher salaries. We need to weigh this against the added cost of recruiting and training new interviewers, and the learning curve effects associated with experience that we and others have observed. This may suggest a reconsideration of the way interviewers are viewed as employees by survey organizations.

All of the results presented thus far in the book suggest that standardization of interviewer behavior during the introductory interaction with sample householders is ineffective. While standardization during the interview process has value for the scientific replicability of the measurement, we see little in the scientific method to argue for standardization during the recruitment of sample persons. Rather, the evidence points to each person taking her or his own route to the participation decision. These routes are influenced by householders' backgrounds and socio-demographic attributes but are influenced by interviewer behavior. The most effective behavior is that mirroring the concerns of the householder. We examine this feature of interviewer behavior in the next chapter.

If training programs could be crafted to provide novice interviewers with a large range of householder concerns, preferably in the natural language of the householders, the trainees could learn to classify these concerns into common themes. Each theme could then be addressed with a different line of arguments for presentation by interviewers. To be effective, interviewers need to provide these arguments quickly in their own words, customized to the terms used by the householder.

Another major finding in this chapter is the marginal effect of self-reported interviewer confidence measures. If more confident interviewers achieve success at greater rates, controlling on their experience, then selecting for high-confidence candidates and training with confidence indicators as outcome measures may be warranted. Selection protocols could take place at intake interviews with job applicants. Trainers might present applicants with a set of scenarios that involve reluctant householders. Each scenario might ask the trainee to invent ways to address the concerns of the householder. The outcome measure would involve judgments on the resourcefulness of the applicant, the perseverance in providing repeated counterarguments. The quality of the counterarguments should not be relevant to the judgments; the number of conversational turns during which the trainee could sustain persuasive behavior would be measured. Those choosing to continue persuasive behavior for a longer period of time would be preferred as trainees.

These are best viewed as speculations based on these empirical results, but there is no doubt that the development of effective training and supervisory protocols remains a challenge to survey researchers. Only with those protocols can the results of this chapter, showing the effects of interviewer tenure and confidence measures, be brought under the control of the designer.

CHAPTER EIGHT

When Interviewers Meet Householders: The Nature of Initial Interactions

8.1 INTRODUCTION

The next two chapters focus on the moments when all of the influences of the survey climate, of the prior experiences of the householder, of the various survey design features, and of the attributes of the interviewer combine to affect the survey participation decision. This chapter describes the interactions between householders and the interviewers who deliver the survey request. This is the behavior at the very bottom of the theoretical framework (Figure 8.1). Then, in Chapter 9 we explore how features of these interactions influence survey cooperation.

Perhaps the most important observation is that these interactions are generally brief. In our experience (see Table 8.2) about 55% of the first face-to-face contacts of the interviewers with householders last between 1 and 5 minutes (up to termination or start of the interview). Each subsequent contact with the same household tends to be even shorter. This may reflect the fact that interviewers need to take more time to introduce themselves and describe the study on early contacts with sample households; later contacts may be directed at finding a suitable time to do the interview or addressing specific concerns about participation; and later contacts are disproportionately with households reluctant to participate. Oksenberg, Coleman, and Cannell (1986) noted that interactions of telephone interviewers and householders tended to be even shorter, with over 50% lasting less than 1 minute. Thus, although the outcome of these interactions is influenced by multiple features of the survey, the interviewer, and the householder, the decision phenomenon is relatively brief.

8.2 THE INITIAL INTERACTION FROM THE HOUSEHOLDER'S PERSPECTIVE

During the initial moments of this interaction, we believe the householder is involved in an active effort to comprehend the interviewer's intent (see Figure 8.2).

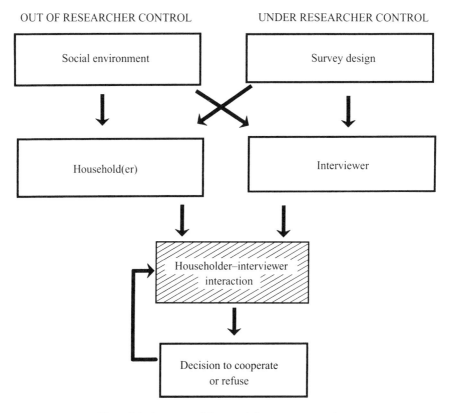

Figure 8.1. A conceptual framework for survey cooperation.

Since in most cases the householder has had no prior contact with the interviewer, all the cues revealing the intent of the interviewer come from components of the interaction themselves. Sometimes this process in face-to-face surveys begins with visual inspection of the interviewer prior to opening the door (through peepholes in doors, through windows). The central question is "What does this person want of us/me?" This process is common to all encounters with strangers, and is a tool persons use to identify their own appropriate behavior. For the face-to-face interview situation, the possible answers include a request for charitable contribution, a sales request, a political petition, a religious petition, a delivery of products, provision of services, a mail delivery, an agent from a social service agency, a meter reader, etc. (see Chapter 2). For the telephone, the candidates are sales persons, credit card offerers, solicitors for charities, or investment agents. Each of these situations might suggest different initial behavior on the part of the householder.

Sometimes initial visual cues provide sufficient information to eliminate large numbers of possible explanations for the visit. A child at the door is unlikely to be delivering mail or express packages. Persons in uniform generally have insignias

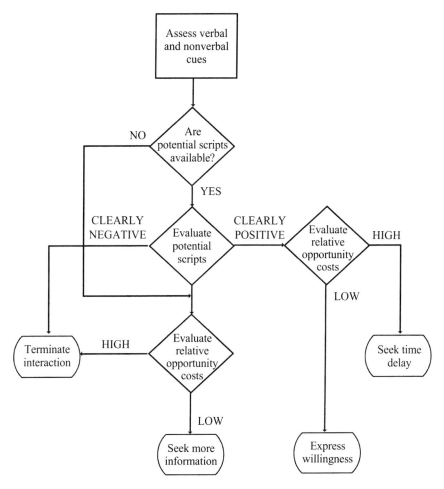

Figure 8.2. Householder strategies for evaluating survey request.

that reveal their institutional affiliation, and thus eliminate all possible scripts except those possibly connected with the institution. An adult without uniform, but holding a clipboard, is often collecting signatures and/or soliciting funds in support of political or charitable activities. An adult male in dirty, torn clothes, knocking on the door late at night, might generate scripts that uniformly suggest large potential costs of the interaction to the householder.

Face-to-face interviewers for some organizations are trained to display evidence of institutional affiliation (e.g., briefcases that are labeled "U.S. Bureau of the Census," nametags that are labeled "University of Michigan"). Such behavior appears to be less frequently true when the institutional affiliation is judged not to be helpful to the survey request (e.g., a market research interviewer).

There appear to be norms of privacy in this society that force strangers at the doorstep to quickly identify the purpose of the interaction. This is probably because none of the long-term social exchange norms apply to strangers. The stranger has not been invited to the home; the stranger has initiated the intrusion into the privacy of the home; the householder owes only "common courtesy" of a brief listening to the stranger's story. Because most interactions are very brief, the householder must generate alternative explanations for the interviewer's contact with the household quite rapidly. The models inherent in cognitive script theory are useful here (see Abelson, 1981). These would assert that the words, behavior, and physical appearance of the interviewer will be used to identify possible "stories" or scripts that are potential explanations of the interviewer's purpose. These scripts are heuristic devices to reduce the cognitive burdens of the householder's deciding how best to behave and how to anticipate the next action on the visitor's part. Depending on the script, past experience with such situations will guide behavior (e.g., if this is a sales encounter, be reluctant or uninterested).

The power of the script chosen by an actor interacting with another is that it is a filter through which each behavior of the "other" is interpreted. Since the same behaviors in different contexts have different meanings, two actors pursuing different scripts in the same interaction can yield interactional problems. The experimental literature in script theory has shown that when subjects are primed to use one script to interpret behavior in a situation, they will fail to recognize and remember features of the situation that are not part of the script (e.g., Bower et al., 1979). Cognitive scripts thus act as filters to the comprehension process, not unlike a schema organized along nontemporal lines. We believe this experimental literature has identified principles of behavior that underlie householder behavior upon first encountering an interviewer.

The relevance of script theory to understanding survey participation is that a householder's choice of the wrong script may prompt her or him to terminate the interaction. If householders perceive a sales call for an unwanted product or service, they may fail to attend to the words of the interviewer describing a different intent. There are several findings in the methodological literature that are consistent with this interpretation. Some involve direct attempts by the interviewer to dispel other scripts. For example, experiments have been conducted on the inclusion of the statement "I'm not selling anything" in telephone survey introductions (Gonzenbach and Jablonski, 1993; Pinkleton et al., 1994). In short, one of the first issues in a contact with a householder is the establishment of the intent of the visit. Any evidence that we can obtain about the process the interviewer and householder use to come to a common understanding of this intent will aid our understanding of the process of survey participation.

Using script theory also permits us to make observations about differences between modes in survey participation. One important difference between telephone and personal visit surveys is the restriction of the telephone to aural cues. For better or worse, the personal visit mode provides the householder and the interviewer with a set of visual cues to help them in their judgments about the other actor. For personal visits, householders often can observe the age, gender, race, dress, carried materials, facial expression, and sometimes the automobile used by the interviewer. All of these provide cues to dispel or support one or more alternative possible scripts even before the interviewer has uttered one word. On the telephone, these cues are stripped away.

The two modes also differ in the set of possible scripts applicable. No physical contact is possible in the telephone mode, so scripts involving immediate physical harm to the householder or the interviewer are eliminated. This might explain the lower response rate differential between urban areas (with higher violent crime rates) and rural areas on telephone versus face-to-face interviews (Groves and Kahn, 1979). Further, some encounters with strangers tend to be personal visits; others occur over the telephone. For example, religious-group solicitations tend to be face-to-face; sales calls seem to be dominantly by telephone. We would hypothesize that the relative frequency of refusals on the telephone because of disinterest in buying a product or service would be higher than in personal visit surveys, other things being equal. Similarly, reasons for refusals in the personal visit mode would entail misinterpretations of other sorts.

In what ways is an interview request similar to those from other door-to-door solicitations, and in what ways is it different? All generally involve the presence of a stranger at one's door. Furthermore, in all cases, such presence is unconnected with past experiences of the householder. In this sense, they are different from visits by meter readers, bill collectors, and the like, where the request involves an element of legal obligation or is brought about by some action (or inaction) on the part of the householder. All are seeking something from the householder—time, commitment (religious or political), information, or money. Generally, none imply the likelihood of a multifaceted relationship (i.e., the interaction is likely to have a narrow purpose; a sales pitch for aluminum windows is not likely to be followed by a invitation to the visitor's home for dinner). Finally, none portend an ongoing relationship. This implies different normative obligations on the part of the householder engaged in the interaction.

The various types of encounters *do* differ in the nature and content of the request. First, they vary in the extent to which they appeal to the common good versus the specific needs of the individual. Calls for charitable contributions and survey requests often involve arguments about the welfare of others. Sales calls and political canvassing appeal to individual interests of the householder. Second, probability sample survey visits are different from most other encounters in that the others generally involve a single contact request. In contrast, the survey interviewer is likely to return if the initial visit is unsuccessful. That is, when the householder terminates an initial contact with an interviewer without consent, the very act of calling back on the household is often *prima facie* evidence that sales scripts are inappropriate to interpret the visits. Finally, the notion of nonreplacability is critical to survey interviewing. For most other contacts, one household or householder is as good as any other (given an unlimited supply of households). However, for probability-based survey interviewing, substitution of households or persons is generally not permissible. This has a profound impact on how the interviewer behaves, an observation which we will explore in more detail later.

We would also argue that over the last few decades, script pools relating to doorstep interactions with unsolicited strangers have become not only more diffuse but also more perverse. For example, 30 years ago, the presence of a salesperson on one's doorstep was not uncommon, while this form of selling has all but disappeared. Similarly, many other solicitations previously conducted in person are being done by telephone or mail. This has reduced, we believe, the number of strangers typically encountered at one's door, and correspondingly reduced the development of (or need for) a well-differentiated and large set of scripts to handle each of these types of encounters. In any case, survey requests are a relatively rare event, and we suspect householders are unlikely to have a specialized script available to deal with such requests. (Perhaps the most pervasive visual image of survey interviewers is that provided in cartoon treatments of surveys, which portray a man in a suit holding a clipboard—an image rarely realized in practice.) Even when the householder has a script evoked by the word "interview," as Suchman and Jordan (1990) note, it might be one of the job interview, the late night talk show interview, the television news magazine interview, or other scripts that may or may not affect consent rates for the survey request.

As the frequency of door-to-door contacts has decreased and crime rates (and corresponding fear of victimization) have increased, unsolicited visits to one's home may increasingly be treated as unusual and possibly even threatening events. The consequence of this is that the natural set of scripts that householders may evoke when an interviewer approaches their door may be less benign than in the past (i.e., householders are more suspicious).

These observations lead to two conclusions. First, householders encountering a survey interviewer are more likely to invoke a "generalized stranger script" or an incorrect script (e.g., salesperson) than to immediately identify the intention of the interviewer correctly. Second, such scripts as may be invoked by householders may be likely to produce an initial reaction of caution or even suspicion.

As portrayed in Figure 8.2, the householders' initial efforts are directed at comprehending and evaluating the intent of the interviewer. The householder attends to all verbal and nonverbal cues, evaluating whether they are informative to eliminating some possible scripts. There *are* cognitive burdens involved in constructing and assessing multiple alternative scripts. Hence, the householder places a premium on rapid identification of which alternative script is appropriate. This isn't always easy. Sometimes, words of the interviewer serve to evoke multiple scripts at once (e.g., "from the University of Michigan," "a study for the U.S. Bureau of Justice Statistics"). Other times, key phrases are used to make judgments of intent. For example, "authorized by Public Law 121-3," in a government survey may be sufficient to convey the notion that compliance is mandatory.

If all of the potential scripts have clearly negative valuations, the householder seeks to terminate the interaction (e.g., "I'm not interested"). If the potential scripts are a mixed lot, some positive, some negative, the householder needs more information. When seeking more information about what script is appropriate is judged too burdensome itself, the householder merely silently chooses a script and delivers an answer based on it (e.g., "I don't want any" or "Sure, come on in."). A specific example of such a high-burden circumstance is when the householder is engaged in some other activity at the time of the contact. Here the cognitive burdens of determining the intent of the interviewer are exacerbated by the opportunity costs of doing so, the inability to continue the prior activities. Sometimes this generates responses such as "I'm too busy to talk to you now" or "Could you call later?" or a host of other similar responses. Such persons may seek to postpone the interaction to another time.

When opportunity costs for seeking more information are low, some householders will question the interviewer about alternative scripts in order to eliminate them (e.g., "What are you selling?"). Such attempts might be especially efficient for scripts that are easily identified or that lead to clear behavioral outcomes. In the above example, once it is known what is being sold, the decision to continue can be made quickly.

Sometimes, all of the possible scripts are positively valued, opportunity costs for the interaction are low, and the householder indicates willingness to continue the interaction into the interview phase.

8.3   CUES FOR JUDGING THE INTENT OF THE INTERVIEWER

When actively judging the intent of the interviewer's call, the householder attends both to the words of the interviewer and other aspects of the contact situation (e.g., the interviewer's accent, dress, demeanor). The chances of the householder choosing the correct script are higher when the words of the interviewer are used in the judgment process. However, those attending to the interviewer's words typically must judge the nature of the request from key words like "survey" or "research project" or other cues that have multiple meanings. Little if any of the information interviewers provide to householders reveals central aspects of the request—the fact that the interviewer will be asking a large set of questions about personal attributes, some of which may require difficult recall or judgment tasks; that the interviewer will document all answers for later study; that the interviewer will completely determine the topics of the conversation; that the interviewer may probe, seeking more complete answers to questions; that the interviewer will refrain from any self-revelations.

Instead, the interviewer may refer to long-term goals of the study, what agency sponsors the study, how the data will be kept confidential, and a host of other auxiliary aspects of the survey request. We find this to be difficult for many survey researchers to understand without some careful thought. To them, the interviewer is asking the householder to become part of a process generating statistical information about a population, to test a hypothesis about human behavior, or to describe the extent of some social phenomenon. The request, to the survey researcher, is to become part of his or her study. To the householder, however, the most immediate and most relevant aspects of the survey request has little to do with population-based information (a product the householder rarely is given). The central features are the grant of time for the interview and the revelation of personal information to a stranger who appears to be interested in asking questions of some unknown sort.

Petty and Cacioppo, in a theoretical model of attitude change, the Elaboration Likelihood Model, have found useful distinctions between "central route" and "peripheral route" cuing to the cognitive processes. The central route is "careful and thoughtful consideration of the true merits of the information presented in support of an advocacy of some position on an issue" (Petty and Cacioppo, 1986, p. 3). The peripheral route is attention to "some simple cue in the situation that induced change without necessitating scrutiny of the central merits of the issue-relevant information" (p. 3). Petty and Cacioppo repeatedly find that attitude change can occur either because the subject is attending to peripheral cues that give credibility, attractiveness, or authority to the speaker or because the subject is attending to central features of the argument and they are powerful enough to shape her or his attitude. Attitude change based on central route processing, decisions based on the merits of the case, appear to be more enduring.

Some of the work of their research program has found that when subjects have low motivation to process the central arguments of a persuasive message, peripheral cues in the situation become more important determinants of their judgments. Further, with low involvement, larger numbers of different arguments presented on the issue yield more change in attitudes than smaller numbers of arguments. Those subjects for whom the topic was salient, who had higher levels of intelligence, who believed they alone were responsible for the judgment, tended to use central route processing.

The relevance to this perspective to the interactions of interviewers and householders is 1) we believe that most householders selected for surveys begin the interaction with an interviewer without a well-formed prior decision regarding participation, 2) most householders begin with a state of low involvement in the topic of the survey, 3) most cues providing by the interviewer appear to be peripheral to the interview request, and 4) some behaviors of householders may be seeking central route information. The low overall involvement rates, following the Petty and Cacioppo theory, would lead away from attention to the central merits of the survey request and toward peripheral cues. The central merits of the survey request in our belief concern the specific interview requested of the householder, not the statistics of the survey writ large. Most arguments presented by interviewers, however, provide information about the larger survey, not the interview requested of the householder. Sometimes, householders do ask, "So what do you want me to do?" or "What does this entail?" or "What do I have to do?"—all calls for a description of the activities actually required of a survey respondent. Interviewers are generally trained to provide a general description of the question-answering process, but not to provide detailed information about specific types of questions, especially not those judged to be sensitive or threatening.

Those making judgments about the intent of the interviewer from nonverbal cues are more likely to evoke erroneous scripts, leading either to quick declinations, to quick agreements to proceed, or to questions seeking clarification from the interviewer. Judgments might be based on observations as simple as noting that the interviewer appeared to be a nonthreatening, interesting, pleasant person; that the interviewer demonstrated devotion to the job by calling on the household in the pouring rain; *or* that the interviewer seemed unsure, tense, unfriendly, or tired.

The householder may not reveal to the interviewer much if any of this comprehension step. Householders may remain silent during the interviewer's delivery of introductory talk about the survey. Alternatively, they may communicate nonverbally their puzzlement, interest, boredom, annoyance, or impatience.

In those cases, when the survey script is correctly evoked by the householder, we believe the four sets of influences in Figure 8.1 come to bear. There are components of the reaction that are rational choices between the burdens of consenting to the survey request (e.g., time spent being interviewed lost to other pursuits) and components that are more based in affective reactions (e.g., whether the interview may involve questions that are embarrassing to the householder).

8.4   INTERACTION FROM THE INTERVIEWER'S PERSPECTIVE

The interviewer is often the dominant actor in the initial moments of contact with a householder. These are the moments when the interviewers identify themselves, describe their affiliation, the sponsor and topic of the survey, the purposes of the survey, the respondent selection procedures, and a variety of other attributes of the survey request. They are in a heightened state of alertness, focused on judging the likely next behavior of the householder and identifying their own appropriate response.

In a series of focus groups over the past few years, we have sought reports of interviewers about their thoughts, affective states, and behaviors during these first few moments of interaction. We chose this method of inquiry because we wanted to inform our quantitative studies with qualitative information about the self-assessed interviewer influences on householders' decisions to participate in surveys. We expected that some interviewers might be quite insightful about which interviewer behaviors seem to affect householder decisions positively and negatively. Many interviewers reported about what they do during various steps of first contact with a household. Experienced interviewers often reported that they adapt their approach to characteristics of the sample unit.

Face-to-face interviewers engage in a continuous search for cues about the attributes of the sample household or the person who answers the door, focusing on those attributes that may be related to one of the basic psychological principles that are hypothesized to facilitate compliance (see Chapter 2). Interviewers note that this search begins upon first visiting the sample neighborhood:

> The interviewer has to be sensitive to the area in which she rings that bell. There are clues around about the people who live in any area: their socio-economic background, their values—often a house can reveal the age of the inhabitants.

> Depending on the neighborhood and ethnic area, I fluctuate my approach.

> I often wear something, a jacket or necklace, which displays the symbol of the college located in our county. People are very proud of our local, growing university and seem to be more cooperative if they can immediately recognize that I am local.

> I try to use a "gimmick" in my attire when visiting HUs. Bright colors, interesting pins, jewelry—nothing somber or overly "professional" or cold looking—fun items of attire like ceramic jewelry, scarfs tied in odd ways. If my [initial drive through of the neighborhood] spots cats, dogs in windows, doors, I make a note and wear something like a cat pin on a coat, etc.

Before making a call on a sample unit, many interviewers often first drive into its neighborhood to check on existence of toys in the yard, dogs, "no solicitation" signs, evidence of whether the unit is occupied during the day, as well as the character of the neighborhood in general. In poor areas, some interviewers choose to drive the family's older car and to dress in a manner more consistent with the neighborhood. In rich neighborhoods, interviewers may dress up. In both cases, the same compliance principle—similarity leads to liking—is engaged, but in different ways.

Which statement an interviewer uses to begin the conversation with the householder is the result of observations about the neighborhood, housing unit, and immediate reactions upon first contact with the person who answers the door.

> I use different techniques depending on the age of the respondent, my initial impression of him or her, the neighborhood, etc.

> From all past interviewing experience, I have found that sizing up a respondent immediately and being able to adjust just as quickly to the situation never fails to get their cooperation; in short, being able to put yourself at their level, be it intellectual or street wise is a must in this business. . . .

> After 3 or 4 minutes you can almost tell what they are going to say, at that time you have to change your attack.

> I give the introduction and listen to what they say. I then respond to them on an individual basis, according to their response. Almost all responses are a little different, and you need an ability to intuitively understand what they are saying. Sometimes, the words they are saying are not the true reasons they are refusing. I attempt to discern what the situation is and respond to it.

The reaction of the householder to the first statement dictates the choice of what second statement to use. With this perspective, all features of the communication are relevant—not only the words used by the interviewer, but the inflection, volume, pacing (see Oksenberg, Coleman, and Cannell, 1986), as well as physical movements and demeanor of the interviewer.

We have termed this phenomenon "tailoring," and it appears to have links to other social-psychological attributes. Figure 8.3 is an illustration of the hypothesized interviewer strategy at work. We utilize the notions of *strategy* and individual *arguments* within each strategy. For example, visual observations of the neighborhood and the sample housing unit may lead the interviewer to choose a strategy for, say, the Consumer Expenditure Survey, appropriate to a low-income person. The inter-

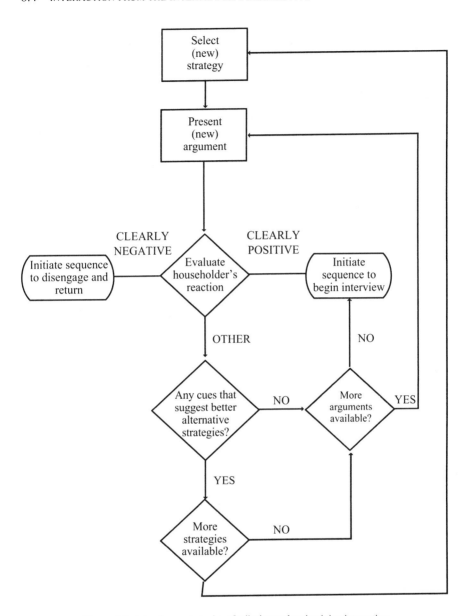

Figure 8.3. Interviewer strategies of tailoring and maintaining interaction.

viewer might introduce the argument in this strategy that the Consumer Price Index, derived from the Consumer Expenditure Survey, affects the rate of change in Social Security, welfare, Medicare, and other transfer programs. There are other elaborating arguments to such a strategy.

Interviewers tell us that they typically have selected one or more possible strategies prior to initiating contact. The first utterances of the interviewer are constituent arguments in the one chosen after the householder's initial response. The successful application of tailoring depends on the ability of the interviewer to evaluate the reaction of the householder to each of the arguments presented. Note that the interviewer's initial goal is to maintain interaction (avoiding pushing for the interview) as long as the potential respondent's reaction remains neutral or noncommittal. An interviewer will continue to present different arguments until such time as the householder is clearly receptive to an interview request, or there are no more arguments to present. For inexperienced interviewers, the latter may occur before the former, forcing the interviewer to (prematurely in some cases) initiate the interview request.

When negative householder reactions to arguments are sensed by the interviewer, evidence exists that the wrong strategy has been chosen, and that an alternative strategy might be more appropriate. Basically, this means that the strategy is not addressing the concerns of the householder. Following the example above, emphasizing the economic role of the CPI is irrelevant to a householder whose real concern is whether the interviewer has called for criminal purposes. Sometimes the cues that interviewers must judge are nonverbal, and thus devoid of rich information. Statements made by the householder generally contain more information about the householder's script choices and their concerns. Householder questions are especially useful because they both reveal concerns and provide the interviewer another conversational turn in which to address the concern.

Telephone surveys limit most possibilities of tailoring prior to the first contact. Typically, the interviewer knows only the general geographical area of the same household. An answering-machine response sometimes gives the interviewer information about the household (e.g., the presence of children) but there is little guarantee of this. For that reason, tailoring opportunities begin for telephone interviewers principally with the first "hello" from a householder.

8.5 EMPIRICAL MEASUREMENT OF INTERACTIONS BETWEEN INTERVIEWERS AND HOUSEHOLDERS

Obtaining verbatim recording of interactions, for example, through tape recording is undeniably the preferred method to capture the content, pacing, inflection, and other potentially important attributes of doorstep interactions. Although taping may itself affect the interaction, most researchers find few observable effects. However, the cost of recording such interactions, transcribing them, and coding them in ways useful to analysis is quite large.

Most of the empirical data we discuss below is not of this ilk. Instead, in order to test some of the theoretical speculations above, we asked face-to-face interviewers to document what happened when they talked with householders, using a structured instrument, described in more detail in Chapter 3. Variants of this instrument were used on several surveys over the past few years, and we present the results from the National Survey of Health and Stress (NSHS) (described in Chapter 3).

The form asked interviewers to document various verbal behaviors of the householder for each contact, and to record various verbal and physical behaviors they performed during the interaction. The specific behaviors chosen were exemplars of the helping and compliance principles discussed in Chapter 2. Interviewers were instructed to complete the form immediately after each contact with members of a sample household with whom they had a conversation regarding the survey. The introductory conversation refers to the time between when the interviewers introduced themselves to the householder and either the time they proceeded with the interview or, failing that, the time the interviewers left the housing unit.

To enrich the analytic utility of the observations about the interaction between interviewers and householders, other data were added: observations about the neighborhood, data on the disposition of every call to the sample household, and data from a self-administered questionnaire of interviewers (also described in Chapter 3). The final dataset created from a combination of these sources contains records for 54,586 calls and 26,888 contacts for the 9,863 eligible sample cases.

In Chapter 3, we review some of the measurement limitations of this interviewer form, and the reader should review that material. In short, in replications of this technique among interviewers in Great Britain there is lower than desirable correspondence between what an interviewer records as having happened during the contact and what a third-party observer records. Further, some of the behaviors documented by interviewers may be based on nonverbal cues provided by the householder, not direct oral communication. Finally, we suspect that there are errors on the part of interviewers in recalling behaviors. We would suspect these errors to induce greater correspondence between documented behavior and the perception by the interviewer of the householder's reaction to the entire visit.

In the analyses that follow, we call attention to interpretations of findings that could reflect systematic recording errors on the part of interviewers. These are contrasted with interpretations guided by the theory reviewed above. The hope is that improvements in the design of contact description measurement can lead to the improvement in the measurement of such data for increasing understanding of causes of survey nonresponse.

8.6   NATURE OF THE HOUSEHOLDER–INTERVIEWER INTERACTION

In this section, using the data from the NSHS, we attempt to describe what happens when interviewers and householders meet in brief contacts forming the setting of a survey request. Based on these data, we first give a brief description of the common features of such conversations—their length, relative frequencies of types of interviewer and householder actions, and variation in actions over successive contacts with the same household. The analysis takes advantage of the large number of contacts to examine how interaction varies among different types of contacts. Based on the perspective outlined above, we expect to find systematic variation in the structure of these conversations.

This leads to an examination of characteristics of householders that are associated with different patterns of behavior. Here socio-demographic attributes, common correlates of survey participation, will be discussed to see whether they themselves shape the nature of the behavior in survey contacts. We address whether interviewers vary in the kind of householder behavior they elicit, controlling on the householder characteristics found predictive of that behavior. This analysis is the empirical check on the theoretical notions that interviewers affect householder behavior in systematic ways. These descriptive analyses provide one of the few glimpses researchers have yet had on thousands of conversations prior to the decisions to participate.

8.6.1 Changes in Interaction over Repeated Contacts with the Same Household

The first question addressed is the level of effort (measured in terms of the number of contacts) required to reach a final disposition on a case. The distribution of contacts is presented in Table 8.1. In 86% of cases, the household roster (household listing) is obtained on the first contact. The roster can be filled out by anyone in the household who possesses the relevant information, but only the selected respondent can provide the interview data. Most cases (over two-thirds) require additional contacts subsequent to the listing to obtain a final disposition. Indeed, it takes slightly more contacts on average to obtain a final disposition after obtaining the roster information (mean, 1.6) than to obtain a listing in the first place (mean, 1.2). Finally, Table 8.1 shows that one-quarter of sample cases are resolved (either with an inter-

Table 8.1. Percentage of contacted sample households requiring different number of contacts to complete two survey tasks[a]

Survey task	Number of contacts						Total (percent)	Mean
	None	1	2	3	4	5 or more		
Contacts to obtain household listing	—	86.2	9.2	2.7	0.9	1.0	100.0%	1.22
Additional contacts to obtain final disposition	32.4	28.7	18.2	9.3	4.5	6.8	100.0%	1.58
Total contacts	—	23.5	31.8	20.7	10.9	13.1	100.0%	2.81
(n)								(9,389)

[a]These are based on all contacts, both those from the field and from the telephone facility.

view or a permanent noninterview) in a single contact with the household, with 76% being resolved in three or fewer contacts. We would expect the results in Table 8.1 to vary by who the eligible respondents were, whether a screening interview was required, what callback scheduling was used, and what mode of data collection was employed.

8.6.2 Characteristics of Contacts

Given that the nature of the interaction (and the nature of the request) may change over contacts, it is informative to examine characteristics of contacts by number of contact. Table 8.2 presents characteristics of contacts separately for the first and each subsequent face-to-face contact on a sample household, based on the contact-description instrument.

The mode of contact tends to change with repeated contacts, from face-to-face mode to telephone mode. It is likely that this is a function of the field administrative rules for NSHS and not generalizable to other surveys. Less than 6% of the contacts use an intercom or telephone for communication in the first contact but over half of the contacts occurring after the first four contacts are by telephone. Most of these calls occur after the household listing is obtained and are seeking contact with the chosen respondent. It is also likely that most telephone contacts have as their goal making an appointment with the chosen respondent for a later face-to-face interview.

Brehm (1993) in his study of face-to-face survey nonresponse makes the argument that quick responses (like "I'm too busy") to interviewer queries have some meaning but are shallow indicators of real motivations for householder behavior. Table 8.2 permits us to see the relative frequency of different statements by the householder. "I'm too busy" and "I'm not interested" are the most common comments, but neither are encountered with great frequency (16.1% and 8.1%, respectively, on the first contact).

Use of stock phrases like "I'm too busy" may reflect miscomprehension about the intent of the interviewer (a choice of an incorrect cognitive script) or householders' rules about never responding to surveys, as well as real time constraints of the householder. As expected, this is the one householder response that does not decline in frequency over successive contacts. With each additional contact, more of those not chronically busy move to "interview" or "refusal" status. The remaining cases eligible for visits in later contacts are disproportionately those who were too busy to act on the survey request in prior visits.

One of the few contact situations typically covered in interviewer training sessions involves how to answer various householder questions about the study. In NSHS contacts, the most common respondent question is "What is the purpose of the survey?", occurring in about a third of the first contacts. The next most common are "How long will the interview take" (27%) and "How was I chosen?" (20%). These questions may be evidence of a more thoughtful consideration of the interview request by the householder; that is, they seek more information with apparent understanding of the nature of the request.

Table 8.2. Percentage of cases contacted exhibiting various attributes, by contact number[a]

	Contact number				
	One	Two	Three	Four	Five+
Duration of contact					
Less than 1 minute	7.6%	19.0%	21.2%	23.0%	24.0%
1 to 5 minutes	55.4	58.6	55.7	56.0	58.3
More than 5 minutes	37.0	22.4	23.2	21.1	17.7
Total	100.0%	100.0%	100.0%	100.0%	100.0%
Mode of contact					
Intercom or telephone	5.8%	28.4%	33.0%	39.3%	52.5%
Closed door	1.6	1.0	1.	0.7%	0.7
Open door, screen door storm door	30.3	16.8	12.9	11.6	10.
Face-to-face, outside	23.2	12.7	9.8	9.4	8.1
Face-to-face, inside	38.1	39.4	40.7	36.2	25.6
Other	1.1	1.7	2.7	2.8	2.6
Total	100.0%	100.0%	100.0%	100.0%	100.0%
Householder comments (multiple mentions)					
a. Too busy	16.1%	10.8%	9.5%	10.2%	12.1%
b. Enjoy surveys	2.2%	1.4%	1.0%	0.5%	0.7%
c. DK topic	3.1%	1.4%	0.9%	0.5%	0.5%
d. Think about it	2.1%	1.0%	0.7%	0.5%	0.6%
e. Don't trust surveys	1.1%	0.9%	0.6%	0.5%	0.2%
f. Waste of time	0.8%	0.6%	0.3%	0.5%	0.3%
g. Surveys help community	1.5%	1.0%	1.0%	0.9%	0.5%
h. Respondent not home	18.8%	10.7%	9.9%	11.9%	14.9%
j. Waste of money	0.4%	0.4%	0.2%	0.3%	0.0%
k. Questions too personal	1.7%	1.5%	1.1%	1.6%	0.6%
m. Not interested	8.2%	7.1%	6.6%	5.7%	5.1%
n. I never do surveys	2.0%	1.4%	1.0%	0.6%	0.7%
o. Government knows everything	0.4%	0.4%	0.2%	0.1%	0.0%
Householder questions (multiple mentions)					
a. Purpose of the survey	32.5%	13.7%	10.4%	8.7%	5.1%
b. Who is sponsor	3.4%	2.3%	1.5%	1.5%	0.9%
c. How was I chosen	19.5%	10.0%	8.9%	6.8%	4.2%
d. How long will it take	26.9%	15.6%	12.2%	9.6%	7.4%
e. Who sees answers	2.4%	2.0%	1.9%	1.8%	1.0%
f. Can I get results	1.8%	2.6%	2.3%	2.1%	1.3%
g. Is there an incentive	0.8%	0.6%	0.4%	0.3%	0.1%
Interviewer statements/actions (multiple mentions)					
a. How results affect respondent	16.9%	10.6%	9.3%	8.2%	6.7%
b. Complete by certain date	2.6%	4.1%	4.7%	5.3%	7.2%
c. Begin asking questions	5.5%	7.4%	9.1%	7.9%	5.6%
d. Interview others in area	18.6%	8.4%	6.0%	4.9%	2.8%
e. Most people enjoy survey	19.2%	12.0%	10.9%	9.5%	7.7%
f. Most people participate	11.8%	7.3%	5.7%	5.1%	3.2%

Table 8.2. *Continued*

	Contact number				
	One	Two	Three	Four	Five+
Interviewer statements/actions (cont.)					
g. Compliment householder	10.3%	8.7%	9.0%	6.6%	4.7%
h. Respondent represents others	28.0%	18.0%	15.5%	14.3%	11.1%
j. Explain random selection	57.6%	25.5%	18.8%	14.7%	9.2%
k. Explain purpose of study	64.6%	31.9%	22.7%	19.4%	12.7%
l. Mention advance letter	77.7%	24.4	14.6%	9.5%	5.1%
m. Mention incentive/gift	53.4%	29.0%	26.0%	22.3%	17.2%
(*n*)	(9,164)	(6,913)	(3,940)	(2,019)	(2,586)

*a*This includes "field" contacts only. All telephone facility contacts are excluded.

As part of introductory comments, interviewers are also typically trained to provide the householder with a brief description of the survey. Describing the study, mentioning the respondent letter and incentive, and explaining the random selection of the household and sample person would more likely occur on the first contact than later contacts. These statement types are required to describe the request and also are answers to some of the respondent questions. This is indeed reflected in Table 8.2.

Table 8.3 uses the categorization of respondent behaviors into "negative," "time delay," and "positive" statements, and "questions." The most common behavior by householders is to ask questions about the survey request. This is true on first contacts and all subsequent contacts. Next most common are "time delay" statements,

Table 8.3. Percentage of contacts exhibiting various householder behaviors, by contact number

	Contact number				
	One	Two	Three	Four	Five+
Negative statements (don't trust surveys, waste of time or money, etc.), one or more mentioned	9.6%	8.6%	7.5%	6.7%	5.8%
Time delay statements (too busy, will think about it, etc.), one or more mentioned	17.2%	11.3%	9.9%	10.5%	12.5%
Positive statements (enjoy surveys, helps community, etc.), one or more mentioned	3.4%	2.2%	1.9%	1.3%	1.1%
Questions asked by householder, one or more questions	50.2%	28.2%	23.8%	19.1%	14.2

leading to later attempts by the interviewer to complete the survey. Direct negative statements are rare (only 10% of first contacts, and fewer in subsequent contacts), but positive statements are even more so. Given the extreme skewness of the latter distribution, this variable is excluded from further analysis.

At this level of aggregation, some evidence exists from other studies with more valid measurements, albeit on nonprobability samples. In similar analyses, Morton-Williams (1991) finds 17% of contacts with an "I'm not interested" comment, with similar rates of time delay statements (19%) and questions about the survey purpose (19%). Hence, the three behaviors were roughly equally likely. Our data show relatively more questions asked about the survey purpose (32% on the first contact), and relatively fewer "I'm not interested" responses (8% on the first contact). Maynard, Schaeffer, and Cradock (1993), in their study of telephone survey contacts ending with declinations, find that 83% terminate early in the interaction, with the first conversational turn of the householder, with statements of the type labeled "negative."

The relative percentages of householders asking questions decline rapidly over successive contacts. Other householder behaviors also decline in their occurrence, but less dramatically than questions. These results are in concert with the finding of shorter interactions for successive contacts with households. A small qualification of these results is that the percentage of householders giving "time delay" statements actually increases somewhat in fourth and higher contacts. This probably reflects the fact that households requiring more contacts prior to final disposition are those that report they don't have time to provide an interview in earlier contacts.

Table 8.2 is useful to get a overview of what happens on different contacts, but it cumulates over many different kinds of contact situations. Some of the important variations in contact situations have been investigated. For example, behaviors in later contacts may be influenced by what happened in prior contacts. To examine this, householder behaviors in the first two contacts of multiple-contact households are classified by a) whether or not a household listing was obtained in the first contact and b) whether or not the householder in the first contact was contacted in the second. Householders not providing a listing in the first contact made negative statements 27.1% of the time, compared to 6.9% for householders providing a listing in the first contact. Negative statements in contact 2 decline to 20.0% for households not listed and to 6.1% for households listed on contact 1. Similar relationships are found for time delay statements and questions.

For households where different householders were contacted in the first two contacts, more negative statements, time delay statements, and questions were elicited than when the same householder was contacted. It appears that the initial interaction is essentially repeated for a householder newly encountered. Generally, these and other results mirror both the relative frequencies of the different behaviors and the overall decline in comments and questions over contacts, although the absolute frequencies of behaviors differ somewhat depending on specific circumstances. Given the conditional nature of later contacts, the remainder of the analyses focus on the first contact only.

We also find that those householders who eventually become final refusals on the survey are more likely to make negative and time delay statements in the first contact than those who acceded to the interview request. Table 8.4 presents these data for the NSHS along with those from the 1990 National Election Study (NES) and AHEAD survey (see Section 3.5.4). However, questions come more frequently on the first contact from those who grant an interview than from those who become refusals (whether on that contact or a later one). We explore the impact of these statements on the final disposition of contacted households in Chapter 9.

8.6.3   Interaction in Telephone Surveys

It is useful to compare some of these results to information from telephone-survey interactions. In another study (Couper and Groves, 1995), we asked interviewers to tape approximately 100 introductory interactions in an RDD telephone survey. The study drew a probability sample of interviewer shifts in a centralized facility and asked interviewers to tape introductions during that shift. (Permission was sought to keep the tape prior to the completion of the call. If the householder refused, the tape was destroyed.) About half of the interactions ended with appointments being made; a quarter are calls that lead to interviews, and the rest ended in refusals.

Because the data came from tapes, we were able to code the number of conversation turns (an uninterrupted utterance by one actor). On average, there were 10.8 turns per interaction, suggesting relatively short interactions. Table 8.5 describes various aspects of householder behavior during their conversational turns. It displays positive, negative, and time delay communication, using the same notions as

Table 8.4. Percentage of first contact exhibiting various householder behaviors for three surveys, by final outcome; overall cooperation rate of surveys

	NSHS	NES	AHEAD
Negative statements			
Interview	5.7%	13.2%	9.5%
Refusal	44.2%	24.7%	42.7%
All	9.6%	15.7%	15.3%
Time-delay statements			
Interview	15.2%	15.8%	8.1%
Refusal	34.7%	32.3%	15.4%
All	17.2%	19.4%	9.4%
Positive statements			
Interview	3.7%	3.3%	3.8%
Refusal	0.8%	0.9%	0.2%
All	3.4%	2.7%	3.1%
Questions			
Interview	52.0%	43.6%	35.1%
Refusal	33.8%	32.3%	19.0%
All	50.2%	41.1%	32.3%
Cooperation rate	89.2%	78.2%	82.6%

Table 8.5. Percentage of conversational turns and full interactions exhibiting various householder behaviors in an RDD telephone survey

Statement type	Turn pairs	Interactions	Mean number of turn pairs per interaction
Positive	0.8%	7.9%	0.09
Negative	5.5	26.7%	0.58
Other informative	8.7	42.6%	0.92
Time delay	15.0	63.4%	1.59
Questions	5.7	34.7%	0.60
Uninformative	64.3	99.0%	6.84
Total	100.0%		
(n)	(1,074)	(101)	(101)

Source: Couper and Groves (1995).

in Table 8.4 (although based on coders' judgments of the intent of the householder's utterance).

The research added two other categories: "other informative" and "uninformative." "Other-informative" statements were those in which some information was volunteered by the householder beyond what was sought by the interviewer. "Uninformative" is the residual category, containing a mix of utterances (e.g., "uh huh," "uhm"). These were categories that the instrument completed by face-to-face interviewers did not include.

As we found in the face-to-face case, positive statements are very rare (on average less than one in a hundred statements, or less than one in every 10 interactions). While the majority of turn pairs can be described as uninformative (i.e., turns in which no information is revealed by the householder that would permit tailoring), almost all (92%) of the interactions contain at least one statement by the householder that could be deemed informative, and an average of 3.8 turn pairs per interaction (or about a third of all turn pairs) provide at least some information relevant to the survey request. There were too few cases to split by call number, as in Table 8.3 for a face-to-face case, but it appears that, compared to Table 8.3, there are fewer questions being asked by householders over the telephone relative to negative and time delay statements begin given. We speculate that this might reflect a reduced level of involvement required by householders over the telephone.

8.6.4 Do Characteristics of the Interaction Vary by Different Social Environments?

We deal with the issue of social environmental impact on survey participation in greater detail in Chapter 6. In this section, we examine the impact of some of these contextual variables on the householder–interviewer interaction. Living in urban areas has often been characterized as productive of brief, superficial interactions with many strangers, a lifestyle that over time produces suspicions about the intent of strangers who seek more lengthy, involved interactions (Franck, 1980). The interaction data for first contacts (Table 8.6) show that negative statements are somewhat more prevalent in central cities of consolidated metropolitan statistical areas (CMSAs), declining from 10% in such areas to 6% in non-MSA areas. Time delay statements are also somewhat higher in central cities of large CMSAs (20% to 15%

Table 8.6. Percentage of first contacts exhibiting various characteristics, by environmental, householder, and interviewer characteristics

	Percentage more than 5 minutes	One or more negative statements	One or more time-delay statements	One or more questions
Social environment				
Central cities of CMSAs	36.9%	9.7%	20.3%	51.9%
Balance of CMSA	41.3%	9.5%	19.9%	53.0%
Other MSA	35.5%	7.9%	14.2%	50.1%
Non-MSA	36.8%	6.0%	15.3%	50.1%
p-value for χ^2	0.44	0.00	0.02	0.88
Physical impediments to access	38.0%	10.3%	20.2%	51.4%
No impediments to access	37.4%	7.6%	16.2%	51.0%
p-value for χ^2	0.78	0.01	0.00	0.85
Householder characteristics				
Male	37.1%	8.9%	16.3%	50.4%
Female	37.8%	7.3%	17.2%	51.5%
p-value for χ^2	0.50	0.01	0.34	0.36
Single-person household	28.0%	9.5%	20.5%	52.9%
Other	39.1%	7.7%	16.2%	50.8%
p-value for χ^2	0.00	0.04	0.01	0.14
30 years or younger	34.1%	4.8%	13.7%	45.9%
31–40 years old	37.9%	7.8%	20.0%	54.6%
41–50 years old	39.1%	11.0%	19.0%	54.7%
51–60 years old	42.5%	13.3%	16.4%	52.2%
61 or older	46.2%	8.7%	6.8%	47.4%
p-value for χ^2	0.00	0.00	0.00	0.00
Interviewer characteristics				
Less than 6 months tenure	35.4%	8.9%	16.0%	55.0%
6 months to 1 year tenure	34.7%	7.9%	18.4%	55.0%
1–5 years tenure	42.7%	8.8%	19.3%	49.4%
5 years or more tenure	35.9%	6.0%	12.4%	44.7%
p-value for χ^2	0.41	0.03	0.00	0.03
1, lowest rating on confidence	32.1%	8.9%	11.8%	41.7%
2 on confidence rating	30.0%	8.1%	17.8%	49.6%
3 on confidence rating	42.5%	8.1%	17.9%	54.9%
4, highest rating on confidence	52.6%	5.9%	13.6%	50.7%
p-value for χ^2	0.00	0.51	0.08	0.02

in non-MSAs). This finding probably reflects both greater reluctance to engage in the interaction for other reasons (i.e., time constraints as an unwarranted excuse) and real differences in available leisure time by level of urbanicity. Time-use studies (Juster and Stafford, 1985; Robinson and Godbey, 1997) do show that residents of large urban areas are less frequently at home, spend more time in transport between home and other activities, etc.

The neighborhood observation forms asked the interviewer to record whether the sample unit had bars on its windows, metal, windowless doors, security-system signs, no-trespassing signs, and a variety of other features. These are viewed as indicators of active attempts by the household to limit contacts with strangers, particularly those with hostile intents toward the household. They may even be indirect evidence of a generalized suspicion of strangers that would lead to reluctance to be interviewed. Table 8.6 shows that householders in such units tend to give somewhat more negative and time delay statements than those in households without those barriers to access.

In contrast to negative and time delay statements, questions appear to be behaviors that are not related to the two social environmental factors examined. There are no important differences by urbanicity or by the presence of physical barriers to accessing the unit. This is consistent with the notion that questions are devices used both by those positively and negatively inclined toward the survey request. We noted this earlier in Section 8.2. In contrast, negative and time delay statements are suspected to be disproportionately evidence of reluctance.

8.6.5 What Householder Characteristics Are Related to Their Behavior in Preinterview Interactions?

The interviewer focus groups taught us to expect different types of interactions in multiperson households, and in such households there was evidence of longer interactions, fewer negative statements, and fewer time delay statements (see Table 8.6). There may be three underlying causes of this—interviewers speaking with multiple household members in a single contact, the likelihood that multiple person households would contain at least one person interested in the contact, and the possibility that those living alone are more reticent to engage in interactions with strangers in general.

Older householders tend to have longer interactions with interviewers than do younger householders, but the nature of that content appears not to be captured by the categories of negative statements, time delay statements, and questions. That is, those 51 or older, despite having proportionately more lengthy contacts with interviewers, provide only slightly more negative statements and fewer time delay statements and questions than do householders in the 31–50 age range. The relative shortness of contacts with householders 30 years or younger is matched by their infrequent use of negative statements, time delay statements, and questions. Interviewer focus groups told us that some of the longer interaction with older respondents may be caused by reticence to engage in the interaction because of fear of strangers and difficulties of the elderly understanding the nature of the request, leading to interviewers giving longer explanations about the survey. Time delay statements are provided most frequently by those age 31–40, which may reflect real time constraints among this group.

There appear to be no large differences between the sexes on the nature of their behaviors during interactions with interviewers. Males show some tendencies toward negative statements (9% to 7%) but otherwise are similar to females in their behavior during contacts.

8.6.6 Do Different Types of Interviewers Produce Different Kinds of Interactions with Householders?

Our theoretical orientation to interviewer differences focuses on those interviewers who have available a variety of techniques to use in household contacts. From focus groups and the past literature (see Chapter 7) two variables measured in interviewer surveys were expected to produce different types of interactions—interviewer tenure and a self-rating on confidence in their abilities to gain householders' cooperation.

These two variables are related to one another, but in a manner that might be counterintuitive. The confidence variable is measured on a four-point scale, with a "4" signifying that the interviewer believes that "With enough effort I can convince even the most reluctant respondent to participate."

Somewhat surprisingly, those interviewers who have less than 6 months experience as an SRC interviewer are most likely to report such high confidence levels. The percentages with high confidence decline with greater experience, except that those with more than 5 years of experience tend to have confidence levels most similar to the new recruits. This implies to us that the confidence measures are affected by the real field experiences of interviewers, and that those experiences tend to depress reported confidence levels.

Does the nature of the interaction on initial contacts reflect those differences, or is the greater success of more experienced interviewers independent of what householders initially said in first contacts? The most distinctive tenure group are those with 5 or more years of interviewing experience. They experience fewer negative statements and fewer time delay statements. They also experience fewer questions. Apparently, through experience, interviewers can learn what the likely concerns of householders would be and shape their approach to those concerns. This would lead to interactions that reduce reluctance and preemptively answer the questions that householders might have.

The interviewers who report highest confidence seem not to differ from others in their receiving negative or time delay statements. Low-confidence interviewers do tend to receive fewer questions from householders. The reason for this result is not clear, and probably can only be clarified with multivariate analysis.

In summary, however, the differences in respondent behaviors over traditional socio-demographic characteristics of householders and over interviewer characteristics are generally not large. The largest variation seems to exist for different householder age groups and different interviewer tenure groups. The social environmental variables show small differences in householder behavior, despite their relationships with the likelihood of survey cooperation.

8.6.7   Multivariate Analysis of Householder Behavior during Contacts

The bivariate results in Table 8.6 will be useful to interviewers and field administrators for anticipating what might occur for different types of households. That is, knowing differences in householder behavior associated with one variable is sometimes sufficient to prepare interviewer strategies. As part of a general theory of survey participation, however, such bivariate results are relatively uninformative. For example, there are more single-person households among the elderly. With both variables being associated with householder behavior, bivariate results do not provide insight into what processes might underlie the findings.

Such questions about the marginal effect of one variable controlling for another is of most interest for two sets of variables key to the theoretical structure above—householder attributes and interviewer attributes. Do interviewers have effects on the nature of householder behavior, controlling for the householder characteristics that are associated with different behaviors? If the answer is "yes," then it is possible that tendencies for certain types of householders to interact in a certain way are neutralized by influences of interviewers.

To gather evidence on the combined effects of variables, two sets of models were fit on first contacts. The first are logistic regression models predicting the likelihood of a householder performing each of the three types of actions, using the environmental and householder attributes in Table 8.6 as predictors. The second are logistic models using those variables and interviewer attributes as predictors. The first set of models produced no important new insights. Only one predictor influence is diminished in the multivariate models—household size. Table 8.6 shows tendencies for single-person households to provide more negative and time delay statements, in a bivariate analysis. Controlling on urbanicity and householder age, however, seems to reduce the empirical impact of household size.

The second set of models added interviewer experience and self-confidence to the predictor set. Following our work on interviewer-level influences (see Chapter 7), a curvilinear effect of interviewer tenure was hypothesized, and natural log of years of experience was used as the predictor. Table 8.7 shows that interviewers appear to have marginal effects, controlling on householder attributes, on the likelihood that negative statements will be made and that questions will be asked. The nature of the effects are interpretable within the larger theoretical perspective outlined above. Experienced interviewers tend to receive fewer negative statements (coefficient of –0.18) and are asked fewer questions by respondents (coefficient of –0.22).

The qualitative investigations of interviewers and the theoretical rationale suggest that experienced interviewers are more adept at reading cues about the household and the householder contacted. In reaction to these cues, the experienced interviewer might alter his or her introduction, addressing concerns of the householder. This behavior, by providing information likely to be salient to the householder, reduces the need for householder questions. In addition, the behavior is likely to inoc-

Table 8.7. Coefficients of logistic models predicting three householder behaviors, by attributes of environment, household, and interviewer

Independent variables	Negative statements	Time-delay statements	Questions
Constant	−2.4**	−2.7**	−0.28
Environment			
Urbanicity:			
Central cities of CMSAs	0.43**	0.28	0.068
Balance of CMSA	0.45**	0.29	0.12
Other MSA	0.30*	−0.073	0.070
Non-MSA	—	—	—
Household			
Physical impediments to access (1 = yes)	0.24*	0.18	−0.019
Male householder (1 = yes)	0.22*	−0.087	−0.038
Householder age:			
30 years or younger	−0.62*	0.78**	−0.089
31–40 years old	−0.12	1.22**	0.26*
41–50 years old	0.24	1.14**	0.26*
51–60 years old	0.49	0.99**	0.18
61 or older	—	—	—
Single-person household (1 = yes)	0.06	0.21*	0.070
Interviewer			
Log (years of experience)	−0.18**	−0.11	−0.22**
Confidence	−0.094	0.030	0.16**

Note: Dependent variable coded 1 = one or more statements made of particular type, 0 = no such statement made.
*$p < 0.05$.
**$p < 0.01$.

ulate the householder against other concerns he or she might have, and thus leads to fewer negative statements.

Interviewer experience does not have effects on time delay statements, as measured in these data. The discussion earlier noted that some time delay statements are proxies for negative statements. If the data could differentiate the time delay statements that were excuses from those that were real reports of limited time availability, the theory would hypothesize that interviewer behaviors would affect the former but not the latter.

The other interviewer attribute that was used is the self-confidence rating. It has measured effects only on questions, with more confident interviewers obtaining more questions, other things being equal. There are no significant effects for negative or time delay statements. This is consistent with the bivariate analyses of Table 8.6.

Finally, in an analysis not presented in Table 8.7, there was an examination of a set of interaction hypotheses. For example, do experienced interviewers act to eliminate the tendency for urban dwellers to provide negative reactions to interviewers?

That is, are they disproportionately successful in reducing negative comments from the urban dwellers? These found no support in the data.

8.7   SUMMARY

Most face-to-face solicitations for a survey interview last less than 5 minutes before termination or the initiation of the interview. During that time the most common comments of householders are that they are too busy to talk, the chosen respondent is not at home, or they are not interested in the interview. Telephone contacts tend to be much shorter. When householders have a question, they often ask about the purpose of the survey, how the household was chosen, and how long the interview will take. Such questions tend to be more prevalent in early contacts than subsequent contacts; comments by householders decline less dramatically in relative frequency over contacts. Interviewer behavior also appears to change over repeated contacts, with interviewers probably describing the nature of the survey in early contacts, but tending to focus on benefits of the survey in later contacts.

The findings are consistent with the overall conceptual perspective of the initial moments of contact between interviewer and householder reviewed at the beginning of the chapter. It is likely that the prelude to the decision to participate in a survey consists of each actor iteratively narrowing possible interpretations of the other's intents. Other things being equal, householders initially approach these encounters using conventional reactions to an approach of a stranger. In urban areas, this leads to somewhat greater negative reactions voiced by them. Observable barriers to interaction with outsiders, like bars on windows, are reliable predictors of more negative statements when householders are contacted. The theory suggests that experienced interviewers can break through these default negative reactions to a stranger by altering their own behavior.

One speculation is that experienced face-to-face interviewers observe characteristics of the housing unit, the household, and the householder to preempt questions on the part of the householder. They provide answers to possible concerns before the householder asks the question. Through this they avoid negative statements and tend to reduce the number of questions posed by householders.

In contrast, self-confident interviewers tend to evoke more questions, controlling on householder attributes and interviewer experience. It is suspected that this arises because they deliberately elicit questions, with the assurance that they will be able to answer them well.

Evidence was obtained that time delay statements are relatively unaffected by urbanicity and interviewer attributes and seem more a function of householder attributes, especially age of householder. This is evidence similar to past research that the majority of requests to call later are reflections of real time constraints of the householder. In this first part of the analysis of the NSHS data, we have focused on what happens when householders speak with interviewers about a request for an interview. The fact that the character of these conversations systematically varies by characteristics of householders can allow interviewers to anticipate what the householder may say. This is what experience teaches interviewers over trial and error. It is anticipated that different interviewer behaviors, in reaction what the householder says, affect the likelihood of survey participation.

In short, from the analyses in this chapter we have learned the following about the nature of contact-level interactions between interviewers and householders:

- Interviewer contacts with householders last a relatively short time, especially on the telephone.
- In face-to-face surveys, first contacts tend to be longer than later contacts, with more questions being asked by householders and more notes of time pressures (reflecting perhaps the unscheduled nature of the first contact).
- In early contacts, interviewers tend to explain the purposes of the survey; in later contacts, they tend to communicate benefits of participating in the survey.
- Householders in telephone contacts exhibit different behavior than in face-to-face contacts, with proportionally more negative comments than questions.
- Householders in subgroups with higher refusal rates (e.g., urban dwellers, males, households with physical impediments to access) tend to proffer more negative comments.
- More experienced interviewers tend to receive fewer negative comments (reflecting the same skills that lead to lower refusal rates; see Chapter 7) but also fewer householder questions (perhaps reflecting the ability to anticipate householder concerns before they are articulated).
- Interviewers with high self-confidence regarding their ability to elicit participation of the sample households tend to have longer contacts with householders, containing more householder questions. This probably reflects their assurance regarding knowledge of the answers to these questions.

The next chapter explores the effect that these interactions may have on survey nonresponse, examining outcomes both at the contact level and the case level (final disposition).

CHAPTER NINE

Influences of Householder–Interviewer Interactions on Survey Cooperation

9.1 INTRODUCTION

We started our empirical study of survey cooperation in Chapter 5, examining characteristics of households and householders. In Chapter 6 we examined measures of the influence of social environmental attributes of the sample household. Some of these would be causes of the participatory decision common to all members of a sample (e.g., experiences of national debates on confidentiality of survey data); others might be shared by those in the same sample cluster (e.g., crime rates). In our search for insight into what causes survey participation this chapter moves to one of the lowest levels of influence, the interaction between interviewer and householder, in order to find systematic variations that might explain why some accept and some refuse survey requests.

The reader should view this chapter as a microscopic view of the survey participation process. We will be searching for proximate causes of the decision regarding survey participation, examining individual actions of the interviewer (e.g., whether they handed to the householder some physical material) and individual utterances of the respondent. Like much of science, this research strategy seeks understanding of a phenomenon by dissecting it into smaller pieces and examining each piece's role in producing the whole. While the perspective offers rich material for inductive theorizing, we also believe it is relevant to practical questions of what survey-design features induce higher participation rates and what statistical model specifications are most appropriate for postsurvey adjustment.

In the last chapter we saw that the interviewer is the dominant actor in the meeting with a householder. We also saw large reductions in the amount of interaction between the two in successive contacts. Further, the nature of the interaction changes, dispensing with preliminary descriptive matter about the survey in early contacts, and focusing on evaluative behaviors of householders in later contacts. This chapter attempts to examine whether data about the nature of the interactions are predictive of outcomes of contacts and of the final disposition of the sample case.

Before we apply the data to such questions we examine in more detail the notions of tailoring and maintaining interaction (see also Chapter 2), which guide our hypotheses in the analysis.

9.2   TAILORING

Successful tailoring requires that a number of conditions be met. First, the interviewer must have a wide variety of techniques or strategies at his or her disposal. In some sense, expert interviewers have access to a large repertoire of cues, phrases, or descriptors corresponding to the survey request. Only then can the appropriate statement for a particular situation be selected.

Second, the interviewer must be a good reader of verbal and nonverbal cues from the householders and their surroundings. The important cues are those indicating the householders' choice of script guiding their behavior, their likes and dislikes, salient issues, time pressures, etc. In short, any verbal or nonverbal behavior that reveals decision factors the householders might use to decide about their participation can be used by the interviewer.

Third, the interviewer must apply the appropriate strategy or technique in response to the cues received. Interviewers must be able to make fast and accurate judgments about the particular script reflected in the householder's initial response to the request, and react accordingly. At times, whether through low motivation to seek alternative scripts, or because of a misjudgment regarding the intent of the interviewer, a script is evoked that is judged inappropriate (from the interviewer's perspective) to the survey request. In such cases, interviewers are often quick to redress the situation by distinguishing their request from other similar situations. More often, though, a script may be used whose meaning is ambiguous to the interviewer. For example, "I'm not interested" may mean "I think you're selling something and I'm not interested in buying," or "I know this is a survey request, but I'm not interested in participating." To be successful in tailoring, an interviewer must be a good reader of such cues, or should find ways of eliciting additional cues that reveal the intentions of the householder.

Fourth, the interaction (whether within a single contact or across multiple contacts) must be of sufficient duration for tailoring to be applied. That is, there is a premium on *maintaining interaction* with the householder, even without apparent progress toward gaining the consent to the survey request. There is some support from training procedures that the "maintaining interaction" model is correct. First, interviewers are typically warned against unintentionally leading the householder into a quick refusal. If the person appears rushed, preoccupied by some activity in the household (e.g., fighting among children), the interviewer should seek another time to contact the unit. A common complaint concerning inexperienced interviewers is that they create many "soft refusals" (i.e., cases easily converted by an experienced interviewer) by pressing the householder into a decision prematurely. Unfortunately, only rarely do interviewer recruits receive training in the multiturn repartee inherent in maximizing the odds of a "yes" over all contacts. Instead, they are trained in stock descriptors of the survey leading to the first question of the interview.

Tailoring need not necessarily occur only within a single contact. Many times contacts are very brief and give the interviewer little opportunity to respond to cues obtained from the potential respondent. Tailoring may take place over a number of contacts with that household, with the interviewer using the knowledge she or he has gained in each successive visit to the household. Tailoring may also occur across sample households. The more an interviewer learns about what is and is not effective with various types of potential respondents encountered, the more effectively requests for participation can be directed at similar others. This implies that the success of interviewer tailoring evolves with experience. Not only have experienced interviewers acquired a wider repertoire of persuasion techniques, but they are also better able to select the most appropriate approach for each situation.

The addition of the concept of tailoring to understanding of survey participation also aids in explaining some of the puzzling empirical findings in the survey-methodological literature. There is a relatively large methodological literature studying the effects of different introductory scripts of telephone interviewers on nonresponse rates. Interviewers were instructed to deliver verbatim versions of different scripts. Over several years the research never yielded a consistent set of guidelines and failures to replicate what occurred. This result is exactly what would be expected if tailoring is a principal influence on the ability of interviewers to gain participation. Tailoring implies that there are few "main effects" of any particular interviewer's utterance. What addresses one respondent's concerns may be irrelevant or antithetical to another's concerns. The lack of consistent findings in the literature stems from differences across populations, topics, interviewers, and survey climates, as well as across what issues became salient to the householder at the time of the contact.

From an analytic viewpoint, therefore, tailoring implies statistical interaction effects involving statements that the interviewer might deliver to the householder. If the householder behaves in one way, there will be positive effects of an interviewer statement; if not, there could be negative or no effects of the statement on the likelihood of participation.

9.3 MAINTAINING INTERACTION

Another concept identified in interviewer focus groups is the premium placed by experienced interviewers on the ability to avoid a termination of the interaction with the householder during initial contacts. The introductory contact of the interviewer and householder is a small conversation. It often begins with the self-identification of the interviewer, contains some descriptive matter about the survey request, and ends with the householder's agreement to proceed with the questioning, a delay decision, or the denial of permission to continue. There are two radically different goals in developing an introductory strategy—maximizing the number of acceptances per time unit (assuming an ongoing supply of contacts) and maximizing the probability of each sample unit accepting.

The first goal is common to some quota sample interviewing. There, the supply of sample cases is far beyond that needed for the desired number of successes. Interviewer behavior should focus on gaining speedy resolution of each case. An acceptance of the survey request is preferred to a denial, but a lengthy, multicontact preliminary to an acceptance can be as damaging to productivity as a denial. The system is driven by number of interviews per time unit.

This goal is shared by salespeople who directly contact households as well. They, by and large, rely on contacting many persons, expecting only a few, as little as 1 in 10 to 1 in 20, to buy their product or service. Productivity with such success rates demands quick resolution of each case.

The second goal, maximizing the probability of obtaining an interview from each sample unit, is the implicit aim of probability sample surveys. The amount of time required to obtain cooperation on each case is of secondary concern. Given this, interviewers are free to apply the "tailoring" over several turns in the contact conversation. How to tailor the appeal to the householder is increasingly revealed as the conversation continues. Hence, other things being equal, the odds of success are increased with the continuation of the conversation. Thus, the interviewer does not maximize the likelihood of obtaining a "yes" answer in any given contact, but minimizes the likelihood of a "no" answer.

We believe the techniques of tailoring and maintaining interaction are used in combination. Maintaining interaction is the means to achieve maximum benefits from tailoring, for the longer the conversation is in progress, the more cues the interviewer will be able to obtain from the householder. However, maintaining interaction is also a compliance-promoting technique in itself, invoking the commitment principle as well as more general norms of social interaction. That is, as the length of the interaction grows, it becomes more difficult for one actor to summarily dismiss the other.

9.4   USEFUL CONCEPTS RELATED TO TAILORING

It appears that the notion of tailoring that we use to understand interviewer behavior is similar to the social psychological concept of "self-monitoring" (Snyder, 1974; Lennox and Wolfe, 1984). "Self-monitoring" refers to the extent that persons "regulate their self-presentation by tailoring their actions in accordance with immediate situational cues." (Lennox and Wolfe, 1984, p. 1349). Researchers in the area find that persons systematically vary in the extent of their self-monitoring; that is, self-monitoring appears to be more a personality trait than situationally induced. Persons high on self-monitoring will be alert to the actions and words of others in an interaction and adjust their behavior based on those actions and words. This behavior seems to us to be similar to that we have observed in interviewers (e.g., see the discussion on self-monitoring in Chapter 7).

Morton-Williams (1993) applies Argyle's notion of "social skills" to interviewer behavior at the time of initial contact with the householder. The social-skills literature notes that most social interaction follows patterns of behavior among the actors that are observable by those external to the interaction, interdependent across actors, and learned over repetitions of similar interactions. "Social skills" is actually a diffuse set of properties of behavior that facilitate the accomplishment of goals in social interaction. Examples of social skills are as vague as being "polite," using the viewpoint of the other actor to acknowledge weaknesses in one's own perspective, adapting to the other actors style of speech, etc. As Argyle (1992) himself puts it,

> In some ways they are like motor skills, such as driving a car, where rapid corrective action is taken when necessary. Those moves made are not social signals; for example, an interviewer asks open-ended questions to get the other to talk more freely. But verbal utterances have to be delivered in the right non-verbal style: indicating warmth and encouragement in this case. It is essential to consider the other's point of view—social influence only works if the right kind of persuasive considerations are presented. (p. 84)

Morton-Williams dissects the survey contact into components that include goals of the actors (which might relate to the burden assessment we believe respondents perform), rules governing behavior, role systems, repertoire of behaviors (set of actions by each actor in search of his or her goals), sequences of episodes (the temporal ordering of actions building toward achievement of goals), stress elements due to conflicting goals, cognitive structure (which appears to relate to shared scripts guiding the actor's behavior), and achievement motivation (how dedicated to her or his goals each actor is). Each of these yield themselves to analysis of alternative interviewer behaviors that may more effectively persuade householders to cooperate with the survey request. Morton-Williams' work is specifically aimed at using social-skills perspectives to raise response rates in surveys, and thus she prescribes interviewer behaviors that attempt to adapt to individual situations, enlarging the interviewer's repertoire of strategies.

Similar concepts are used in the literature on direct sales. The notion of adaptive selling, first introduced by Weitz and colleagues (e.g., Spiro and Weitz, 1990; Weitz, 1981; Weitz, Sujan, and Sujan, 1986) appears to closely resemble that of tailoring. For example, Lassk *et al.* (1992, p. 613), note that "psychological adaptiveness in the sales area is the ability to correctly comprehend verbal and nonverbal communications and then transform this interpretation into a persuasive communication." This has led to a line of research attempting to identify stable characteristics of successful salespeople (including self-monitoring) and training in adaptive sales behaviors (see, e.g., Goolsby, Lagace, and Boorom, 1992; Levy and Sharma, 1994; Tanner, 1994).

If the notion of tailoring is akin to that of adaptive selling, why don't survey researchers apply what is known from the personal selling field to the interview situation and train interviewers to be effective salespeople? There are several reasons for this.

First, survey interviewers already experience much greater "success" than sales-people, routinely getting cooperation rates that exceed 80%, whereas the average success rate in direct sales (cold calls) is closer to 10% (see Oakes, 1990). Some sales approaches actually recognize this. Surveys are at times used to disguise sales approaches, so much so that the acronyms SUGGING and FRUGGING (selling and fundraising under the guise of surveys, respectively) have become common par-lance in the survey industry.

Second, as we have already noted, the sales script is likely to evoke an initially negative (or at best neutral) reaction from the householder. Interviewers tell us that they take great pains *not* to resemble salespeople in their demeanor and actions. By deliberately distancing them from the salesperson image, interviewers may be able to break the householder's reaction to a presumed sales approach and lead the householder to greater attention to the real purpose of the interviewer's visit.

Third, the nature of the request and the goals of the actors in the two interac-tions differ greatly. Whereas selling is a process whose success "depends on the salesperson properly identifying and satisfying the needs of the customer" (Szymanski, 1988, p. 65), in some sense the householder has no possible need for a survey interview. The focus of the survey interaction is often on the needs of the survey organization, government agency, or larger society. While not emphasizing direct benefits of participation to the householder, the survey interviewer is often attempting to minimize the costs of doing so. In other words, while sharing some concepts in common (e.g., script theory, self-monitoring, psychological adaptive-ness), selling and interviewing are two fundamentally different activities. This does not mean that survey research has nothing to learn from the personal selling literature; rather, we believe these approaches cannot be applied uncritically to sur-veys.

9.5   PAST RESEARCH ON INTERVIEWER–HOUSEHOLDER INTERACTION AFFECTING COOPERATION

Past research on the interaction between interviewer and householder has been dominated by studies of effects of altering what the interviewer does and says dur-ing the survey introduction. These include the effect of phrases that attempt to com-municate the social utility of the survey (Dillman, Gallegos, and Frey, 1976), to de-scribe the nature of the respondent task (O'Neil, 1979), to evoke norms of reciprocation or influences of liking (Cialdini, 1984), and to dispel erroneous scripts ("I'm not selling anything" in Gonzenbach and Jablonski, 1993; Pinkleton *et al.,* 1994). Most of these ignore all householder behavior except the final outcome regarding participation in the survey.

Two research efforts have studied in more detail behaviors of both actors. The pathbreaking work of Morton-Williams (1991, 1993) used tape recordings of doorstep introductions by interviewers to identify effective interviewing behaviors. Morton-Williams utilized notions of social skills to describe the effectiveness of in-terviewer behaviors that adapted to the individual circumstances of different householders. Many of the concepts in social skills fit nicely with the script-theoretic notions, together with the concepts of tailoring and maintaining interaction.

In applying social-skills concepts to the contact situation, Morton-Williams reports findings quite consistent with the theoretical structure of Figures 8.1–8.3. Because her work is oriented toward improving the social skills of interviewers, she focuses on the behavior of the interviewer during the contact situation. She notes that the initial "goals" of the householder are "to find out what precisely the stranger on the doorstep wants," (1993, p. 6) (what we called the "script identification"), that one "rule" interviewers use is to "be responsive to the situation" (p. 7) (our notion of tailoring), and that "accurate observations of the contacted person" are essential (reading cues from the householder to guide tailoring).

Maynard, Schaeffer, and Cradock (1993, 1995) have analyzed transcriptions of first contacts of interviewers and householders on telephone surveys. They use conversational-analytic techniques to identify consistent features of the interaction and to classify types of contacts. They set about studying properties of telephone contacts that end in the householder declining the request of the interviewer. They note, as implied by our discussion above, that some householders attempt to narrow down the set of alternative scripts applicable to the interaction by delaying their decision for some time during the interviewer's introduction. For these persons a declination comes later than for others. ". . . [B]y timing the start of withdrawal later, call recipients show themselves not to be hostile to any or all encounters involving unknown and/or organizational callers, but to specific ones." (p. 16) They label some declinations as "expressive" when they are preceded by householder questions, and note that this occurs when interviewers fail to address the questions adequately. Much of their analysis studies the actions of the interviewers after initial declinations are provided by the householder.

9.6 PREDICTING THE OUTCOME OF CONTACTS USING CHARACTERISTICS OF THE INTERACTION

Although the last chapter presented results that are consistent with our theoretical perspective, the major question has not yet been tested—do these observations capture systematic influences on the likelihood of participation in the survey? From a purely predictive perspective, does the nature of the interaction provide new information about the likelihood of consent to the interview, separate from the socio-demographic variables commonly used? Do the observations add predictive value to the kind of records of prior call results normally kept by survey interviewers? From a theoretical perspective, is there supporting evidence that tailoring of interviewer behavior to householder characteristics increases response propensities?

We address questions like these in two steps. First, we look at outcomes of individual contacts on sample households. We measure the influence of householder and interactional characteristics on contact-level outcomes, after controlling for the effects of prior contacts. In that portion of the chapter we dissect the decision process into successive contacts on sample households, to examine whether the theoretical perspective above is informative about the likelihood of a refusal on a specific contact. Second, we aggregate the data to the level of the sample case and examine whether the nature of the interaction between householder and interviewer, as reflected in these observations, is informative about the likelihood that the household eventually yielded an interview.

Our theoretical perspective suggests that the variables measuring the nature of the interaction will reduce the measured effects of socio-demographic variables in predictive models of contact- and case-level outcome. Much of the effect of the socio-demographic attributes arises because they act to shape the nature of the interaction with the interviewer. When we identify those behaviors themselves, the power of the more fixed attributes of the householder should diminish in importance.

9.6.1   Testing the Influence of Interviewer–Householder Interaction on Contact-Level Outcomes

The process of survey participation often requires several contacts by the interviewer, prior to obtaining an interview or a final refusal. Thus, we first examine how the nature of the interaction between householder and interviewer affects the outcome of individual contacts. Then we examine whether the character of those interactions affects the final disposition of the case.

Under the principle of maintaining interaction, customized appeals for householder participation require the interviewers to have sufficiently long conversations with the householders over time so that they can observe the unique concerns of the householder. "Success" by the interviewer on any individual contact is either an interview or some other result that permits future interaction with the sample household. In this section we chose, therefore, to examine the likelihood that an interview, an appointment, or other outcome occurs, relative to a refusal. We do this separately for first, second, third, and fourth contacts with sample households. We have thus dissected the survey activities pertaining to a sample household into smaller pieces, looking for systematic influences on householder reluctance or cooperation within those pieces.

9.6.2   Socio-Demographic Correlates of Contact-Level Outcomes

Because we are interested in placing our findings in the context of the past literature on socio-demographic correlates of survey participation, Table 9.1 presents the results of logistic regressions predicting the outcome of the first four contacts on cases, using only such socio-demographic indicators. The predictors in the models are those used in Chapter 5, the result of theoretical motivation above.

The reader will note that the case bases for the regressions decline with the contact number. Approximately 2,100 cases obtained their final disposition in the first contact, 1,230 in the second, etc. Although some households required more than four contacts to obtain a final disposition, their numbers are too small for useful statistical analysis.

Table 9.1. Coefficients of logistic model coefficients predicting contact-level success versus refusal, by attributes of householder and environment (standard errors in parentheses)

Independent variables	Contact number			
	1	2	3	4
Constant	3.07**	2.36**	2.06**	1.99**
	(0.18)	(0.16)	(0.23)	(0.26)
Environment				
Urbanicity				
Central cities of CMSAs	−0.29	−0.036	0.28	0.11
	(0.18)	(0.18)	(0.22)	(0.25)
Balance of CMSA	0.14	−0.013	0.12	0.11
	(0.16)	(0.16)	(0.19)	(0.18)
Other MSA	0.11	0.20	0.16	0.13
	(0.16)	(0.14)	(0.18)	(0.18)
Non-MSA	—	—	—	—
Household				
Physical impediments to access (1 = yes)	−0.44**	−0.28**	−0.20	−0.18
	(0.11)	(0.10)	(0.14)	(0.24)
Male householder (1 = yes)	−0.22*	−0.18*	0.043	−0.20
	(0.10)	(0.088)	(0.15)	(0.17)
Householder age				
30 years or younger	−0.20	−0.082	0.14	0.30
	(0.17)	(0.12)	(0.18)	(0.25)
31–40 years old	0.15	−0.16	0.48*	0.65*
	(0.19)	(0.14)	(0.18)	(0.29)
41–50 years old	−0.11	0.078	0.23	0.03
	(0.20)	(0.13)	(0.18)	(0.26)
51 or older				
Single-person household (1 = yes)	−0.29*	−0.045	−0.24	0.18
	(0.12)	(0.15)	(0.20)	(0.32)
(*n*)	(9,101)	(6,977)	(3,647)	(1,758)

Note: dependent variables coded as 1 = success (interview, appointment, other), 0 = refusal.
*p < 0.05.
**p < 0.01.

Table 9.1 shows that the demographic variables tend to be predictive of the likelihood of a positive result on contact 1, but are less so for contacts 2–4. The important predictors from the past literature on final disposition are indeed important also for the *first* contact—refusals are higher among units with barriers to access, males, and those who live alone. On *later* contacts these variables are less important, with uniformly smaller coefficients (accompanied by higher standard errors because of the smaller case bases).

For none of the contacts do the urbanicity variables have measurable impact. Urbanicity is one of the most frequently cited correlates of response propensity, but has no statistically significant effect on the outcome of individual contacts in the NSHS. (In separate analyses, we find significant urbanicity differences on final dispositions.) The smaller effects on the contact level could arise from interviewers retreating, whenever possible, in the face of some resistance, regardless of the residential environment.

We interpret these results as implying that, with no indication of householder behavior, socio-demographic correlates commonly found in the past literature *do* separate householders by likelihood of cooperation. However, among those requiring more than one contact to obtain a final disposition, the socio-demographic variables lose their predictive power. This occurs for two reasons:

a. Householders in some of the demographic categories with higher response propensities disproportionately give interviews in the first contact and thus fall out of the active set of the sample. This reduces the statistical power to detect the marginal effects of these demographic variables on later contacts.

b. For later contacts behavioral attributes of prior contacts are important indicators of the likelihood of cooperation. This says that regardless of (a), the second, third, and fourth contacts are affected by the specific events of the first contact. These events are the background brought by both the interviewer and the householder to the later contacts.

9.6.3   Interviewer-Level Effects

Our theoretical perspective led us to expect that, controlling on householder attributes, experienced, confident interviewers would tend to obtain positive outcomes at the individual contact level. There is some evidence for this on the first contact (see Table 9.2), but these relationships, too, diminish in size and statistical significance over later contacts.

We are surprised by the negative effect of interviewer confidence in contact 4, but note the highly selective nature of these cases. Further, as we noted in Chapter 8, interviewers with the least experience (less than 6 months) had high levels of confidence relative to more experienced interviewers. It may be that the confidence measure is not reflecting the self-assurance than comes from command of persuasive skills but some spirit of enthusiasm that is not useful for refusal conversions.

In interpreting this finding it is also important to note that the data do not arise from a design with interpenetrated interviewer assignments; that is, interviewers have not been randomized to sample households. The assignment procedures can affect these results; in Chapter 7 we note that experienced interviewers are often asked to persuade householders to cooperate after they have already refused another interviewer. In that analysis, when statistical controls were added for such reassigned cases, more experienced interviewers were found to increase the likelihood of cooperation over less experienced interviewers. Restricting the present analyses to contacts by field interviewers only, the number of interviewer switches on adjacent contacts is extremely small (about 3% on average across pairs of contacts) for

Table 9.2. Coefficients of logistic models predicting contact-level success versus refusal, by householder, environment, and interviewer attributes (standard errors in parentheses)

Independent variables	Contact number			
	1	2	3	4
Constant	2.61**	2.08**	1.93**	2.50**
	(0.28)	(0.24)	(0.35)	(0.31)
Environment				
Urbanicity				
Central cities of CMSAs	−0.28	0.018	0.34	0.20
	(0.19)	(0.18)	(0.23)	(0.24)
Balance of CMSA	−0.14	0.066	0.20	0.17
	(0.17)	(0.16)	(0.19)	(0.18)
Other MSA	0.13	0.26	0.28	0.11
	(0.18)	(0.15)	(0.17)	(0.17)
Non-MSA	—	—	—	—
Household				
Physical impediments to access (1 = yes)	−0.44**	−0.27**	−0.21	−0.15
	(0.11)	(0.099)	(0.14)	(0.25)
Male householder (1 = yes)	−0.25*	−0.18*	0.040	−0.25
	(0.091)	(0.084)	(0.15)	(0.17)
Householder age				
30 years or younger	−0.24	−0.15	0.05	0.26
	(0.17)	(0.12)	(0.18)	(0.25)
31–40 years old	0.12	0.12	0.43*	0.62
	(0.19)	(0.14)	(0.18)	(0.30)
41–50 years old	−0.11	0.016	0.19	0.079
	(0.21)	(0.14)	(0.19)	(0.27)
51 or older	—	—	—	—
Single-person household (1 = yes)	−0.31*	−0.065	−0.27	0.16
	(0.12)	(0.15)	(0.20)	(0.32)
Interviewer				
Log(years worked)	0.12	0.042	−0.043	0.14
	(0.07)	(0.085)	(0.084)	(0.098)
Confidence	0.15*	0.095	0.064	−0.28**
	(0.06)	(0.068)	(0.083)	(0.10)
(*n*)	(8,837)	(6,769)	(3,503)	(1,687)

Note: dependent variables coded as 1 = success (interview, appointment, other), 0 = refusal.
*$p < 0.05$.
**$p < 0.01$.

the NSHS data. Adding an indicator for such switches, however, adds little impact to the interviewer experience measure.

Since we have shown that householder demographic variables lose their predictive power for cases requiring multiple contacts, the next logical analytic step is to examine whether the character of the interaction between interview and householder predicts the outcome of these contacts.

9.6.4   How the Interaction between Householder and Interviewer Affects Contact-Level Outcomes

We examine in two separate steps whether the contact description form captured behaviors that affect contact level outcomes:

a. We test whether, controlling on the recorded outcome of the prior contact, the documented householder behavior informs us about the likelihood of a success on the current contact. This is important to the entire effort to measure interaction-level attributes. If interviewers can routinely make observations about the contact-level interaction that are informative about the likelihood of later cooperation, then there is hope both for field administrative use of these data (adjusting field follow-up effort based on the information) and for use in postsurvey nonresponse adjustment.

b. We test whether changes in what the interviewer does from one contact to the next are related to the likelihood of success on a contact. This is a weak indicator of between-contact tailoring. For example, if interviewers are tailoring their behavior in the second contact to the householder cues in the first contact, we should expect more changes between the two contacts in what they do, relative to those who aren't tailoring.

Table 9.3 presents the results of logistic regressions by contact number, using the same dependent variables as Table 9.1 and 9.2, but using predictors reflecting the nature of the interaction. For example, the first column uses the characteristics of the interaction in contact 1 and contact 2 to predict the outcome of contact 2. The dependent variable is coded as a binary variable with positive outcome (interview or appointment) modeled relative to a negative outcome (refusal). In all models we control for the outcome of the prior contact, modeled as three dummy variables representing the 2×2 table of refusal/no refusal by whether a household listing had already been obtained. Earlier analyses revealed the importance of treating these two variables as interactive rather than main effects.

One of the key analytic questions generated by the conceptual framework above is: "Does the character of the interaction in contact 1 color the outcome of contact 2, beyond the effects of the coded outcome of contact 1?" This question is interesting both theoretically and practically. Theoretically, the recorded disposition of contact 1 reflects the announced decision of the householder at the time of contact 1. Thus, if there are marginal effects of the character of first contact interaction, we have evidence that the disposition codes typically used insufficiently describe the householder's reaction. Practically, response rates might be explained by whether interviewers attend to the nature of the prior interaction in preparing for the next contact with the sample household. This has implications for the use of call notes by interviewers to guide their actions on later contacts.

Table 9.3 shows that controlling for the outcome of the first contact, householder behavior during the prior contact is informative about the chances of success of the current contact. Specifically, the logistic regression coefficient for negative state-

Table 9.3. Coefficients of logistic models predicting contact-level success versus refusal on different contacts, by characteristics of prior contact (standard errors in parentheses)

Independent variables	Model predicting outcome of:		
	Contact 2	Contact 3	Contact 4
Constant	0.029	0.67**	0.49
	(0.18)	(0.24)	(0.25)
Outcome of prior contact			
Listing, refusal	—	—	—
Listing, no refusal	3.09**	2.44**	2.37**
	(0.19)	(0.24)	(0.27)
No listing, no refusal	2.52**	1.76**	1.72**
	(0.23)	(0.34)	(0.41)
No listing, refusal	0.26	−0.35	−0.14
	(0.21)	(0.28)	(0.54)
Householder behavior			
in prior contact			
Negative statements (1 = yes)	−1.12**	−1.21**	−1.19**
	(0.14)	(0.18)	(0.31)
Time-delay statements (1 = yes)	−0.49**	−0.37	−0.47*
	(0.13)	(0.19)	(0.22)
Questions (1 = yes)	0.19	0.36*	0.32
	(0.098)	(0.15)	(0.27)
(*n*)	(7,046)	(3,674)	(1,774)

Note: dependent variables coded as 1 = success (interview, appointment, other), 0 = refusal.
*$p < 0.05$.
**$p < 0.01$.

ments is negative and larger than that for time-delay statements, suggesting that the former has a greater negative impact on the probability of gaining cooperation in the current contact. Householders who ask questions in the prior contact, controlling for the outcome of that contact, are somewhat more likely to produce a positive outcome in the current contact (reaching statistically significant levels in contact 3).

The strength of the householder behaviors as predictors does not radically change over successive contacts. (The larger standard errors for coefficients reflect diminished case bases.) There is some evidence that householder questions in later contacts become even more powerful predictors of later cooperation (a coefficient moving from 0.19 to 0.32), but replication is needed to have much assurance in this result. This offers a stark contrast to the declining importance of householder sociodemographic attributes over successive contacts.

In short, Table 9.3 answers the first question posed by finding that the likelihood of a positive outcome in the current contact is informed both by the character of the earlier interaction and the stated outcome of the prior contact. This means that the outcome codes provide summary guidance about the likelihood of future behaviors at the sample household, but that the nature of the interaction provides independently informative information. Prediction of the outcome of the current contact from merely knowing the outcome of the prior contact is less powerful.

The second analytic question we pose is whether the concept of tailoring finds empirical support in the data. The concept of tailoring notes that successful interviewers alter their behavior in reaction to cues provided by the householder. This alteration is aimed at addressing specific concerns of the householders, as suggested by their verbal and nonverbal behavior. As a crude indicator of tailoring, we created measures of whether interviewer behavior changed across two adjacent contacts. Unfortunately, the data provide only a weak indicator of tailoring. It is not a measure of altered behavior in direct response to householder behavior, but instead, a measure of how much interviewers change what they do over successive contacts. This is expressed as $1 - (I^c \cap I^{c-1})/(I^c \cup I^{c-1})$, where I^c is the different interviewer actions performed in the cth contact. This indicator of tailoring is also grossly measured in terms of broad categories of interviewer actions (see Table 8.2) and does not capture the nuances of variation in arguments within a particular strategy (see Figure 8.3).

We saw earlier that interviewers observe smaller numbers of new cues from the householders over successive contacts. Interviewers are most active in contact 1 and contact 2, providing a variety of information and arguments for cooperation to the householder. The frequency of such behavior in later contacts is lower. Given this, we speculated that *changes* in interviewer behavior were less likely to be measurable positive influences on later contact-level outcomes.

Table 9.4 shows that, controlling for characteristics of the prior contact, the tailoring indicator has effects in the right direction but does not attain traditional levels of statistical significance ($p = 0.066$) for the outcome of the second contact. Alternative versions of the tailoring measure (e.g., collapsing the measure into binary indicators or using counts of behaviors rather than ratios) *do* reach traditional levels of significance ($p < 0.05$).

We interpret this finding to mean that change in interviewer behavior between contacts may influence the likelihood of a positive outcome in the next contact. Emphasizing different aspects of the survey avoids refusals in the second contact, regardless of what was the outcome of the first contact.

No effects of tailoring are evident in models for later contacts in Table 9.4. This is consistent with the expectation that the greatest opportunities for tailoring arise in early contacts, but also with the fact that the power of tests is much reduced by fewer observations.

Thus, to summarize the models on contact-level outcomes, the typical socio-demographic variables partially explain the outcome of the initial contact, but are not useful in predicting the outcome of subsequent contacts. Further, simple interviewer observations of what the householder does in one contact are useful in predicting the outcome of the following contact. Finally, tailoring might increase the likelihood of success in early contacts with a sample household.

Table 9.4. Coefficients of logistic models predicting contact-level success versus refusal on different contacts, by characteristics of prior contact and tailoring indicator (standard errors in parentheses)

	Model predicting outcome of:		
	Contact 2	Contact 3	Contact 4
Constant	−0.14	0.74*	0.59*
	(0.23)	(0.27)	(0.26)
Outcome of prior contact			
Listing, refusal	—	—	—
Listing, no refusal	3.08**	2.44**	2.36**
	(0.19)	(0.24)	(0.27)
No listing, no refusal	2.53**	1.76**	1.70**
	(0.23)	(0.34)	(0.41)
No listing, refusal	0.28	−0.34	−0.15
	(0.21)	(0.28)	(0.54)
Householder behavior			
in prior contact			
Negative statements (1 = yes)	−1.12**	−1.21**	−1.20**
	(0.14)	(0.18)	(0.26)
Time-delay statements (1 = yes)	−0.49**	−0.37	−0.46*
	(0.13)	(0.19)	(0.22)
Questions (1 = yes)	0.18	0.37*	0.34
	(0.10)	(0.15)	(0.28)
Indicator of tailoring (change of behavior)	0.28	−0.10	−0.15
	(0.15)	(0.18)	(0.18)
(*n*)	(7,046)	(3,674)	(1,774)

Note: dependent variables coded as 1 = success (interview, appointment, other), 0 = refusal.
*$p < 0.05$.
**$p < 0.01$.

9.7 EFFECTS OF INTERVIEWER–HOUSEHOLDER INTERACTION ON THE FINAL DISPOSITION OF SAMPLE HOUSEHOLDS

This section presents the second step in answering the question about how interviewer–householder interaction affects response propensities. It turns our focus to that more typical of the literature on survey nonresponse—the likelihood of obtaining an interview among those sample cases contacted. Of key interest is whether the measures of interviewer and householder interaction are informative about the likelihood of ultimate cooperation, controlling on socio-demographic attributes of the householder.

To accomplish this, we use a household-level, not a contact-level analysis, limit our analysis to cases with more than one contact, and jointly examine the effects of socio-demographic and ecological attributes of the householder, householder behavior, interviewer attributes, and tailoring. To avoid any ambiguity of causal order, the householder behavior variables reflect behavior in contacts prior to the one yielding the final disposition. For example, the "negative statements" variable equals 1 for cases where a householder made any negative statements in contacts prior to the last one; 0, otherwise. The tailoring indicator measures changes in interviewer behavior only between the first and second contact. (Using some other summary indicator was rejected because it would fail to be a simple analogue of the indicators used on the contact-level models.) The base model is Model 1 of Table 9.5, including only the socio-demographic predictors. Model 2 tests the marginal effect of householder behaviors. Model 3 tests the further marginal effect of interviewer variables.

We remind the reader that the NSHS data used here show somewhat different patterns of correlates with cooperation than do the decennial match data in Chapters 5 and 6. This arises for three reasons: 1) the predictors available from the data set are different from those of the match data (the model specifications are different); 2) NSHS has a restricted age range of eligibility relative to the match data sets; and 3) Table 9.3 presents models estimated only on those cases with multiple contacts (this affects the coefficients for the single-person households, for example).

Model 1 shows the expected result of lower cooperation among male and older householders, age, and among those with barriers preventing easy contact with the household. Although the urbanicity variables show decreased cooperation in urban areas, as expected, the effects are not statistically significant. The hypothesized lower cooperation of persons living alone also are not found. Finally, the marginal effects of interviewer tenure and self-confidence are negligible. In short, the base model shows most of the expected effects of householder socio-demographic attributes, but not those of interviewer attributes.

Model 2 shows that controls on the householder behavior diminish the effect of their socio-demographic attributes and improve the overall fit of the model relative to model 1. For example, although householder age was informative of the likelihood of an interview in model 1, it is less so in the presence of controls on householder behavior. In other words, among persons who provide negative statements, time delay statements, or ask questions at the same rate, there are few differences in cooperation across age groups. Similar statements can be made for different urbanicity groups, housing units with some barrier to access, male householders, and single-person households. All of these results are evidence that the interviewer observations about what happens during contacts are independently informative about the ultimate likelihood of cooperation.

Model 3 adds to these variables another central tenet of the theory—interviewer tailoring, as measured by changes in interviewer behavior between the first and second contact. This indicator has weaker predictive power for the final disposition than it had for the result of the second contact (first column of Table 9.4), and it fails to achieve statistical significance at traditional levels. Its effect is in the right direction; changes in interviewer behavior between the first and second contact improves the likelihood that the household will eventually grant an interview.

In short, the models in Table 9.5 show that traditional socio-demographic corre-

Table 9.5. Coefficients of logistic models predicting final cooperation, by environment, householder, and interaction predictors, multicontact cases only (standard errors in parentheses)

Independent variables	Model 1	Model 2	Model 3
Constant	1.85**	2.76**	2.68**
	(0.27)	(0.29)	(0.29)
Environment			
Urbanicity			
Central cities of CMSAs	−0.28	−0.25	−0.27
	(0.15)	(0.16)	(0.16)
Balance of CMSA	−0.22	−0.047	0.010
	(0.15)	(0.16)	(0.16)
Other MSA	0.11	0.22	0.20
	(0.14)	(0.15)	(0.16)
Non-MSA	—	—	—
Household			
Physical impediments to access (1 = yes)	−0.25*	−0.19	−0.20
	(0.11)	(0.13)	(0.14)
Male householder (1 = yes)	−0.28**	−0.20*	−0.20*
	(0.80)	(0.10)	(0.10)
Householder Age			
30 years or younger	0.51**	0.20	0.23
	(0.16)	(0.19)	(0.19)
31–40 years old	0.68**	0.44*	0.48**
	(0.17)	(0.18)	(0.18)
41–50 years old	0.35*	0.17	0.24
	(0.17)	(0.18)	(0.18)
51 or older	—	—	—
Single-person household (1 = yes)	−0.033	0.0045	0.0044
	(0.14)	(0.16)	(0.16)
Interviewer			
Log(years worked)	−0.049	−0.14	−0.14
	(0.071)	(0.086)	(0.084)
Confidence	0.088	0.043	0.036
	(0.067)	(0.078)	(0.077)
Householder behavior			
Negative statements		−2.45**	−2.47**
		(0.13)	(0.13)
Time-delay statements		−0.67**	−0.68**
		(0.099)	(0.10)
Questions		0.77**	0.76**
		(0.098)	(0.097)
Tailoring indicator (change of behavior between contact 1 and 2)			0.19
			(0.15)
(*n*)	(6,835)	(6,625)	(6,447)

Note: Dependent variable coded 1 = interview, 0 = refusal.
*$p < 0.05$.
**$p < 0.01$.

lates of response propensity partially reflect the different kinds of behavior they groups exhibit when they interact with interviewers. These behaviors are more proximate causes of the decision to participate in the survey, and rather crude measures available from the contact descriptions are sufficient to measure them. Such crude measurement is not sufficient to reflect interviewer tailoring, however, and the data do not themselves support the marginal effects of tailoring, controlling on householder behavior. We believe, however, that tailoring remains an influence on final disposition and assert that a purified measure of it, reflecting interviewer effectiveness in adapting their behavior to respondent concerns, would provide empirical support for the theory.

9.8 SUMMARY

We have found results largely consistent with the theoretical propositions outlined earlier. Socio-demographic and psychological predispositions of householders are useful proxy indicators to their likely behavior in the first contact with an interviewer. These attributes are associated with behavioral tendencies that are consistent over time. The indicators, however, are fallible; they are correlates, not causes of the survey participatory behavior we study.

Once the first contact is completed, the behaviors manifested in that contact are much more powerful indicators of the likelihood of survey participation in any specific later contact with the household as well as of the likelihood of participation occurring over all future contacts.

Further, the nature of the householder behavior provides systematic indications of response propensity. Negative statements at any prior contact bode poorly for ultimate participation. But negative statements are relatively rare, and they are by no means perfect predictors of negative outcomes. In that sense, they are clearly not endogenous to the decision to participate. Interviewer behavior in subsequent contacts can eliminate their influence.

Questions from the householder, on the other hand, bode well for the likelihood of their participation in future contacts. Time delay statements fall in between the two other behaviors. These may contain a mix of genuine time constraints and polite proxies for refusals.

We think there are several reasons for questions predicting later positive outcomes: a) this is evidence that the householder is attending to the conversation enough that he or she is motivated to seek clarification, b) the householder's question forces the continuation of the conversation for at least one more turn, and c) the question provides rich cues to interviewers that are stimuli to tailoring behavior on the part of the interviewer. We believe this finding to be contrary to the impression of many inexperienced interviewers, who react to householder questions as signs of reluctance.

We believe the results of these analyses, measured relatively crudely across contacts here, would extend to behavior within a given contact. We expect that within-contact tailoring should be a more powerful influence on the decision to participate, as it immediately addresses the concerns and objections raised by the householder.

As we note in Figure 2.1, the decision to cooperate or refuse is shaped by the interactions between householders and interviewers. The data at our disposal, albeit providing weak indicators of the richness of these interactions, nevertheless support the notion that such interactions are informative of the process of reaching a decision about participation in a survey.

In short, we have learned the following from the analyses in this chapter:

- Negative and time delay statements from householders in one contact portend lower cooperation in later contacts and lower cooperation rates at the final disposition of the sample case.

- When a householder poses questions about the survey request in one contact, it forecasts higher likelihood of eventual cooperation with the survey request at some later contact. This may arise both because it signals some minimal level of interest on the part of the householder and also because it provides a vehicle for interviewer tailoring of behavior to the concerns of the householder.

- The type of utterances provided by householders are informative about the likelihood of eventual cooperation, beyond the information contained in call-level result codes typically used by survey organizations.

- The content of the conversation (i.e., negative statements, time delay statements, or questions) signal differences in cooperation rates among age groups and across units with and without physical impediments to their access.

Thus, in some sense, in contact conversations there exist traces of evidence of the impact of the influences at other levels (e.g., the environment, the household) that, if observed and interpreted appropriately, can be used by talented interviewers to alter the cooperation propensities of householders.

9.9 PRACTICAL IMPLICATIONS FOR SURVEY IMPLEMENTATION

When householders encounter interviewers for the first time they face two burdens—first, determining what is the intent of the visit and, second, deciding whether they should consent to the request of the interviewer. The first task, which we believe is best understood as a problem of selecting the best cognitive script, forces many householders to ask questions of the interviewer. This is no doubt due to the relative rarity of survey interview requests *and* the tendencies for interviewers to be trained to describe the survey and its goals and not the interview task itself. The amount of respondent behavior in the first contact usually is greater than that in later contacts, mainly because that first contact has all the burden of determining the purpose of the interviewer's visit.

The content of the interaction between interviewers and householders offers more information about the likelihood of future cooperation than do the result codes assigned to individual visits to the household. For example, respondents who ask questions in one contact but end it with a refusal are better candidates for conversion to interviews than are interim refusals who exhibited no such curiosity. With attention to the development of more reliable, cheap indicators of such householder behavior, response propensity models that used as predictors such indicators might be able to identify subsets of the sample with homogeneous response likelihoods (see Chapter 11). That is, the true response propensities of final refusals who asked questions of the interviewer may be more similar to interviewed cases who did so than to other refusals who asked no such questions, other things being equal.

Interviewers can act to change the base tendency of a householder to participate, even after they have exhibited one of the behaviors above. Altering the kind of information they provide to the respondent, emphasizing one aspect of the design or intent of the survey over others, can influence householders to participate. We believe that the most effective changes in interviewer behavior are those shaped by real concerns revealed by householders. This notion of tailoring has the power to change the calculus of the decision making on the part of the householder.

The finding that the content of the interaction is independently predictive of outcomes is also important for practical survey work because it largely supports the practice of encouraging interviewers to make notes about the nature of the interaction during the contact. Unfortunately, interviewers are typically given few guidelines about what is important and unimportant to document. Senior interviewers sometimes say that on refusal conversion efforts it is more important for them to learn what the householder said in prior visits than what the interviewer said. What this empirical work suggests is that some householder and interviewer behaviors are more important than others. Further work might focus on what respondent and interviewer behaviors are most predictive of behavior in the next contact.

The findings of this chapter also imply that interviewer training regimens might be structured to give trainees skills in tailoring. We suggest that the training might have four steps: 1) the assembly of householder concerns about the request, using the native nomenclature of the population; 2) development of alternative kinds of information relevant to those concerns; 3) training of interviewers to classify householder comments into different categories; and 4) the training of interviewers to provide, quickly and in words appropriate to an individual householder, responses to householder comments and questions.

This suggested process has some similarity to the practice in telephone and field surveys of providing interviewers with common questions (and their answers) posed by householders and stock answers they might provide. It differs, however, in two important respects. First, it provides interviewers with examples of both questions and concerns, which might not manifest themselves in verbal questions. For example, time delay statements suggest concerns that the interviewer should be armed to address, through explicit apologies for uninvited intrusion into the home at an inconvenient time and manifesting flexibility at setting a next contact at a more convenient time. Second, it emphasizes diagnosis and tailored replies, delivered in quick response to the displayed behavior. It is common for inexperienced interviewers to generate quick but soft refusals because they did not quickly give effective responses to a specific concern of the householder. Thus, the training proposed must work on speed of classifying a comment into an appropriate category and of quickly delivering an appropriately phrased response to a comment.

The training technique that seems well suited to these needs is repeated simulation of householder behaviors with interviewers asked to respond to them. Initial training might concentrate on classifying behaviors into different categories of householder concerns about the survey request. Later training would grade both appropriateness of classification and speed and wording of response. Speed, we believe, is an effective measure of how skilled the interviewer has become in tailoring.

The training might end with a "flight test" of common concerns raised, with classification and speed graded by trainers. Interviewers who fell in the lower ranges of performance might be given remedial training prior to beginning production work.

CHAPTER TEN

How Survey Design Features Affect Participation

10.1 INTRODUCTION

In this chapter we turn to the final set of influences in our model (Figure 10.1), that of the survey design. We save this for last because this chapter is based largely on a review of existing literature, rather than original analyses of the data we have at our disposal.

The theory of survey participation we espouse asserts that survey designers have two ways of manipulating survey participation rates. The first is the behavior of interviewers during the initial contacts with the sample household, which we discussed in Chapter 8. Training of interviewers normally has some component to teach them how to describe the importance of the survey, how to answer questions about the study, and how to motivate the householder to participate. In Chapter 8 we saw how interviewers vary in their use of these behaviors.

The second set of influences the designer can manipulate are aspects of the survey task and the nature of the relationship constructed with the sample household. These can be separated into design features used to increase the rate of contacting sample households and those used to increase cooperation of those contacted. They include advance contact with sample households by telephone or mail, repeated calls on the unit, offers of incentives to householders, mode of data collection, and a host of other features.

The ideal data set to investigate the power of these design features would contain observations on cases from many different surveys. The surveys would vary experimentally on each of the design features in question. Unfortunately, no such data set exists, nor does the decennial match data set available to us offer much opportunity to study empirically effects of such features. Instead, in this chapter we examine various data sets available to us and also use past survey methodological research to gain some insight into both why each influence exerts its effects and how the influences combine to produce effects on survey participation.

OUT OF RESEARCHER CONTROL   UNDER RESEARCHER CONTROL

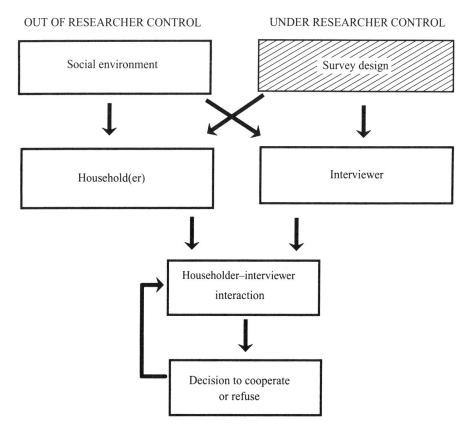

Figure 10.1. A conceptual framework for survey cooperation.

10.2   THE BALANCE OF COST, TIMELINESS, MEASUREMENT, AND SURVEY ERRORS

When survey designers begin their work they face thousands of individual decisions about different features of the survey. In addition to choosing the sampling frame, inventing a sample design, developing and pretesting alternative questions, and choosing a mode of data collection, they must choose whether to send an advance letter, alerting the sample unit to the interviewer's call; whether to offer cash or other incentives to householders; when and how often to attempt contact with sample households; whether to attempt to call again on households that refused to provide an interview, and so on. Most of these decisions have potential cost and time implications.

Given a fixed amount of money to conduct the survey, therefore, the designer must choose whether efforts to achieve a higher response rate are more important than fully pretesting the questionnaire, drawing a larger sample size, or using more in-depth measurement of key concepts in the survey. Similarly, increasing the effort to achieve high response rates may cause a delay in the completion of the data collection and thus affect the timeliness of initial estimates from the survey.

One trade-off decision has been the source of great speculation among survey researchers for some years. Many have suspected that respondents who consent to an interview only after intense persuasion may provide only superficial attention to the respondent task. In short, they may have sufficient motivation to provide the interview (perhaps to avoid continuation of unpleasant encounters with the interviewer), but insufficient motivation to listen thoroughly to the questions, to search their memories carefully for the appropriate answer, and to provide candid replies. This describes a trade-off between nonresponse error and measurement error properties of a survey.

The hypothesis of higher measurement error among reluctant respondents deserves more careful thought. The underlying premise of the hypothesis is that lack of desire to grant the interview would continue after a commitment is given to provide the interview. In essence, the respondent would have passed a threshold that separates refusals from acceptances, but would fail to attain a level of motivation to attend carefully to the respondent tasks.

If such a theory were true, we would expect that the differences between reluctant and willing respondents would be a function of the level of reluctance and the difficulty of the response task for an individual item. For example, a question asking about whether they owned or were renting the housing unit may be easier to answer accurately than not. In contrast, a question about the number of times the respondent has visited a doctor in the last year may require a level of cognitive processing too burdensome for the marginally engaged respondent, and yield either over- or underreporting of events.

10.3 SURVEY DESIGN FEATURES AFFECTING LIKELIHOOD OF CONTACT OF SAMPLE HOUSEHOLDS

In Chapter 4 we noted that surveys with many calls on sample households, at different times of day, on different days of the week, and over an extended period of time generally achieve higher response rates. Sometimes the survey designer controls these features of the survey implementation. Other times, the survey design fixes the length of the data collection period (e.g., the ten-day interviewing period of the Current Population Survey) and assigns to each interviewer a certain number of sample cases, but does not specify the number of calls to make or the time spacing between them. If the workload is large relative to the length of the data collection period, such survey designs in essence limit the number of calls and variation in time of the calls. This section reviews what the field knows about the effects of number of calls on cases, length of survey period, interviewer workload, and interviewers' attempts to observe characteristics of households relevant to response propensity.

10.3.1   Number and Schedule of Callbacks

In Chapter 4 we discussed the ability to contact a sample household as a function of the number of calls made on the household. Some survey designs have specific rules for the calling patterns on households and thus bring this under design control. Face-to-face interviewers are instructed to make a designated number of calls at designated times. Some telephone surveys use software in computer-assisted interviewing systems to enforce such rules. There are, for example, surveys that specify that the interviewer make four call attempts over a number of days in order to make first contact with a household, and then to obtain a final disposition with the household within four additional calls. Cases that cannot attain the desired outcome by the chosen call number are coded as noninterviews.

The imposition of a calling rule is generally not made to increase the likelihood of contact or cooperation, but rather to reduce the average amount of effort to obtain contact or cooperation. Thus, as Weeks, Kulka, and Pierson (1987) illustrate, the imposition of a calling pattern in a centralized telephone-interviewing facility was able to increase the percentage of numbers reached on the first call by about twenty percentage points by making calls on weekday evenings versus weekday mornings.

10.3.2   Length of Data Collection Period

Some surveys measure a phenomenon that exists only for a brief moment in time. Surveys about the public's initial reaction to a world event (e.g., military action of the United States against another country) often are interested in measuring attitudes formed on the basis of first news, separately from those formed on the basis of more comprehensive reports. They often choose survey periods of a small number of days. Some surveys have set reporting times such that the amount of time between the events being reported and the publication date of survey estimates is small. For example, the Current Population Survey seeks reports of behavior during the week including the 12th of the month, and reports the findings from the survey of 60,000 households by the first Friday of the next month, leaving about 10 days for data collection.

At the extremes of lengths of data collection it is clear what might happen to contact rates, other things being equal. If a survey were limited to a few hours in a single evening (as is done in some political attitude polls), all those households who were away from home on that evening remain uncontacted, regardless of efforts made by the interviewers. On the other hand, all households are likely to have someone at home at some point during a survey data collection period lasting several months.

The surveys we matched to the decennial census have data collection periods that vary from 10 days to several months (see Chapter 3, Table 3.1). However, the noncontact rates do not vary systematically by these lengths. One possible explanation is that calling patterns of the surveys vary with length and that the vast majority of households are at home in a 10 day period.

Thus, the important questions about the length of data collection period are:

What percentages of households cannot be reached for each length of a data collection period? How short does a data collection period have to become for its length itself to produce lower contact rates?

In an analysis of National Election Study data, Brehm (1993) describes the probability of determining whether a selected address or telephone number is nonsample (nonresidential unit or nonworking number) with each passing day that a sample number receives calls in an 18 day survey period. More than 80% are determined within 4 days, and more than 90% of the nonsample cases are determined within a week. He also notes that after the first 7 days, nearly 80% of the interviews had been taken, and each additional day yields less than a 5 percentage point increase in the cumulative number of interviews taken. He estimates that increasing the survey period from 11 to 18 days might result in a 5 percentage point increase in response rates.

Another useful approach to this question examines a very high contact rate survey conducted in a small number of days. The Current Population Survey (CPS) is one such survey, with a data collection period of 10 days and a contact rate of 97`%. Unfortunately, the data available on the study do not permit one to separate the contact step from the interview step. In using these data, we are forced to assume that interviewers are contacting and interviewing cases as soon as possible in the 10 day field period. That is, they are working at maximal productivity.

Figure 10.2 presents the percentage of cases completed for each of the 10 days in the CPS interviewing period. The data come from the June, 1997, CPS and are re-

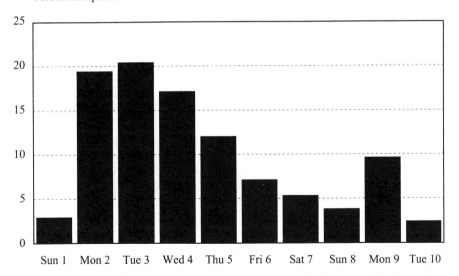

Figure 10.2. Percentage of CPS first-month sample cases completed by day, June, 1997. *Source:* Harris-Kojetin (personal communication).

stricted to cases in their first month in the CPS sample (the same set we use in the decennial match). Interviewers are usually instructed not to make calls on Sunday for these new cases. The figure shows that by day 3 (Tuesday), about 43% of the cases were resolved, and by the end of one week (Saturday), about 79% have been resolved. Given that noncontacted sample cases are kept active until the end of the field period, this gives an idea of the likely contact rates given maximal effort in a short period of time.

10.3.3 Interviewer Workload

Another limiting factor on the efforts of interviewers to reach cases is the size of their workload—on average, how many cases they have been assigned to complete during the data collection period. This attribute of a survey affects the number of calls interviewers make on cases, their ability to choose times of days and days of weeks that will maximize contactability, and the freedom of the interviewer to give attention to reluctant respondents. The size of workload that exceeds interviewers' capacities is probably a complex function of the mode of data collection, the length of the interview, the compactness of the sample segments (in face-to-face surveys), the nature of the study population, and the number of days of the data collection period.

From our perspective, in the causal chain of contacting a case, workloads affect the level of effort (number and timing of calls) applied to contacting and obtaining cooperation from each sample case.

10.3.4 Gathering Information About The Sample Household to Guide Future Calls

Face-to-face interviewers often attempt to obtain information about a sample household prior to making an initial call. Interviewers tell us that they often drive by the sample housing unit, observe the neighborhood, and attempt to discern characteristics of the sample household. They look for bicycles in the yard, the presence of cars in the driveway or in front of the house, signs of someone being present, the state of disrepair of the unit, evidence of concerns about crime (bars on windows), and evidence of avoidance of stranger contact ("no soliciting" signs).

This search for information is obtained to guide their calling patterns on the sample household. It is also conducted to make decisions about their personal safety in calling on the household in the evening. The behavior is the first example of attempts to tailor their approach to the sample unit to characteristics of the unit.

10.4 SURVEY DESIGN FEATURES AFFECTING COOPERATION

Most of the research literature on reducing nonresponse is focused not on contacting sample households but on gaining their cooperation after they have been contacted. This section reviews the major survey-design features that have been found to affect cooperation.

10.4.1 Agency of Data Collection

It appears that one of the heuristics householders use in deciding about whether to grant a survey request concerns the actor making the request. In survey situations, the interviewer is one of these actors, but the interviewer is also explicitly acting as an agent for the survey sponsor. The sample person's knowledge about the sponsor, as well as their attitudes toward the sponsor, can be used as a short-cut for judging the desirability of granting the interview. If the organization indirectly acts to benefit the target persons or their group, then they might be more positively disposed to the request. If the householders know nothing about the sponsor, they may focus more on the interviewer's characteristics to aid in their judgements.

There is a small research literature on this issue. Presser, Blair, and Triplett (1992) conducted a split-sample experiment using an RDD telephone survey in the District of Columbia about a mayoral primary in 1990. For one-half of the sample the interviewers stated that the study sponsor was a university; for the other, a local newspaper. A random adult respondent was selected in each household using the last-birthday method. There was no statistically significant difference in response rates (65% with university affiliation; 61% with newspaper affiliation).

In a similar RDD study in Denver, Everett and Everett (1989) experimentally varied sponsorship by a university and a commercial research agency. The dependent variable examined was the rate of refusal, given contact (that portion of the participation process that is probably more sensitive to experimental manipulation). They found significantly higher rates of refusals for the commercial-agency affiliation (among male interviewers, an 8% refusal versus a 5% refusal; for females, a 6% refusal versus a 4% refusal).

There is a similar but older experiment in a face-to-face survey, manipulating affiliation of the interviewer. In 1976, a split-sample national area probability survey was conducted jointly by U.S. Census Bureau interviewers and Michigan Survey Research Center interviewers. The reader should note that this experiment actually used different interviewer corps for the two affiliations, not merely having the same interviewers describe their affiliation differently. Thus the experiment measures the combined effects of affiliation and any differences in training and behavior of interviewers in the two organizations. Sample segments were split randomly and assigned to both organizations. The Michigan Survey Research Center interviewers obtained a higher refusal rate (13%) than that of the U.S. Census Bureau interviewers (6%) (see National Academy of Sciences, 1979).

These findings are consistent with the theoretical reasoning that sponsoring institutions that are perceived to offer benefits to the target person may generate more cooperation than would other organizations. Some of this explains the historically higher response rates of government statistical agencies than of commercial organizations (de Heer and Israels, 1992; Goyder, 1987; Wiseman and McDonald, 1979).

It also underlies the practice in some surveys of seeking co-sponsorship or endorsement of a survey effort by professional organizations or any other group that would be respected by sample persons. Thus, it is common for surveys of physicians to seek endorsements from medical associations or hospital associations, in order to borrow legitimacy from those institutions. The value of such manipulations is dependent on the strength of positive affect toward the chosen institutions *and* to the homogeneity of affect within the population. It is possible in heterogeneous populations that one sponsorship will act to increase response rates for one subgroup (e.g., a trade union for blue-collar workers) but another for another subgroup (e.g., management of a company for white-collar workers). For example, in an expenditure-diary survey in the Chicago area, Ferber and Sudman (1974) found that sponsorship by a government agency increased response rates in the suburbs, whereas sponsorship by a university increased rates in the city.

10.4.2   Advance Warning of the Survey Request

It has become commonplace in face-to-face surveys around the world to mail a letter to the sample household that alerts the unit to the upcoming call. From a theoretical perspective, this design feature can be used to manipulate a variety of influences known to affect survey participation:

(a) Using official stationery of a sponsoring agency legitimized in seeking the survey information communicates that the survey is so authorized.

(b) Having the letter signed by a person in authority might magnify that effect.

(c) Personalizing the name and address of the sample household and the salutation of the letter can convey the seriousness of purpose of the endeavor, relative to mass mailings.

(d) Reviewing the social benefits of the survey can communicate, "independently" of the interviewer, the survey's contribution to the common good.

(e) Including gifts or incentives in the letter can evoke norms of reciprocation and focus the attention of the sample household on the benefits of responding rather than the burdens of responding.

(f) By alerting the household to an upcoming call by the interviewer, the letter can be consistent with norms of politeness that unannounced calls of salespersons and evangelists (or even criminals and scam artists) typically violate.

(g) By providing information about the nature of the survey interview, the letter can educate the sample household with regard to the task being requested.

(h) By reviewing the confidentiality provisions of the survey, it can allay fears of undesired distribution of private information.

Finally, perhaps for all of these reasons, interviewers *believe* that the advance letters enhance their own confidence in seeking the interview request at the first contact with the sample household.

For example, Figure 10.3 contains the advance letter used in the Current Population Survey. The letter follows many of the theoretical features described above, appearing on official Department of Commerce stationery, being signed by the director of the Census Bureau, reviewing the social benefits of the survey, announcing the upcoming visit by an interviewer, and emphasizing the voluntary nature of the request. The letter concludes with, "Your cooperation will be a distinct service to our country," invoking norms of civic duty (Cialdini, 1990; see also Groves, Cialdini, and Couper, 1992). Finally, the CPS letter provides answers to commonly asked questions about the survey to address any doubts before the interviewer arrives.

Luppes (1994) conducted a content analysis of advance letters used by government statistical agencies for consumer expenditure surveys in Europe. His study gives us a notion of common practice. Most letters had 5 to 6 paragraphs, with each paragraph averaging 4 or 5 sentences. All but one of the letters were one page long. The content of the letters was diverse across different organizations. Many devoted content to the goals of the survey and to direct benefits of the study. Other common topics included a description of the survey organization and provisions for the confidentiality of data provided. Most offered a telephone number contact for more information. Almost all contained an announcement of an upcoming call by an interviewer. It is our impression, from informal inspection of other letters, that this kind of content is common across surveys. In another study on the content of advance letters, White and Freeth (1996) asked survey respondents to critique the advance letter they had received and found that they preferred short letters explaining the purpose of the survey and the sorts of information to be collected.

The literature on advance letters is a relatively small one but has addressed both the net effect of letters on cooperation rates and the marginal effects of the various features listed above. The Traugott *et al.* (1987) studies on advance letters in telephone surveys used a dual-frame sample design, mailing advance letters to some cases in the list portion of the sample. In the first experiment, the advance letter from a university survey research center raised response rates from 56% to 69%; in the second experiment, from 69% to 78%. The authors concluded that although the letters increased response rate, the cost efficiency of advance letters depended on other survey costs. They also noted that in telephone surveys, letters could be sent only to sample households with listed phone numbers and mailable addresses on their listings. In a similar dual-frame sample design, Pennell (1990) found that the effect of advance letters on cooperation rates in the 1988 New York Reproductive Health Survey was small and positive, but not statistically significant. He suggests the effect of the advance letter may have been muted by respondents' interest in the topic of the survey and that some respondents who were supposed to have received the letter may not have been exposed to it.

Another telephone survey experiment on advance letters, by Dillman, Gallegos, and Frey (1976), manipulated the content of the advance letter, varying the amount of information about the study and inducements to respond. The control group was a "no-letter" condition. Three letters varied the amount of information about the study and sets of reasons supporting participation in the survey (low reward, medi-

CPS-263(L) LOS ANGELES
(04-97)

UNITED STATES DEPARTMENT OF COMMERCE
Bureau of the Census
Washington, DC 20233-0001

OFFICE OF THE DIRECTOR

FROM THE DIRECTOR
BUREAU OF THE CENSUS

You may have read in the newspaper – – or heard on the radio or television – – the official Government figures on total employment and unemployment issued each month. The Census Bureau obtains these figures, as well as information about persons not in the labor force, from the Current Population Survey (CPS). This information, which we collect for the Bureau of Labor Statistics, provides vital up-to-date estimates of the number of persons working, the number who are unemployed, and many other related facts. Occasionally, we ask additional questions on education, health, family income, housing, and other important subjects.

A Census Bureau representative, who will show an official identification card, will call on you during the week in which the 19th of the month falls. The representative will ask questions concerning the ages, employment status, and occupations of the members of your household, as well as other related information. By law, Census Bureau employees hold all information you give in strict confidence. Your answers will be used only for statistical purposes in a manner in which no information about you as an individual can be identified.

We have selected your address and about 48,000 others throughout the United States for this survey. Because this is a sample survey, your answers represent not only yourself and your household, but also hundreds of other households like yours. For this reason, your participation in this voluntary survey is extremely important to ensure the completeness and accuracy of the final results. Although there are no penalties for failure to answer any question, each unanswered question lessens the accuracy of the final data. Your cooperation will be a distinct service to our country.

On the other side of this letter are answers to questions which participants ask most frequently about this survey.

Thank you for your cooperation.

Sincerely,

Martha Farnsworth Riche

Further information
may be obtained from:

Regional Director
Bureau of the Census
Los Angeles Regional Office
15350 Sherman Way, Suite 300
Van Nuys, CA   91406-4224

Telephone: 818–904–6393

Figure 10.3. Advance letter for the Current Population Survey.

Who uses this information? What is this survey all about? In a country as big as ours and one that changes so rapidly between decennial censuses, people in government, business, and other groups need up-to-date facts in order to plan efficient and adequate programs. It is important to know how many people are working or out of work (to help direct programs which would contribute to an expanding economy and provide new jobs), how many children will be attending school (to plan for schools and the training of an adequate number of teachers), how many new families are forming (to plan for adequate housing to meet their needs), and so on. Occasionally, we may combine data from the CPS with data from other Government agencies to provide a comprehensive set of summary information about employment, income, and participation in various Government programs. The CPS is one of the most important and timely sources of information used to make such plans.

How was I selected for this survey? Actually, we selected your address, rather than you personally for this survey. Each month, we scientifically select about 15,000 groups of addresses to represent the United States. Each of the address groups contains about four housing units and altogether result in about 48,000 interviewed households each month. If you should move away while your address is still in the survey, we would interview the family that moves in.

How many times will I be contacted and how long will it take? Our representatives contact occupants of a selected dwelling eight times -- 4 months in one year and the same 4 months in the following year. Our representatives contacted this address four times last year and your address is scheduled for four more interviews. In addition, we contact a small number of households twice during one of the 8 months to ensure the validity of our statistics and verify that our representatives are doing the best job possible. After the eight interviews are completed for a household, it is not contacted again regarding the CPS. On average, an interview should take about 10 to 15 minutes. Your interview may be somewhat shorter or longer than this depending on such things as the number of adults in your household or the type of questions asked in a given month. If you have any comments about this survey or recommendations for reducing its length, send them to the Associate Director for Administration/Comptroller, ATTN: Paperwork Reduction Project (0607-0049), Room 3104, FB 3, Bureau of the Census, Washington, DC 20233.

What protection do I have? Is this survey authorized by law? All information individuals give to the Census Bureau is held in the strictest confidence by law (Title 13, United States Code, Section 9). Our representative has taken an oath to this effect and is subject to a jail penalty and a fine if he/she discloses any information survey respondents give him/her. Title 13, United States Code, Section 182, and Title 29, United States Code, Sections 1 through 9, authorize the collection of most of the information we request in this survey. In addition, portions of the survey in any 1 month may be authorized by one of the following: Title 7, United States Code, Sections 1621–1627; Title 38, United States Code, Section 219; and Public Laws 89-10, 92-318, and 93-380. In some months, the survey may contain questions authorized under laws other than those cited; further information concerning the authority for any particular portion of the survey can be obtained from the representative who contacts your household. The Office of Management and Budget Control number for CPS is 0607-0049. Without this number we would not be able to conduct this survey.

Why do you include me? I'm retired. Some retired persons may feel that their activities are not important to this type of survey and wonder why we include them. In order to have an accurate picture of the entire population, it is necessary to include persons in all age groups. Our experience with interviewing retired persons shows that many of them are actually participants in the labor force because they work part time or are looking for work. This information, along with data on other subjects such as income, health, and housing, assists in the measurement of the economic condition of the elderly population as a whole.

Figure 10.3. (*continued*).

um-reward, and high-reward letters). The three letters also varied in their length as a function of the level of information, with the "high-reward" letter requiring two pages. As Table 10.1 shows, the two lower-reward treatments achieved higher response rates than the no-letter condition, but the high-reward letter did not (our computations of the standard error of the differences, assuming simple random sampling, and $\alpha = 0.05$). This study may illustrate the combined effects of letter length and content. Length of the letter could be interpreted as communicating importance of the communication, but it may also reduce the readership of the letter because of the burden of reading a longer text. A follow-up experiment that tests a

Table 10.1. Percentage of cases with various outcomes by advance letter treatment group

Outcome	No letter	Low-reward letter	Medium-reward letter	High-reward letter
Completion	85.0%	91.7%	93.4%	89.4%
Partial completion	1.1%	1.1%	0.7%	0.7%
Refusal	13.9%	7.2%	5.9%	9.9%
Total	100.0%	100.0%	100.0%	100.0%
(*n*)	(280)	(278)	(287)	(274)

Source: Dillman, Gallegos, and Frey (1976). Reprinted with permission of University of Chicago Press © American Association for Public Opinion Research.

letter condition with high reward but of equal length to the lower-reward condition would be interesting to conduct.

Another study allows us to measure the effect of differential authority on the power of advance letters. Brunner and Carroll (1969) conducted a split-sample experiment, with one-third of the sample receiving advance letters on stationery from a market research organization, one-third, letters, from a university; and one-third, no letter. Those receiving the letter from the university achieved response rates 30 percentage points higher than the "no letter" condition. The letter from the market-research organization decreased response rates by 6 percentage points relative to the "no letter" condition. This study shows the potential backfire effects when the advance letter communicates aspects of the survey request that may be unattractive to the sample households. In this case, one *post hoc* hypothesis is that the sample households judged that the intent of the commercial survey was not as beneficial to them or to the common good as that of the university based survey.

Some summary observations about advance letters are in order. First, one limitation of letters is that their effect is dependent on unknown behavior in the household. Most sampling frames in household surveys in the United States (this is not true in countries using population registers as frames) do not have names of persons on the frame. Advance letters are sent to the sample address, with the letter addressed to "Householder" or "Occupant" or "Head of Household." If the targeted respondent in the household is different from the person who normally opens such mail, then the effect of the letter is diminished. Couper, Mathiowetz, and Singer (1995) show that in about half of the households, one person sorts the mail prior to reading, and over 60% throw away some mail without opening it. Letters addressed to individuals who are targeted as respondents do not suffer from such problems.

Second, for a letter to have its full effect, a household must read the advance letters. Households with uniformly low literacy rates or households without members who can read the language in which the letter is written are not subjected to the influence of the letter. Others may open the letter, inspect it briefly to determine whether it is a bill or an offer of some benefit to the household. If it is not, they discard it.

Third, several backfire effects of letters have been hypothesized. It has been argued that advance warning of the survey request merely solidifies opposition to the request, permitting sample households to prepare their refusals for the time of the first contact. Although this may be true for a subset of cases, the experimental effects appearing in the scientific literature appear to counter this argument on the whole. The experimental effects support the notion that sponsorship and the content of the letters are important influences on the nature of their effects on cooperation. Another potential backfire effect may arise if a long period of time passes between sending the advance letter and attempting contact. For example, Pennell (1990) found the cooperation rate among those who had received the advance letter four weeks prior to be 10 percentage points lower than those who received no letter at all.

10.4.3   Respondent Incentives

Another common design tool to improve participation is an incentive—giving the sample unit something judged by the sponsor of the survey to be valued by the unit. Most incentives take the form of money, either cash, checks, or money orders. Some incentives are objects that have cash value—gift certificates, coffee mugs, books, calculators, jewelry, tool kits, kitchen magnets, American flags, pens, medallions, and a host of other things.

The literature shows that incentives appear to increase overall response rates. In an empirical analysis of response rates across surveys Goyder (1987) shows that controlling on a large set of other survey design features (including length of the survey, effort to obtain response, sponsor, topic) incentives tend to lead to higher response rates. Further, larger incentives are generally associated with higher response rates. For example, a field test of the National Adult Literacy Survey, a personal visit survey in which interviewers revealed the incentive after a household roster was obtained, tested a $20 and $35 incentive, paid by check at the completion of the interview (Berlin *et al.*, 1992). The study had three components—a screener interview, a background questionnaire, and an exercise measuring ability to read and comprehend text.

As Table 10.2 shows, higher response rates were obtained with incentives than without incentives. Looking at the "overall" response rate, the percentage of per-

Table 10.2. Response rates for different respondent tasks by incentive group, National Adult Literacy Survey Field Test

Respondent task	Incentive group		
	$0	$20	$35
Screener	87.4%	87.7%	90.0%
Background	78.6%	82.7%	84.4%
Exercise	92.8%	97.9%	98.0%
Overall	63.8%	71.0%	74.4%

Source: Berlin *et al.* (1992).

sons completing all three components, the $20 incentive group has 7 percentage point higher response rates than the no incentive group. The $35 incentive group achieves a still higher rate but not greatly larger than the $20 group.

This study replicates the findings of others. It is true that the larger incentives lead to higher response rates, but it appears that the effect itself diminishes as the amount rises.

An interesting question is whether it is possible that increasing incentives could actually lead to reduced cooperation. There has been speculation that a sufficiently large incentive may inappropriately direct the householder's attention to assessing the burden of the task. With sufficient attention to this aspect of the task, versus issues of the importance of the survey or the pleasure of the interaction with the interviewer, reluctance could be generated. There is slight evidence for this from a study by James and Bolstein (1992). The study offered replicate samples checks of $5, $10, $20, or $40. Instead of increasing response rates with incentive amount, the rates were 67%, 67%, 79%, and 69%, respectively. That is, there was evidence of a decline in response rate moving from an $20 incentive to a $40 incentive. Theoretically, this effect should be a function of the nature of the respondent task being sought. In the James and Bolstein work, a one-page (two-sided) questionnaire with 14 questions was mailed to the respondents. Relative to most surveys this is a small task request.

Theoretical Constructs Relevant to Incentive Effects. There is some debate about what principles underlie the effects of incentives. From one viewpoint, that of social exchange theory, an incentive is an act of kindness bestowed on the householder by the survey organization. This kindness evokes the norm of reciprocity. That is, in accepting the gift, the householder is obliged by social norms to respond in kind. One way to return the favor is provided when the survey request is made. With this viewpoint, the value of the incentive is both a function of its perceived cash value but also of the apparent thoughtfulness of the giver, the genuineness of the sentiment accompanying its provision, and the absence of its attachment to any *quid pro quo*. Thus, if the householder interprets the incentive as a precursor to a disproportionately large request, its effects might be nullified.

A second interpretation of incentives is much more rationally based. The incentive is a benefit of responding that can be evaluated in the context of amount of effort and opportunity costs attached to the survey request. These efforts and costs might be measured relative to the perceived value of the householder's time. Thus, low-income householders might react to incentives more dramatically than high-income householders. The decision calculus in this perspective is a direct cost–benefit tradeoff. At the extreme, the incentive is an offer of payment for services, and the householder evaluates whether his or her time should be given at that price.

From a practical viewpoint, one decision a survey designer must make regarding incentives is whether to give them to all sample units prior to the survey request or to make the incentive conditional on their providing the interview. Although the literature could be stronger on this point, there is a consensus that prepaid incentives appear to induce cooperation more effectively than promised incentives. For example, in an experiment linked to the National Medical Care Expenditures Survey, Berk *et al.* (1987) varied the timing of a $5 incentive to sample person in the second wave. One-third of the sample received the incentive prior to the request, one-third were promised the incentive if a self-administered questionnaire were completed, and one-third were not alerted to the incentive at all prior to the questionnaire (but were sent payment later). As shown in Table 10.3, in both modes of data collection, prepaid incentives led to increased response rates over no incentives or promised incentives.

It is appealing to think of this common finding as support for the social exchange interpretation of incentive effects. Even if true in some cases, this effect may not be present in unusually complex and lengthy surveys, or for ones that impinge on the time of householders in major ways. In these cases, when the householder is clearly informed about a burdensome task, a more economic exchange approach, balancing the burden with remuneration to respondents, may be operative. In such cases, we might expect promised payment to perform more like prepaid incentives.

A Note About Ethics. Many researchers are most concerned about the short-run problem of obtaining high cooperation rates for their current survey. If the chosen survey design features are inadequate to achieving a desired response rate, investigators sometimes offer higher incentives to reluctant respondents, hoping that money will change their minds. This practice clearly offers greater reward for the reluctant, and evokes concerns that the easily compliant are disadvantaged by this design. The counterargument is based on the logic that the compliant would not have cooperated unless they believed that the interview would be sufficiently beneficial to them (without the need for an incentive). That is, opportunity costs and bases of burden judgements vary across individuals and thus only some require external rewards (incentives) to enter a compliant state. It is unfair, in this argument, to give incentives to those whose situation provides sufficient rewards to respond without incentives. One telling question to ask in such circumstances is, "How would the easily compliant behave if they were informed of the design decision to give incentives only to the reluctant?"

Differential Effect of Incentives. One piece of counterevidence for the social exchange interpretation of incentives would be larger effects of incentives for lower-

Table 10.3. Response rates by mode and incentive treatment group, National Medical Expenditure Survey

Mode	Treatment group		
	No mention	Promised payment	Prepayment
Telephone	50%	50%	62%
Personal visit	75%	67%	84%
Total	66%	60%	73%

Source: Berk *et al.* (1987).

income households. This would be evidence that the benefits of the incentive were being weighed with the metric of one's "price" per unit time—does the incentive compare favorably with what I am worth?

Is there evidence for larger effects of incentives among lower-income groups? Negative evidence was found by several investigators (Kanuk and Berenson, 1975; Miller, Kennedy, and Bryant, 1972; Willimack *et al.*, 1995; Goetz, Ryler, and Cook, 1984). All these studies, except the Willimack *et al.* one, involved promised incentives. These found that there were no differences among either education or income groups or geographically defined groups that vary on socioeconomic status. Conversely, a diverse set of surveys varying by mode, nature of incentives, and sponsorship found the expected larger effects of incentives in poorer households (Nederhof, 1983; Ferber and Sudman, 1974; James and Bolstein, 1990). It is interesting to note that most of these latter studies used prepaid incentives, not promised incentives. The *post hoc* speculation that comes to mind here is that lower socioeconomic groups react more strongly to incentives only when they are prepaid.

To summarize the research on incentives, there is ample evidence that incentives act to increase cooperation with survey requests. There is some evidence that they act to reduce the number of calls required on sample cases, thus acting to reduce interviewer effort and costs. Larger incentives lead to higher cooperation, although at decreasing rates as incentives rise. There is some evidence of backfire effects, loss of cooperation with an incentive that appears to be unusually large relative to the burden of the request. Despite the common belief that incentives are more effective on low-income groups than higher-income groups, the research literature is rather ambiguous, with studies showing diverse effects.

10.4.4   Interviewer Incentives

Survey designers sometimes offer incentives directly to interviewers for each completed interview. It is common that this feature is introduced when most of the remaining cases have been contacted and express reluctance to provide the interview. This use of incentives implicitly assumes that interviewers are not doing all that is desirable to obtain the cooperation of sample units.

Interviewer incentives are merely a version of piece-rate or bonus-payment schemes for employees. There are a variety of interviewer incentive schemes that can be constructed. Incentives can be offered for each completed interview or for certain sets (every five or ten interviews); incentives can be offered for achieving targets on average hours of work to produce an interview; incentives can be offered to individual interviewers or to teams of interviewers (the latter being more common in centralized telephone interviewing facilities); incentives can be in-kind (parties for interviewers, compensatory time) or cash; incentives can be paid immediately upon completion of the designated task or at the end of the survey period.

To our knowledge, there are no published experiments measuring the marginal effect of these incentives on overall response rates. We speculate that short-run effects might be heavily dependent on other features of the survey design. For example, if the survey task is a short, simple one, then interviewers may emphasize these features in refusal conversion to obtain cooperation. However, in such cases it is likely that the cooperation rate might already be quite high at the time interviewer incentives would be introduced. Hence, the marginal effects of incentives might be small. Similarly, if incentives were introduced into a survey data collection period of several months, after several rounds of refusal conversion efforts, it is unlikely that large marginal increases in response rates would result.

10.4.5 Respondent Rules

Central to our theory of survey participation is the observation that different house-holders are affected by different influences on their survey decision. The obvious practical implication of that observation is that the respondent rule—who is chosen to report the survey information—will partially determine the response rates obtained.

The common alternative respondent rules are:

(a) a household informant among those members judging themselves capable of responding
(b) the member most knowledgeable on the survey topic
(c) a randomly selected person (usually from adult members of the household)
(d) the household head (or person in whose name the housing unit is owned or rented)
(e) individual response from each eligible member of the household
(f) a designated respondent, sometimes followed by the use of a proxy respondent upon reluctance or inability of the designated respondent to provide the information

These respondent rules vary on the degree to which they limit exposure of the recruitment protocol to persons likely to avoid contact with strangers or likely to find the survey interview unattractive for some other reason. Allowing anyone to respond or permitting proxy reporting provides the survey with a substitution pool from which to choose another respondent.

Using a knowledgeable informant rule can produce a higher or lower response rate, depending on the topic. In an experiment within a telephone survey, a "knowl-edgeable adult" respondent rule obtained an 81% response rate versus a random respondent rule's 75% response rate (Cannell et al., 1987). In this study the knowl-edge sought concerned the health status and behaviors of all household members. That person was often a female, who played the role of the "health monitor" of the household (i.e., that person who makes doctor and dental appointments and cares for the sick of the household). If the knowledge sought was held by another in the household, it is possible that the comparison would have differed. Similarly, proxy respondent rules increase response rates to the extent that the original respondent rule disproportionately identifies reluctant respondents. On the National Health Interview Survey, proxy reporting is permitted for the core questionnaire (asked of all household members), but often not for supplements (asked about random adults in the household). The completion rate for the core is generally 5–10 percentage points higher than for the supplemental questionnaire (Botman and Thornberry, 1992).

Clearly, the largest literature concerning nonresponse effects of respondent rules focuses on different ways of selecting a "random" respondent. An early method, the "Kish" method, involves a listing of all eligible persons and then the application of some random selection device. There are variants of this, not requiring a full rostering of the household (see Groves and Kahn, 1979). There also exist selection schemes that ask one question to identify the respondent, "Which adult had the most recent birthday in the household?" or "Which adult will have the next birthday in the household?" These schemes are used to select a respondent without requiring an elaborate listing of all persons. Most do not implement a randomized selection scheme. While some studies have found the next/last birthday method to be a useful selection scheme, other evaluations of the method (Forsman, 1993; Oldendick *et al.*, 1988; Salmon and Nichols, 1983) have shown that there is a tendency for females to be identified disproportionately as being the designated person. The *post hoc* hypothesized mechanism for this appears to be a tendency for the phone answerer to identify as the householder having the birthday designated. Some survey organizations appear to believe that the request of the household roster is a more important source of nonresponse than do others.

10.4.6   Topic Saliency

A basic finding from the social psychological literature on helping behavior is that if subjects feel that they are qualified to help someone in need, their likelihood of helping is greater (Schwartz and Clausen, 1970). This is relevant to the hypothesis that surveys asking about information of great interest to the sample persons generally obtain higher response rates than those asking about low-saliency information.

In practice, it appears that survey designers believe this hypothesis to be true. It is common for interviewers to introduce a survey by noting that the topic (e.g., access to health services) is likely to be of great interest to them. Conversely, when topics are likely to be of little interest (e.g., television watching for nonviewers of television) the interviewers may describe the goals of the survey in terms that are vague, nonspecific to the uninteresting area.

There are interpretations of these practices from cognitive perspectives. It is likely that topic saliency also acts to improve comprehension by the respondent of the introductory comments of the interviewer. When the householders have knowledge of the topic, they can relate the words of the interviewer to past experiences and to semantic memories relevant to the task. Further, if the householders have often rehearsed the recall of information relevant to the survey topic, they are likely to judge the burden of the interview to be lower.

From a more rational choice perspective, the perceived benefits of the survey may be enhanced with surveys on salient topics. Interviewers in the Consumer Expenditure Survey, one component of the U.S. Consumer Price Index, report that they alter their approach for different householders. When they encounter a retired householder, whose Social Security payments are linked to the value of the Consumer Price Index, they will explicitly note that the survey is used to effect changes in their payments. They believe this observation of saliency acts to improve cooperation.

There are three kinds of evidence in the literature about the role of saliency. First is the typical finding of achieved response rates by topic of the survey. In the decennial match the Consumer Expenditure Survey achieves lower response rates (90.5%) than the National Health Interview Survey (96.0%). That difference may be due to a lower level of interest in expenditures than in health issues. However, it may also be due to the fact that the Consumer Expenditure Survey is a much longer interview on average, and requires either accessing financial records of the households or difficult estimation tasks. However, it is common to find consistent differences in response rates by topic (e.g., health surveys with higher response rates than political surveys).

The second kind of evidence is the relationship between the statements made by householders during contacts with interviewers and their likelihood of participating in the survey. For example, Couper (1997) notes that persons who express a lack of interest in the topic of the National Election Survey household interview tend not to grant the interview. Table 10.4 shows that the response rate of those reporting a lack of interest is 8 percentage points lower than those who do not provide such a response. Comparing this 8 percentage point difference to the effects of other statements, for example, those responding that they were "too busy," reminds us that there are other reasons for not participating. The table, however, shows an effect in the right direction. The paper also notes that these statements of noninterest are also related to real differences in reported knowledge and participation in electoral affairs.

The third kind of evidence regarding the power of topic saliency is differential sensitivity to incentive effects by topic saliency. An interesting example of this is the

Table 10.4. Cooperation rate by household utterances during contacts, 1990 National Election Study

Type of utterance	Cooperation rate	(*n*)
Total	78.2%	(2,555)
"Too busy"		
Yes	61.9%	(766)
No	85.1%	(1,789)
"Not interested"		
Yes	71.8%	(510)
No	79.8%	(2,045)
"How long will it take?"		
Yes	89.0%	(824)
No	73.0%	(1,731)

Source: Couper (1997), adapted from Table 1.

work of Baumgartner and Rathbun (1997). In a study about electricity usage, they found that an incentive had no measurable effect on participation among persons who had volunteered for an electric rate program, but had large effects among program nonvolunteers. These results suggest that survey designers should consider the saliency of the topic to the sample population before deciding whether to use incentives.

10.4.7   Psychological Burdens of the Interview

A rational cost–benefit analysis of a survey request might involve assessments of the burden of the interview for the respondent, among other things. Burden is probably a multidimensional concept here, consisting of the amount of time devoted to the task, the cognitive burdens of comprehending the questionnaire script and searching memory for answers, and any psychological burdens of revealing embarrassing personal attributes to the interviewer (see Bradburn, 1978). Obviously, the most easily measurable of these attributes is the length of time that the interview would require of the respondent. It is that measure of burden that is most often used by designers and regulators of surveys (e.g., the Office of Management and Budget, for U.S. Federal surveys).

Let's dissect the notion of the length of interview as an influence on the decision to participate into different components.

Householder Knowledge of the Length of the Interview. First, the respondent must know about the length of the interview for it to have direct effects on the survey participation decision. In most U.S. Federally sponsored surveys, advance letters sent to the sample household contain some mention of the length of the interview. Many times this mention is not a prominent feature of the letter and could be easily missed by the reader.

Interviewers report that they generally do not emphasize the length of the interview in initial comments with the sample householder. Most interviewers are trained to give an honest estimate of the length of time, when asked by the householder. As we see in Chapter 8, about 27% of the householders ask this during the initial doorstep interaction. That result suggests that many decisions about participation are made in ignorance of the length of the interview in face-to-face and probably also in telephone interviews. In many telephone and face-to-face surveys, the length of the interview is unknown to the householder at the time of the interviewer. We would expect, therefore, that interviewer-assisted modes would tend to show smaller response rate differences by length of interview than self-administered modes.

How could we assemble evidence about the effect of length of interview? The best circumstances would be a survey for which length was unambiguously known and where other influences on survey participation were minimal. The cleanest examples of this are from mailed self-administered surveys. For these surveys, if the householder opens the envelope and sees the questionnaire, we can more readily assume that the length of the questionnaire will be obvious and *might* be used to judge the length of time it would take to complete the questionnaire. These designs eliminate the interpersonal influence of the interviewer; however, they remain subject to the effects of advance letters, format of the questionnaire, page size, etc.—all of the other attributes that might be used to help assess the burden of cooperation. There is a small literature on length of questionnaire, generally measured in number of pages, and achieved response rates, nicely summarized by Bogen (1996). On mail surveys, the findings from experimental studies are mixed, with the tendency to show higher response rates with shorter questionnaires. In self-administered surveys, the effect seems to be sensitive to the format of the questionnaire (see Champion and Sear, 1969); in interviewer-administered surveys there is even less consensus on the issue. Indeed, Champion and Sear demonstrate that a questionnaire with more pages, formatted with more white space, can generate higher response rates than one with fewer pages but less white space.

Effects on Interviewer Behavior of Different Lengths of Interview. Another way that length of interview can affect cooperation rates is through interviewer behavior. Interviewers report in focus groups that they feel greater reluctance to seek long interviews than short interviews. This reluctance might stem both from perceptions that such requests would violate norms of etiquette but also from expectations that such requests generate higher refusal rates. Hence, even if respondents are not explicitly told about the length of interview, interviewer behaviors may change with longer interview requests in a way to generate more refusals. That is, surveys with long interviews produce lower cooperation rates because interviewers behave in ways that generate more final refusals than is true for shorter surveys.

For evidence on this line of reasoning, we examine interviewer-assisted surveys. We seek those that experimentally varied the length of interview and explicitly announced the length of the interview to householders during the introductory script. Frankel and Sharp (1981) did so for a survey about housing conditions, explicitly requesting 25 minutes for an interview versus 75 minutes. Requests for the longer interview generated slightly higher refusal rates (30%) than the shorter (27%). These differences were not statistically significant at traditional levels. In a similar experiment using telephone surveys, Collins *et al.* (1988) found that a 40 minute request generated a 14% refusal rate versus a 20 minute request generating a 9% refusal rate. These are relatively small differences and require us also to speculate on the effect of hearing the news about the length of interview from the interviewer. They, however, confound the effects of interviewer behavior and real length differences. Even the combined effects seem small.

Burden as Cognitive Demand or Psychological Threat. The topics and measurement techniques of surveys often determine the how difficult it will be for a respondent to provide answers to a questionnaire. There is no experimental literature varying the topic of the interview for different populations. However, it is common to find higher response rates for health surveys, relative to surveys about political attitudes and relative to surveys on consumer expenditures. We find it attractive to view these differences as arising from perceived burdens of answering questions on topics about which the respondent knows he is ignorant.

Burden as a Function of Opportunity Costs. Long interviews are burdensome to those householders who have pressing demands from their household obligations, their work demands, or social affairs. Conversely, long interviews may be of minimal burden to householders contacted at a time when those alternative uses of time are not present. Interviewers are filled with stories about some respondents (often lonely and isolated) enjoying the interaction with them so fully that the interviewer has difficulty extricating themselves from these homes.

Summary. Bogen (1996) summarizes the literature by noting that ". . . the relationship between interview length and nonresponse is more often positive than not, but it is surprisingly weak and inconsistent." Our theoretical perspective would lead us to expect this result, because the studies vary on other influences on participation (e.g., persuasion efforts by interviewers, topic saliency). For example, if a survey topic is very salient to the sample, it is likely that longer questionnaires would have little effect. If the interviewer is pleasant and the householder can look forward to an engaging conversation, similar results might apply. Or, conversely, if a short questionnaire asks both obscure and embarrassing questions, burden could be high. That is, length is not a good indicator of burden in many surveys. When the underlying cause of reluctance is the psychological threat posed by discussion topics in the survey, then the nonresponse can be inherently nonignorable. That is, the sample persons' attributes on the survey variables are determining their reluctance to participate.

10.4.8 Interviewer Behavior Guidelines

There have been several attempts in the research literature to experimentaly vary interviewer behavior during contact with sample households. Most of these studies were conducted in telephone surveys, after practice had revealed lower than desired response rates in that mode.

The early experiments attempted to script the interviewer's behavior, focusing on conveying to the phone answerer the importance of the study, the legitimacy of the sponsoring organization, and the small psychological or time burden of the interview task. For example, O'Neil *et al.* (1979) experimentally varied the introductions of interviewers on three dimensions—whether the phone answerers were given a detailed versus cursory explanation of the survey, whether or not the phone answerers were given positive feedback on cooperative answers they supplied, and whether or not the phone answerer was asked two questions illustrative of the type of questions in the survey. The experiment failed to produce interpretable variation in response rates. Dillman, Gallegos, and Frey (1976) experimentally varied whether or not the interviewer made reference to the name of the householder, whether or not an offer of the study results was made, and whether or not the sponsorship was mentioned. Again, there were no significant differences found.

The null results from the early published literature are probably matched by scores of other null findings that were never published. Given the theoretical perspective we now take, with the central role of interviewer tailoring, it is understandable why the results have been obtained. More recently, Morton-Williams (1991), in a field experiment, scripted some interviewers' behavior during contacts and left others to convey specified information in a manner and order they found useful to the particular sample case. The fourteen interviewers using the scripted procedure attained a 59% response rate; those not scripted, a 76% response rate.

A more recent version of the attempts to manipulate interviewer behavior focuses on an attempt to prevent the phone answerer from misinterpreting the call, through choosing the incorrect script. Gonzenbach and Jablonski (1993) experimentally varied whether interviewers said, "We are not selling any product or service," as part of their introduction. They found that those provided that information obtained a response rate about 6 percentage points higher than those who did not. In a replication, Pinkleton et al. (1994) found the inclusion of the statement "I'm not selling anything" did not significantly affect response rates. However, the authors of this study hypothesized that an introduction that more clearly delineates a survey from a telemarketing call might positively impact response rates. Tuckel and Shukers (1997) go so far as to suggest that the persuasive power of the opening statement may have as much to do with the non-sales statement as with the statements about the benefits of participation.

These findings need careful interpretation. It is a common result of the attitude change literature (McGuire, 1985) that a persuasive argument is more effective if it refutes counterarguments prior to their articulation by the subject. However, weak rebuttals to counterarguments not yet considered by the subject can make salient the counterargument. Thus, for those phone answerers who suspect a sales call, the interviewer's statement may be effective. For those not evoking that script, the interviewer's statement may raise suspicion that some deception is taking place, and detrimental effects of the statement might occur.

10.4.9 Follow-up Procedures

If the initial contact reveals that the householder is reluctant to cooperate with the survey protocol, there are a variety of design features that are commonly applied.

Interviewer Switches. A common tool in field administration is the reassignment of reluctant cases to more senior interviewers or to supervisors. These more experienced staff recontact the case, usually making formal reference to the prior visit. A common behavior in these visits is to explicitly ask of the householder what their concerns might be about survey participation. This question can be asked more easily by someone who was not privy to the original conversation. By asking it, the new interviewers can gain a summary of the causes of reluctance that they can then address. This alone should lead to some success because of reassignment. Use of more senior interviewers at this stage magnifies the effect because they generally have more skill at tailoring.

Persuasion Letters. After the first contact with a reluctant respondent, some designs send a letter to that person directly, reiterating some of the material of the advance letter (if any) and emphasizing the importance of complete measurement of the sample. These letters are attempts to communicate directly the authority and legitimacy of the survey sponsor in seeing the information, and the importance of the information to the larger society or the respondent's welfare itself. Sometimes the letters mention that an interviewer will be calling again to seek a reconsideration of the householder's decision. These letters work when the personalization of the correspondence conveys the seriousness of the survey endeavor and the arguments are persuasive to the householder.

Mode switches. In telephone and mail surveys, it is common to attempt a different mode of contact with reluctant respondents, if the survey budget can support the greater expense. For example, some telephone surveys recontact reluctant respondents by face-to-face visits; some mail surveys use telephone contact as a follow-up to nonreturns. These mode switches move in the direction of using the persuasive powers of interviewers more directly, attempting to tailor the approach of the survey to the concerns of the reluctant sample person.

10.5   SUMMARY

By and large, much of survey practice involving nonresponse evolved through trial and error of alternative procedures. For this reason, it should be possible to link common practice (which survives this natural selection process) to one or more of the theoretical influences that guides this book. We have seen above that these links can be readily made for most design features.

The weakness in common practice, and in the research literature, appears to lie in the absence of attention to diversity within household populations. For example, the value of an advance letter for an illiterate household is likely to be lower than for others. The attraction of incentives may vary across persons. The opportunity costs of a long interview may be higher for some than for others.

The ability to tailor design features to individual characteristics of sample households and persons is limited by the nature of sampling frames normally used in household surveys. This prevents much customization of approach to individual households. It is likely that the customization step, if ever to be feasible, would require interviewer action. One approach to this problem is the creation of an information gathering step prior to the contact at which the interview is requested. The purpose of this visit would be to classify the sample unit into a category that would be given a distinct set of treatments. This contact would introduce the interviewer to the household, not to request an interview, but to obtain indicators of possible questions and concerns that would be raised by the household about a survey sponsored by the organization. These concerns would identify what type of presurvey information should be sent to the household and what regimen of approach would be set up for contact attempts. If feasible, "at-home" times could be obtained at the visit or the respondent might be identified so that mailed material might be personalized.

Clearly, the counterargument to such a plan is the relative costs of obtaining such information. The tailoring attempts must be successful enough that they would reduce repeated callbacks on households that proved reluctant to the traditional approach. The motivation of the design, however, is the perspective that future gains in survey participation will require interviewers to address the individual concerns of sample households, at the earliest opportunity in the contact process as possible.

CHAPTER ELEVEN

Practical Survey Design
Acknowledging Nonresponse

11.1 INTRODUCTION

Theoretical principles underlying household survey participation would have little utility to the field if they did not help to identify preferred procedures for the design, collection, and analysis of survey data. The chapters that preceded this one have attempted to coalesce a series of theories about human behavior that are relevant to the process of gaining the participation of a sample of households in a survey, to understand differences between nonresponse arising from failure to contact the sample household and that due to refusals, and to link various design features of surveys to underlying influences on human decision-making regarding survey participation. This chapter applies those lessons to alternative survey design features.

In the previous chapter we reviewed the literature on length of the data collection period, respondent rules, incentives, interviewer training, advance letters, interviewer workloads, interviewer switches, and changes of mode. These are tools developed and used by survey researchers to reduce nonresponse. Most of them have been designed to reduce overall nonresponse rates. There is little research on how different subpopulations react to the different design features or how one feature interacts with another. Further, there is little direct focus on how the design features might affect nonresponse error versus nonresponse rates.

There are three relevant deductions from the theoretical perspective on survey participation presented in this book. The first deduction is that nonresponse rates of different subpopulations may be sensitive to different stimuli. For example, individual householders bring to the survey request different backgrounds that cannot be effectively addressed with a single, undifferentiated script of an interviewer. Since most survey samples provide little useful information about sample households prior to the contact attempts, an adaptive protocol for eliciting participation is warranted. Ideally, the researcher would be informed during the data collection process of any imbalances in the participation of sample persons that would cause nonresponse error in key survey statistics. This information could be used for selective targeting of nonresponse reduction efforts, using techniques appropriate to the groups disproportionately nonrespondent. The basic need is that the investigator should collect information prerequisite to reducing nonresponse error and apply the information to tailoring approaches to the disproportionately omitted groups.

For a trivial example, we have repeatedly shown that the pervasive differences in survey response rates between urban and rural areas can be substantially explained by differences in housing structures and household compositions. The researcher should be armed with information on response rate differences between these two groups as the data collection proceeds. Then the researcher faces the question of whether attempts to reduce the higher nonresponse rates of the urban areas are cost efficient, relative to the bias-reduction possibilities offered by a postsurvey adjustment using urban status as a factor and the costs involved in postsurvey adjustment. This chapter focuses most directly on the use of information about nonresponse tendencies that could be used toward that end.

The second key deduction from a strong theory of survey participation is that nonresponse will not always produce errors in survey statistics. For example, the tendency for persons who live alone to refuse surveys may have little import to a survey measuring food-intake patterns. It could have large impacts on the estimation of the proportion of households with children. This implies that mindless maximization of response rates has no theoretical support. It *is* true that since

$$\bar{y}_r = \bar{y}_n + (m/n)(\bar{y}_r - \bar{y}_m)$$

the lower the nonresponse rate (m/n), the lower the nonresponse error, on a linear statistic, *other things being equal*. We've seen, however, that some subgroups form larger portions of the nonrespondents in high response rate surveys than in low response rate surveys. Hence, statistics related to those subgroups might have larger $(\bar{y}_r - \bar{y}_m)$ terms in high response rate surveys. Theoretically, it should be preferable to bring nonresponse rates to a more consistent level across subgroups than to indiscriminantly reduce nonresponse. The latter alternative will more quickly, easily, and cheaply reduce the overall nonresponse rate, but do so more for groups that are the easiest to locate and persuade to participate.

The third deduction from the theory stems from the sensitivity of the saliency of different attributes of the survey request to situational factors. The theory notes that many survey design features have their effect because they make favorable consequences of the interview more salient to the householder prior to their decision. However, any stimulus presented during the survey request can interact with individual characteristics of the householder. Tailoring is an antidote to unanticipated negative reactions to the approach by the interviewer. Most of these features are assigned in hopes that the nonresponse problem facing the survey will be minimized by implementing the design features. Yet rarely do any of them singly or all of them in combination eliminate nonresponse.

We conclude from the above that reduction of nonresponse error will still demand effective postsurvey adjustment schemes. Much remains to be done to improve the specification of postsurvey adjustment models. This was an original motivation of the research in this book and the empirical findings reinforce the impressions of great opportunities in this area. The greatest lacunae lie in the absence of measurement tools to provide predictor variables for adjustment models. The statistical tools for adjustment are numerous and the field actively creates new tools. The weakness lies in availability of effective measures to use in the models. In other words, it is not the estimation of the models, but their specification that needs research.

This chapter describes a set of procedures that are motivated by the research in this book, leading to possibly more effective nonresponse reduction and nonresponse-adjustment procedures. It is organized about the traditional design and execution steps in a household survey. Figure 11.1 illustrates the key steps of the survey design, collection, and analysis process. Consistent with the perspective of this book, the bold lines note which of the steps can be used to improve nonresponse reduction and adjustment. This chapter is organized about these key steps.

11.2   SELECTION OF SAMPLING FRAMES

The traditional literature on sampling frames focuses on the coverage properties of alternative frames (e.g., Wright and Tsao, 1983). That is, coverage error is the dom-

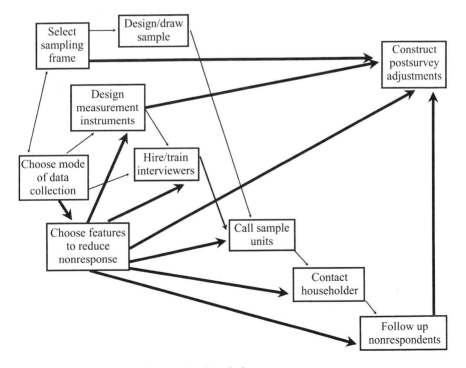

Figure 11.1. Steps in the survey process.

inant concern regarding frame selection and use. This attends to the relative proportion and types of population units not covered by the frame. For example, area frames are susceptible to the failure to include all units in small multiunit structures. Telephone frames include multiple entries for units with more than one telephone number and totally omit households without telephones. Knowing that nonresponse is an inevitable outcome of a survey process forces attention to whether alternative frames might vary in their nonresponse reduction or adjustment possibilities.

For example, in telephone surveys in the United States, two frames are popular; one, the frame of listed telephone numbers, and another, the frame of all possible telephone numbers in legal area code, prefix (the first six digits of a U.S. telephone number) combinations (sometimes called the "RDD" frame). The first frame covers a smaller portion of the telephone household population than the second (about 70% versus about 95%). It enjoys, however, much lower cost of data collection because of the smaller portion of nonresidential numbers in the frame. Further, it might be more attractive from the perspective of minimizing nonresponse error in overall estimates. For example, with the street address of the telephone household, links between the number and social environmental data linked to addresses can be made. This means that the frame can potentially contain a richer set of data on all sample cases (both respondent and nonrespondent cases) than could be true of a sample from the RDD frame. These qualities of the listed frame need to be evaluated in the context of the lower coverage rate of the frame.

The data on the listed frame might include structure type, population density of housing, socioeconomic status indicators, and racial-composition measures. All of these would arise from the most recent block-level estimates from the Census Bureau. In addition, it would be feasible to append aggregate data from other data sources, using address ranges. These might include more-current age, gender, data from drivers' license data bases, city–county data book (county level) data for crime, employment, and other indicators. The utility of these data is dependent on two factors: the homogeneity of these attributes among households in the same block, and the predictive value of these attributes for the key survey variables. These two attributes would vary by survey topic and item from the auxiliary data sets.

If a variable were quite homogeneous within neighborhoods for a target population (e.g., socioeconomic status indicators) *and* related to the dependent variables, then it would be a candidate for monitoring during the data collection period. If large imbalances in response rates for different levels of the variable were evident during the data collection, then special efforts might be targeted to the low response levels. If these failed to correct the imbalance, then the variable would be useful for weighting class adjustments in the estimation phase or in other postsurvey adjustments for nonresponse.

Another way to reduce nonresponse error through selection of a sampling frame is to attempt enrichment of the frame during its development. In face-to-face surveys, observations can be made by the staff listing addresses in secondary sampling areas. These observations might involve variables we found to be useful predictors of nonresponse—multiunit structures, physical impediments to accessing a unit, evidence of presence of children. In telephone frames this might include matching of listed units to census and other geocoded information (e.g., proportion of multiunit structures, neighborhood-level socioeconomic data). With advances in geographical information systems and statistical mapping, the availability of data for this purpose is increasing.

When a designer must decide between a rich frame offering incomplete coverage and a lean frame offering complete coverage, a tradeoff of coverage for nonresponse error is implicitly involved. Sometimes, past studies of frames can be used to assess the coverage error levels. For example, many major national U.S. federal government household surveys contain measures on whether the sample household has a telephone. Secondary analysis of differences between telephone and nontelephone households on key statistics of the planned survey may assist in frame selection decisions.

Multiple frame designs were developed for situations where use of a frame with inadequate coverage was advised for cost or other reasons, but when coverage desires forced use of multiple resources. For example, there are studies using mixes of telephone and area frames (e.g., Lepkowski and Groves, 1983; Hartley, 1974). There are other examples using list frames for some portion of the sample and area frames for others (e.g., AHEAD, Survey of Consumer Finances). Thus, the statistical tools exist to blend multiple frames into a single survey statistic. Their proposed use for nonresponse error reduction merely adds a new rationale for their use.

11.3   CHOICE OF MODE OF DATA COLLECTION

The next step in executing a household survey is choosing whether the survey will be conducted by face-to-face methods, telephone methods, or mail self-administered methods. Our research speaks most directly to the face-to-face mode. However, because so many household surveys are conducted by telephone, we think it important to summarize the implications of this research for telephone surveys. This requires comments on potential mismatches as well as likely applicability.

With regard to nonresponse due to contact, the face-to-face mode and the telephone mode differ on key resources. First, since area frame addresses are often listed by survey staff prior to the contact attempts, there is some chance to collect information useful to contact at that time. There is no functionally equivalent step in telephone surveys based on RDD samples. Hence, the interviewer is armed only with information available at the time of sampling—an exchange name and, potentially, census or other ecological data for the exchange or zip code groups. Further, on each call the face-to-face interviewer has the chance to observe other properties of the sample unit, and sometimes to question neighbors about when the sample household is likely to be at home. In short, auxiliary information can potentially inform the interviewer's judgement of the best time to call next in face-to-face surveys. Answering machines come closest in the telephone mode to a precontact source of information. Some of their recordings reveal whether the household consists of unrelated roommates, whether there are children in the home, whether there is an elderly respondent, whether the household is reluctant to reveal anything but their telephone number. From time to time, this information might be useful to an interviewer in customizing the next approach to the household.

Answering machines also illustrate an impediment to access in the telephone mode. We saw in Chapter 4 how strong an effect on face-to-face contactability were locked apartment buildings, gates, and other features of housing units. The telephone mode has its analogues. There is large diversity in how survey organizations instruct interviewers to deal with answering machines; some leave short messages on the tape, others merely terminate the call and schedule the next call for another time slot.

There are other access impediments in telephone surveys, and we expect as the technology evolves that new ones will be invented and marketed to those households who value their privacy. The current technologies include caller ID, which allows the household to discern whether they recognize the phone number of the caller prior to answering the telephone. In addition, some areas permit subscribers to block calls from various numbers to their home number, thus directly restricting access to a subset of all numbers. The first two features, answering machines and caller ID, may lead to interviewer calls not being answered at times. We expect that these numbers will require more effort to obtain contact, however. It would be wise to record the existence of an answering machine on call records of sample units, in order to guide more intensive calling to these numbers and possibly to use the data for postsurvey adjustment.

In Chapter 5 we examined a set of household-level influences on cooperation in face-to-face surveys. We have little contrasting evidence about the situation in telephone surveys. What does exist might suggest that elderly respondents may show less compliance with survey requests on the telephone than in person (Cannell *et al.*, 1987; Sykes and Collins, 1988). If this is true, we suspect that some blend of two reasons may apply—problems of hearing in the restricted audio medium, and cohort effects of the current group of elderly respondents that make them reticent to use the telephone for long distance calls from strangers. These hypotheses need more study, with proper experimental controls. Other than age differences, there might be differences because of contrasting roles between the "phone answerer" and the "door answerer." That is, some study is worthwhile of whether one household member tends to answer the phone or the door, and whether the same person does both. If not, then the first person who encounters the interviewer in the two modes would vary, with potential effects on cooperation.

Finally, there may be different household influences on telephone surveys than face-to-face surveys because of common uses of the mode. For example, it appears likely in the United States that there are many more sales calls by telephone than by face-to-face mode. Hence, the default cognitive script used by a householder encountering a telephone interviewer is that the caller is selling something. This might lead to a greater challenge to telephone interviewers for communicating their purpose.

However, the other household-level influences (e.g., size of household, presence of children, etc.) we would suspect apply equally well to telephone surveys as to face-to-face surveys. They might be viewed as socio-demographic indicators of psychological predispositions to requests for information from strangers.

In Chapter 6 we reviewed social environmental influences on cooperation. We found that the strong effects of urbanicity on face-to-face response can be substantially explained by characteristics of urban households. Our experience has been that urbanicity effects are smaller in telephone surveys. One reason for this was mentioned above—the ability to successfully reach residents of locked apartment buildings and other units by telephone. Similarly, we would expect lower effects of fear of crime, although it would be likely that some fear of scam artists would persist in the telephone mode.

When we turn our attention to the effects of interviewer attributes and the nature of the interaction with the householder, there are rather stark differences between the two modes. First, the telephone strips away revelation of interviewer attributes that are communicated visually (e.g., through dress, eyes, demeanor). All cues to help the householder choose a cognitive script come through the audio medium. They are thus restricted to gender (usually) and age (sometimes). Research has shown that pace, volume, and inflection might be used by householders to judge the intent of the interviewer (Oksenberg, Coleman, and Cannell, 1986). In addition, in the United States race of the interviewer is sometimes inferred by accent. We'd expect small marginal impact of these differences, except on survey topics directly related to those attributes.

Perhaps more important differences lie at the interaction level. We have noted that telephone communication between strangers appears to be guided by different norms than is face-to-face communication. Telephone interactions involving survey requests tend to be shorter than those in face-to-face modes (Oksenberg, Coleman, and Cannell, 1986; Groves and Kahn, 1979). More householders hang up the phone in the middle of a telephone interviewer's introduction than close the door while the face-to-face interviewer is talking. This heavily restricts the ability of the interviewers to tailor their remarks based on cues they perceived from the householder's behavior. In the telephone mode, many more efforts might be required by the interviewer to extend the interaction in order to acquire such cues. In short, the challenge of tailoring is probably greater in telephone than in face-to-face surveys.

Finally, we note that two other aspects of telephone surveys move in the direction of greater nonresponse error. As we noted above, the use of incentives, advance letters, and other cooperation enhancements is restricted on the telephone. Telephone interviewers cannot display identification or written materials that convey the authority of the organization they represent to collect the survey data. Further, the inability to observe characteristics of sample units restricts the ability to collect data potentially useful for nonresponse adjustment.

Most of the contrasts appear to illustrate disadvantages of the telephone mode, and we expect this underlies its generally lower response rates compared to face-to-face surveys. However, approaches suggested in this book merit further study. First, as we noted above, richer telephone frames, containing data useful for tailoring and for postsurvey adjustment, should be sought. Second, consideration might be given to separating the initial contact from the survey request—an introductory call that attempts to establish the legitimacy of the interviewer. This call would be totally oriented to distinguishing the interviewer from sales callers and from transient "mass marketing" contacts. A later call would make the request, after addressing any concerns the householder might have. Obviously, much developmental work is required to test the merit of these ideas, and they have cost implications for telephone surveys. However, the influences we have uncovered affecting survey cooperation appear to require different approaches than are routine at this time.

11.4 DESIGN OF MEASUREMENT INSTRUMENTS

Much of the survey methodological literature is devoted to the construction of survey instruments (see Sudman and Bradburn, 1974; Sudman, Bradburn, and Schwarz, 1996). Almost all of these efforts are devoted to question structure, wording, and order for oral or visual presentation to a respondent. Little of the work is devoted to observations that could be obtained by interviewers about the sample household, the targeted householder, the living arrangements of the sample household, the social ecology of the neighborhood, etc. While interviewers have traditionally been viewed as the agents of the researcher, there is much room for improving their function as active observers of influences on response propensities.

Based on the research represented in this book, these observational powers of the interviewer are one key to reducing nonresponse rates and building statistical adjustments to reduce nonresponse error, given an achieved response rate. This implies improved nonresponse handling capabilities for those modes offering more observational powers. Unfortunately, mail self-administered questionnaires offer very restrictive means of observation; all measurement must take place through householder actions. Similarly, the telephone as a survey tool restricts almost all auxiliary data to those from the audio medium, during electronic connection with the sample telephone number. In contrast, the face-to-face mode provides observational data through visual inspection prior to, during, and after the contact with the sample units.

To repeat, the value of these observations is in guiding the customization of the survey request to the household and/or collecting variables useful as predictors in postsurvey adjustment models. With such purposes in mind, the weaknesses of the mail and telephone modes force attention to deliberate efforts to collect useful data on the sample prior to or after the request.

For example, "foot in the door" techniques have support from the psychological concept of consistency. The finding that acceding to a small request produces higher likelihood of agreeing to a later, larger request, suggests that contacts with sample households prior to the survey may be worth investigating. Such contacts could attempt collection of auxiliary data, if they were judged to be less intrusive or sensitive than the survey data themselves. For example, elderly householders can sometimes be discerned by the quality of voice.

The field has much to do to improve the richness of observational data on the full sample. There are, however, some basic features of observation that have such value that they should be routinely captured and analyzed. This section reviews these features, ordered by the major influences of our conceptual framework.

11.4.1 Measures of the Social Environment

Some characteristics of the social environment, as we found in Chapter 6, are often known from the sampling frame (e.g., urbanicity). Others, in the United States, can be appended to the sampling-frame record through matching with public records. If the block identification from the last decennial census is known, then block-level measures of structure type, population density, persons per room, and owner or rental status can be obtained *as of the last decennial census.*

When staff from the survey organization are sent to list housing units in sample segments they can collect other ecological data. These could include the number of units in the structure of the sample unit and sometimes the owner/renter status of the neighborhood.

On listed telephone frames, the zip code and county can often be used to link with other data sources. With county identification of the sample segment, other data like police reported crime rates can be obtained. Further, on all U.S. telephone numbers the area code and prefix identify a telephone exchange, an area of land served by the prefix. The number of prefixes in an exchange (knowable from the sampling frame) provides a useful measure of urbanicity.

11.4.2 Measures About the Interviewer

We learned in this book that interviewers' tenure and confidence appear to affect the levels of cooperation they obtain in surveys. We have conceptualized interviewer experience as containing both time and organizational variety as key dimensions. We seem to find more support for length of employment as a critical dimension. Thus, we would suggest that the number of months or years of active survey interviewing experience would be a valuable measure to collect and retain in survey records. This could be measured on the job applications as well as being supplemented by administrative records of the employing organization, as the interviewer increased in job tenure in the same organization.

Theoretically, the value of tenure as an interviewer for nonresponse purposes is a function of the number of encounters the interviewer has with sample persons in recruiting them as respondents. For personnel with part-time interviewing jobs, this might be taken into consideration. Indeed, a preferable measure might be the number of encounters they have had with sample persons, seeking their participation in some survey.

The confidence measure we used (see Chapter 7, Section 7.7) captured self-reported assurance that the interviewer could persuade others to participate in a survey. Despite the useful performance of this item in the analyses, we have not spent sufficient time in developing batteries of items for interviewers. We suspect that further basic work would be useful to improve interviewer-level measurement of expectational states. Theoretically, the best measure would be one reflecting the interviewer's expectational state at the time of a contact with the sample unit, although such measurement would prove difficult.

11.4.3 Measures about the Sample Housing Unit

When the interviewer visits a sample unit, various features of the unit relevant to accessibility by strangers are visible. These include locked gates to the property, "no trespassing" or other signs limiting access, bars on windows and external doors, and multiple locks on the external door. These can be measured in a simple checklist provided to the interviewer.

At the same time, the interviewer can make observations regarding the presence of children: toys left outside on the property, small bicycles and tricycles, swing sets, or car seats in automobiles parked in the driveway. Visits during the day sometimes provide higher odds of observing the presence of children; hence, they are valued for that purpose.

In the telephone mode, two relevant household-level variables are observable—the listed status of the telephone number and the existence of some technological impediments to access. The listed status of the telephone number can be discerned through various commercial services that will match a given RDD telephone sample with directory files. Unlisted numbers are mixes of new residents to the exchange, those with intermittent telephone service, and those who requested unlisted status. It is probably not an ideal mix with regard to homogeneity on response propensities, and if there were ways to separate those who requested unlisted status from others, they would be useful.

The existence of answering machines can be determined on most noncontact calls, depending on whether the machine is in service. This would be a useful indicator of difficulty of gaining contact and, for some users of answering machines, the use of the machine to screen calls, rather than to capture calls received while out of the home. In addition to the existence of the machine, documentation on whether calls were ever made without evidence of the machine activated might separate types of users of the machines, who vary on their cooperation propensities.

Other features of telephone service offer enhanced privacy protection (e.g., caller ID, and call blocking, at this writing). In some areas, the caller to such numbers can discern such services if they enter special codes prior to the call. Such information might be another useful indicator of steps taken by the household to limit contact to outsiders.

11.4.4 Measures Concerning Contact-Level Events

In Chapters 8 and 9 very basic measures of what the householder said during a contact were shown to be useful predictors of later cooperation rates. For example, Chapter 9 shows that when householders ask questions of the interviewer, there are higher odds of obtaining the interview through repeated callbacks. Those who seek some delay in the timing of the interaction tend to have higher odds of providing an interview than those who provide negative reactions to the survey request.

There clearly remains much work to do to explore alternative observations during the interaction with interviewers and to improve the reliability of those measures (see Campanelli, Sturgis, and Purdon, 1997). We would suggest some basic measurement development work, using both tape recording and interviewer recording of events, to gain further insights into what types of information are well collected using interviewer observation and what types are poorly measured.

We are especially interested in attempting to expand the measurement to observational measures on interviewer behavior itself. This would be central to attempting measurement of many of the tailoring concepts described in the book. However, we are not sanguine that interviewers, who succeed by focusing intently on the householder behavior, can provide accurate documentation on their own reactions to individual householder behaviors.

In this area of measurement, centralized telephone interviewing might offer better capabilities. The routine monitoring of telephone interviews, while usually focusing on the question–answer behavior of the interview, can be used to measure the preinterview conversations of interviewers and householders. We have taped such interactions—a tool acceptable for methodological studies but ill-suited for production work. We could imagine monitoring systems that attempt to measure the first few moments of each interaction, but know of no facility now doing so.

11.4.5   Measures Concerning the Result of Calls on Sample Households

These consist primarily of documentation of the time and day of a call attempt on a sample unit, the outcome of the call (typically using some disposition coding scheme), the interviewer who made the call, and written notes about the nature of events during the call. This measurement is useful in order to a) determine when the next call should be made, b) diagnose whether the next contact should alter the arguments presented by the interviewer to tailor them to concerns raised in a previous contact, and c) to decide whether a different interviewer might be more successful in calling the case. These are critical data for gauging the best next steps for contacting the sample household and for guiding other nonresponse rate reduction protocols. They can also be useful in postsurvey adjustment models, especially to separate those with limited time at home from those with little willingness to cooperate.

11.4.6   Observable Correlates of Survey Measures

Successful adjustment for nonresponse in surveys stems from identifying and observing subgroups in the full sample that are homogeneous on the survey variables (i.e., within a subgroup, respondents and nonrespondents resemble one another on the survey variables). For some surveys, interviewers could be used to assist in observing attributes of the housing unit and householders that may be proxy indicators of key survey variables.

Each survey presents its own opportunities and challenges. Some examples might communicate the types of observations we have in mind. In a study of disability measures, observations of assistive devices outside and around the unit might be useful (e.g., ramps, railings). In a study of the elderly (as we exhibit in Section 11.7 below) interviewers recorded whether the householder gave the comment, "I'm too old" or some other self-deprecating remark when the interviewer requested survey participation. In a study of assets and wealth, efforts could be made to determine the owner/renter status of the housing unit and tax assessments through property record checks. In a study of political attitudes, utterances by the householder about interest in politics can be used (e.g., Couper, 1997). In a study of health, observations that householders are in the hospital, in bed, or "not up to the interview" might be indicators useful as proxies to their health status. In surveys of socioeconomic status, the size or estimated value of the housing units can be used.

Obviously, surveys that measure household characteristics yield themselves to this treatment better than do those measuring personal characteristics. Surveys of observable attributes are easier than surveys of internalized states. Further, the standards of observable measurement often must be lower than those of direct questions posed to a compliant respondent. The goal must be to replace assumptions of complete homogeneity of nonrespondents with some sorting of respondents and nonrespondents into subgroups useful for adjustment.

11.5   SELECTION AND TRAINING OF INTERVIEWERS

The research in this book supports the belief that experienced interviewers are more adept at gaining participation than inexperienced interviewers. From the theories we espouse, we suspect that the increased effectiveness results from trial and error in developing tailoring skills, not explicit training effects. We conclude that based on both experience with interviewer training regimens across time and organizations and a recent study of interviewer training materials (Miller-Steiger and Groves, 1997). Most training protocols (80% of those studied) contain material on how to answer explicit questions of the respondent, but many fewer (28%) had lessons involving attempting to diagnose less explicit concerns of the householder. Many materials note that the interviewer should listen carefully to the householders and to answer their questions, but appear to give little attention to training on maintaining the interaction with the householders and tailoring behavior to their concerns.

The theories and empirical results presented in this book motivate a rethinking of interviewer training procedures. Some of the ideas suggested by Morton-Williams (1993) are worthy of examination. With a focus on the notion or tailoring more specifically, three steps in interviewer behavior seem important in a training protocol: 1) recognition of the meaning of cues regarding householder concerns regarding the survey request, 2) knowledge of relevant information about the survey in order to address those concerns, and 3) effective delivery of that information in a manner comprehensible to the householder. As we note in the commentary on the first-contact protocol, this training should give the interviewer skills in helping the householder choose the correct cognitive script (i.e., a survey request) for the interviewer's call, to aid in generating householder talk (to provide cues regarding concerns), to improve diagnostic skills in householder concerns, in developing a wide repertoire of counterarguments, and in customizing the delivery to individual householders.

Finally, using measures on interviewers in postsurvey adjustment will improve the prediction of response propensity (as shown below in Section 11.7). We have shown that tenure and confidence measures are useful, but much work remains to be done developing other useful measures motivated by the theories. For example, these might be measures from standardized measurements of the ability to tailor arguments to a variety of concerns potentially raised by householders. Such measures could be explanatory of a component of variance in response propensity associated with interviewers. Such standardized measurement fits the two-pronged goal of all the new observations of this chapter—potential for use in both nonresponse reduction and nonresponse adjustment. The scores on interviewers could be used to reduce variation, by providing remedial training to those scoring at lower levels. That remedial training might use exemplars from the performance of interviewers achieving high response rates.

Failing success on increasing scores through training, the scores become candidate predictors in adjustment models. To the extent that those with low scores consistently fail to gain the participation of the same types of sample units, the predictors will be valuable.

11.6   CALL ATTEMPTS ON SAMPLE UNITS

Our contribution to understanding nonresponse arising from noncontact is principally in distinguishing the correlates of noncontact from those of refusal. These are the role of physical impediments to accessing the housing unit, multiunit structures, household size in at-home patterns, and ages of householders. The remedies for noncontact nonresponse are, we believe, more straightforward.

The indicators of at-home rates can be used to schedule calling patterns on sample households. For example, our results indicate that households with physical impediments to access should be called first. They average more calls to first contact, in order for the interviewer to negotiate the gatekeeping features of the unit. Similarly, in urban areas, multiunit structures tend to require more calls to first contact.

We note that the same forward-scheduling applies to other modes of data collection. Since unlisted numbers appear to have lower response rates and higher burdens for contact and persuasion than listed numbers (Traugott, Groves, and Lepkowski, 1987), they should be the first sample units called in a telephone survey, in order to provide more time for calls and refusal conversion attempts. When an answering machine is encountered, more frequent calling might be warranted, other things being equal, to attempt to locate a time when the household is not using the machine to trap calls.

11.7   THE FIRST-CONTACT PROTOCOL

The results of the research imply that the first contact with a sample household might best be used as an initial sorting of the sample into two groups—the enthusiastic and compliant versus all others. The first group will communicate willingness to proceed with the interview in an unproblematic way. This is a group of nonnegligible size. In NSHS (see Chapter 4) approximately 35% of all the sample provided an interview on the first contact. We suspect a good portion of those required little persuasion from interviewers. For these, interviewers might proceed only when explicit behaviors by the householder indicate interest in the interview.

The second group of householders consists of those who are curious but tentative, those who are uninterested initially, those who select incorrect scripts (e.g., fearing criminal victimization, perceiving a sales approach), and those who are hostile to any survey request. For all these the theories suggest that the job of the interviewer is to maximize the number of conversational turns in order to gain information useful in diagnosis of a) whether an incorrect cognitive script was chosen by the householder, and b) if the appropriate script were chosen, any issues that must be addressed to clarify the purpose of the interviewer's call prior on the direct request for the interview. This guidance focuses the attention on maintaining interaction with the householder in order to tailor the interviewer behavior to the householder. In turn, it reduces the pressure on interviewers to deliver interview requests quickly (and thus to enter into persuasive communication prior to knowing the concerns of the householder).

This protocol on the first contact is consistent with the training of some interviewers to "retreat" upon the first sign of hostility by the householder. The only alteration suggested from the theory would be to obtain as much evidence as possible during such visits of concerns on the part of the householder. This information can then be used in the next approach to the household, whether by the same interviewer or a different one more closely matched to the characteristics or concerns of the householder.

We speculate that having interviewers record notes about householder behavior is probably more important than their notes about interviewer behavior. In successive calls, the more information that interviewers have about the householder the better. Often this is rich qualitative information but difficult to summarize. Refusal conversion interviewers might spend some time digesting such material in preparation for calls, but profit from the knowledge in constructing their approach. Further, the use of such information in postsurvey adjustment requires a coding step in order to make it analytically useful. In short, much work remains to create efficient ways to record this kind of material.

We note that the telephone mode eliminates visual observation during contact attempts. Useful aural cues arise only after connection has been made, from answering machine tapes or from direct contact with a householder. Further, the problem of maintaining interaction with an unknown householder appears to be greater in the telephone medium than the face-to-face medium. Many refusals merely hang up the telephone while the interviewers deliver the initial greeting. We infer that these judgments are made very superficially, based on unknown voices, especially those with obvious "agent" scripts. "Agent" scripts are those that sound as if the caller is acting as the agent of another and has been trained to deliver the others' words as specified.

We have some evidence that the challenge of tailoring is perforce much higher on the telephone. Each conversational turn is more precious and thus interviewers would tend to make premature judgements about householder concerns. This, combined with the quick terminations, lead to lower response rates on the telephone. This reinforces the notion that learned scripts delivered in a monotone may fail to maintain interaction with the householder, because such behavior resembles other unwanted telephone intrusions. Further, since the density of cues is lower, interviewers need to be especially adept at detecting and reacting to householders' behavior informative of their concerns.

11.8   EFFORTS AT NONRESPONSE REDUCTION AFTER THE FIRST CONTACT

In addition to changes in the behavior of interviewers at the first contact, the theories suggest changes in the follow-up protocol for surveys, when the first contact with the householder does not yield an interview. The utility of follow-up activities is a direct function of how much information was collected about reluctant or busy householders during the initial contact. With no information recovered from the first contact, the interviewer is dependent on the ability to tailor during the next contact based on immediate cues. With some information, the interviewer can customize the first introductory comments during the next contact, as well as anticipate householder behavior.

It is common for a householder letter to be sent following a failed initial contact. We would speculate, based on theory and our empirical results, that letters sent to householders after an unsuccessful first contact would be more successful when the letter acknowledged the householder's comments, expressed an understanding for their legitimacy, and then provided counterarguments tailored to them. Explicit concessions could be stated (e.g., " Since you are very busy, we are quite willing to visit at a time when it is convenient for you"). Some organizations are already doing this.

This follow-up activity can be the responsibility of individual interviewers in dispersed data collection designs, in which interviewers are assigned a set of cases to handle from initial contact through to final disposition. Another level of organization within the survey project is necessary to guide the disproportionate allocation of follow-up resources. Especially in computer assisted surveys, a central data base containing administrative data on cases could be used to monitor the completion rates of various subgroups. All the data collected on the social environment of sample units and on the units themselves could be routinely examined to discern imbalances in the distribution of respondents and nonrespondents on the observed variables. With severe imbalances, the decision maker has the option of reining in follow-up efforts in subgroups of the sample with relatively high response rates and targeting subgroups with disproportionately low response rates. Clearly, this decision would result in lower overall response rates, with the hope of lower nonresponse error related to those subgrouping variables.

Finally, the notion of tailoring works to explain the circumstances under which mixed-mode surveys succeed, especially those that start with modes requiring much more householder active cooperation and end with those requiring minimal reactive acquiesence. Double sampling or two-phase sampling schemes were suggested early by survey statisticians (Deming, 1953; Birnbaum and Sirken, 1950; Hansen and Hurwitz, 1958), as a way to concentrate resources on a probability sample of nonrespondents after the main survey was finished. If the sample were completely measured, the design offers an unbiased estimate of the nonresponse error on measured variables (albeit with higher sampling variance). Based on theory, we would speculate that double-sample designs would be most successful at this goal within designs where tailoring was necessarily constricted during the main study. For example, if mode switches (e.g., from personal visit to telephone) were not possible during the main study, the double sample using the second mode will disproportionately cover those persons comfortable only with the second mode. If reluctance is a feature of one householder and the double sample relaxes the respondent rule to permit another informant (e.g., in household-level measurement), then the double sample can benefit from that change.

11.9   POSTSURVEY ADJUSTMENTS FOR UNIT NONRESPONSE

Despite the best efforts during the data collection period aimed at reducing nonresponse rates, in almost all surveys some sample units remain unmeasured. The sections above argue that those remaining should be more fully the result of targeted efforts to reduce the $(\bar{y}_r - \bar{y}_m)$ terms. It should be expected, however, that despite creative and expensive strategies to measure elusive groups, subgroups will vary in nonresponse rates. The final tool lies in the estimation stage—the use of postsurvey adjustment to reduce nonresponse error.

The survey researcher has imputation, selection bias modeling, and case-weighting adjustments as tools for adjustment for unit nonresponse (see Kalton, 1983; Little and Rubin, 1987; Lessler and Kalsbeek, 1992). Hot-deck imputation creates a data set with complete data for all nonrespondent cases, often by substituting the data from a respondent case that resembles the nonrespondent on certain known variables. The full data vector of the respondent is then duplicated for the nonrespondent case. Imputation is more often used for item missing data adjustment, through the use of predictive models for the missing datum, based on data available on the same (or similar) cases. Heckman's (1979) solution to unit nonresponse was the use of a response-propensity equation specified as a probit or logit function. A density function transformation was used as an added predictor in the structural equation, to account for the "selection bias." This technique has been found to be somewhat sensitive to normality assumptions of the joint distribution of error terms in the selection and structural models (Winship and Mare, 1992). All of these techniques eliminate the bias of nonresponse, under certain assumptions. The fact that such assumptions must be met implies that these methods are based on models of the processes producing cooperation with a survey. In other words, postsurvey adjustments for nonresponse entail implicit or explicit theories of survey participation.

We believe that model specification in adjustment approaches might be informed by a theory of survey participation like that described in this book. A full understanding of decisions to participate in surveys would guide the choice of procedures to keep nonresponse at acceptably low levels, and would help in the allocation of resources for such efforts. Such a theory would also guide assumptions inherent in statistical adjustment of survey data.

First, given the results of our research, we expect that well-specified response propensity models must have different functional forms for noncontact and refusal components of nonresponse. The tendency of the literature to lump these phenomena together is no doubt one reason for the failure to replicate findings across studies. That is, a survey with a high noncontact and low refusal rate should theoretically have a different set of correlates of overall nonresponse than one with a low noncontact rate and high refusal rate.

We have begun to explore the use of the theory reviewed above to construct response propensity models, reflecting the two-step process of survey participation—first, contact with the household; then, a decision on participation, given contact.

As the first stage of the response propensity model, we posit a logistic model for the probability of contact:

$$Pr(C_i = 1 | z_i) = 1 + \exp(-z_i' \beta)$$

where $C_i = 1$ if the individual i has been contacted and zero otherwise; z_i is a vector of individual and block-level covariates and β is a vector of regression coefficients. The unknown β parameters are estimated by maximizing the weighted log-likelihood:

$$\Sigma w_i [C_i z_i' \beta - \log[1 + \exp(z_i' \beta)]] / \Sigma w_i$$

The weight w_i is the sampling or selection weight. The summation in the above expression is over all sampled individuals. The contacted individuals are then assigned

$$w_i^* = w_i [1 + \exp(-z_i' \hat{\beta})]^{-1}$$

weights to make them representative of the population. This updated weight is then used in the estimation of the parameters in the next stage of the model.

In the second stage, conditional on contact, that is $C_i = 1$, the result of the attempted interview is modeled through a second logistic model:

$$Pr(R_i = 1 | x_i, C_i = 1) = 1 + \exp(-x_i' \gamma)$$

where $R_i = 1$ if the individual i is a respondent and zero otherwise, x_i is a vector of individual-, block-, and interviewer-level correlates, and is the corresponding regression coefficients. The unknown γ coefficients are estimated by maximizing the weighted log-likelihood:

$$\Sigma w_i^*[R_i x_i'\gamma - \log[1+\exp(x_i'\gamma)]]/\Sigma w_i^*$$

The summation in the above expression is over all the respondents. Consequently, the final weight is derived as $w_i^{**} = w_i^*/\hat{\theta}_i$ where

$$\hat{\theta}_i = [1 +\exp(-x_i'\hat{\gamma})]$$

In our implementation, all the relevant numerical integrations were carried out using Gaussian quadrature. Special software was developed using GAUSS programing language.

The practical problem for the survey investigators is identifying and measuring the causes of participation in their studies. From the conceptual structure we have found helpful, we would assert that there is a large and separate set of causes of survey contact versus cooperation that are independent of the survey topic. These would be ignorable sources, after appropriate adjustments. Other causes, those linked to the survey topic or to some survey design features, may be productive of nonignorable nonresponse.

If survey researchers chose to make postsurvey adjustment part of consideration at the survey design phase, they would need to consider both general causes of survey participation and the survey-specific causes. Identifying the general causes could utilize a conceptual framework like that above. Identifying the causes linked to the survey topic (and thus likely to be nonignorable) requires particular hypothesis generation concerning the population studied and its knowledge, beliefs, and attitudes toward the survey questions.

11.9.1   Examples of Two-Stage Propensity Model Adjustments

We used the above approach on data from the first wave of the Survey of Asset and Health Dynamics of the Oldest Old (AHEAD) and the National Survey of Health and Stress (NSHS). These two surveys are described in some detail in Chapter 3. This section describes the various data sets that were assembled to help construct response propensity models for the two surveys.

Sample Frame Information. Both samples were linked to 1990 decennial census block-level data, using the census block designation. This supplied several variables—percent minority population in the block, percent occupied units that were rented, percent single-person units, median value of owner-occupied units, median rent in rental units, urbanicity, percent multiunit structures, percent of units vacant, mean number of persons per unit, and persons per square mile. In addition, the AHEAD sampling frame contains a designation of type of frame (area frame versus

HCFA), a designation of the stratum of the case (e.g., Florida, Hispanic or Black oversamples), and a measure of the sample person's age (from the earlier HRS screener or the HCFA files).

Observational Measures. In the NSHS and in AHEAD cases with personal visits, interviewers recorded observations about the neighborhood, the housing unit, and other observable household characteristics. We include a measure of physical impediments to accessing the unit.

Contact Observations. As is described in Chapter 3, interviewers completed a series of contact description items, recording details of the interaction they had during the first contact with the household.

Interviewer Questionnaire. During training for the surveys, all interviewers completed a questionnaire designed to elicit their attitudes toward seeking cooperation from sample persons, expectations about the difficulty of this task, and attitudes and experiences in surveys of the elderly. Interviewer administrative records, containing demographic information and experience-related measures, were also appended to the data file. The interviewer tenure and confidence measures came from these efforts.

11.9.2 Traditional Postsurvey Adjustment for Nonresponse

Traditional approaches to nonresponse adjustment make use of variables available on the frame or collected during screening. The adjustment used for AHEAD was constructed by assigning households to adjustment cells based on census division (10 regional groupings of states), MSA/non-MSA status, age group (70–79 and 80+) and racial composition of the neighborhood (non-Black/non-Hispanic, Black and Hispanic) based on census tract information. A total of 79 adjustment cells were formed, and the reciprocal of the weighted response rate (based on selection weights) used as the adjustment factor. The final weights include the selection weights, the nonresponse adjustment, and a poststratification factor. The latter is not used in our analyses, to permit comparison of alternative nonresponse adjustment approaches to the traditional approach used for AHEAD.

11.9.3 Modeling the Probability of Contact

As in Chapter 4, a logistic model was fit that included proxy variables for the at-home patterns of the sample household and the extent and diversity of interviewer attempts to contact the sample person. As is to be expected in most survey work, the data resources were somewhat different in the two surveys.

The first set of variables to consider in any propensity model should be survey design features related to contact rates. In AHEAD, some variables in Table 11.1 reflect various design features related to contact rates. The first is a sampling frame variable, whether the sample person was selected from an area probability sample

Table 11.1. Coefficients of logistic models predicting contact versus noncontact, for AHEAD and NSHS (stage 1 model)

Predictor	AHEAD	NSHS
Design features		
HCFA Frame	−0.25*	
Total calls	−0.037*	
Face-to-face attempt	0.41*	
Social environment		
Urbanicity:		
City		−0.81*
Suburb		−0.49*
Town		0.082
Rural area		—
Log(population density)		−0.044*
% Rental on block		−0.0078*
% Single-family units	0.015	−0.0023
Persons/unit in block	−0.39*	
% over 70 yrs.	−0.013	
% Minority		−0.0055*
% Multiunit housing		−0.0034
Median value (owned)		−0.040*
Median rent (rented)		0.0005
Housing unit		
Physical impediments to access		−0.43*
Sample person		
Estimated age	−0.054*	

*$p < 0.05$.

screening, with household roster and telephone numbers available, or whether from an administrative record frame, from the Health Care Financing Administration (HCFA), with records associated with those receiving medical care benefits available to all persons 65 years or older. The latter frame contained a relatively recent address but no telephone number. Sample cases from the HCFA frame were released later in the field period, thus limiting the number of calls possible to these cases. The coefficients in the table show that this negatively affected contactability. Hence, we also included the number of calls made on the unit (and found the expected lower contact rate for those with many calls). Finally, this was a mixed-mode survey, permitting either telephone or face-to-face attempts to contact (with face-to-face attempts associated with higher contact propensities). The model in Table 11.1 shows that face-to-face attempts increased the chance of contacts in general. There were no design features used in the NSHS contact model.

The second set of variables are social environmental characteristics, including characteristics of the census blocks, urbanicity classifications, etc. Here, as one sees in Table 11.1, the predictors of contact in AHEAD are different from those in NSHS. We believe that the differences arise from two sources—the fact that

AHEAD measured only older persons and the fact that the survey used both telephone and face-to-face modes to attempt contacts. This may explain the lack of urbanicity effects in AHEAD on contactability, for example. In contrast, the NSHS
model shows the same pattern of negative effects on contactability that we have
seen in several chapters of this book.

The third set of variables are sample-unit attributes, for which only NSHS shows
an effect for the physical impediments measure. The AHEAD survey, with its ability to surmount the physical impediments through telephone calls, does not show
this effect, and hence the variable is not included in the model.

Finally, since the AHEAD sampling frame was household listings from an earlier
survey and the HCFA listings, the estimated age of the sample person was available
on the frame. Thus, a person-level attribute was available, included in the contact
model, and found to show that older persons were less likely to be contacted, partially because of health problems.

As we described earlier, the models in Table 11.1 permit the first stage of an adjustment strategy. We take the expected value for each contacted case in the sample,
under the models, transform it into an estimated probability of contact, and then
weight the contacted cases using the reciprocal of that probability. In some sense,
this produces a weighted data set representing all sampled respondent cases. We
then use these case weights to estimate the stage 2 (cooperation) propensity model.

11.9.4   Modeling Cooperation, Given Contact

The predictor variables for the second-stage propensity equations are organized into
similar levels of influences, with two additional levels to those used in the contact
models. The social environmental level and the sample unit level are supplemented
by indicators of the interviewer-level influences and influences at the contact interactional level (Table 11.2).

Urbanicity influences regarding cooperation apply to both surveys, with lower
cooperation in large urban areas, controlling on all other attributes. (Given our results in earlier chapters, we suspect that some of these effects reflect household-level differences correlated with residential setting, because we have no rich indicators
of household characteristics.)

The physical impediments measure has measured effects on the cooperation rate
among contacts in the NSHS data. This is a model distinct from those in the previous chapters of this book. Given the data in the match study, we were not able to
measure the effects of impediments to access on likelihood of cooperation once the
impediments were surmounted. We suspect this negative coefficient in the table reflects the tendency for less-cooperative persons to live in units with such restricted
access. (For example, "no trespassing" signs portend both difficulty of access and
cooperation.)

Both cooperation models utilize the observations about the interviewer–householder interaction on the first contact. As expected, negative comments have the
highest negative influence to nonresponse, with time delay second. Questions being
asked are positive influences in both data sets.

Table 11.2. Coefficients of logistic models predicting interview versus noninterview among contacted units, AHEAD and NSHS (stage 2 model)

Predictor	AHEAD	NSHS
Social environment		
Urbanicity	−0.37*	−0.44*
Log(population density)	−0.11*	
Percent Minority		0.0030
Percent Vacant units		0.0072
Median value (owner)		−0.023*
Median rent (renters)		−0.0005*
Persons per room in block		−0.21*
Interviewer		
Interviewer age	0.020*	
Female interviewer	0.30*	
Interviewer self-confidence		0.29*
Sample unit		
Physical impediments to access		−0.25*
Contact interaction		
Negative statements	−2.05*	−3.11*
Time-delay statements	−0.61*	−1.15*
Questions	1.21*	1.35*
Positive statements	1.62*	
"Old age" comments	−0.32*	

$*p < 0.05$.

The second-stage model, predicting participation given contact, is the stage we have chosen to introduce variables that are jointly predictive of the participation decision as well as potentially predictive of a set of key survey variables. In the AHEAD study, interviewers also recorded an observation potentially predictive of some of the survey measures on health—a statement by the householders that they were "too old" to do the interview or to be relevant to the survey. Such a comment portends lower probabilities of interviewing the person.

Since the survey interview contains many variables associated with perceived limitations due to advanced age, this statement might be correlated with values on those variables. Hence, it is included to improve homogeneity on the survey variables, conditional on propensities, not necessarily to enhance the fit of the propensity model.

There may be some concern from a causal modeling perspective that the inclusion of these interactional variables creates biased estimates of their coefficients because they are endogenous to the outcome of the case. That is, because a sample person has decided to refuse to be interviewed, he or she will issue a negative statement early in the interaction as a result. That is, the causality of the situation is the opposite of that specified in the model—a refusal causes a negative statement. There are two comments on this perspective: a) first, the propensity models are not causal models, but predictive models; our goal is predicting overall likelihood of a particular outcome, and b) these statements are not fully endogenous because both this data set and others show that even persons who make negative statements eventually participate in the survey, partially in reaction to informative and persuasive interviewer behaviors that follow the initial negative statements and precede the final decision regarding participation.

At this point in the analysis we can partially assess the wisdom of the two-stage modeling of propensities. It is clear that with this example, different models apply to the contact process than to the process of obtaining participation, given contact. Indeed, different models apply to "other noninterviews" than to "refusals." This supports the notion that propensity models should reflect the separate processes in their specification.

11.9.5   Effects of Weighting on Survey Statistics

This section compares how two different postsurvey adjustment schemes affect simple statistics, like means and proportions in the two surveys. The two adjustment schemes are the traditional weighting class adjustment procedure and the two stage fixed effect model adjustments.

Table 11.3 compares the two adjusted means to the unadjusted mean for several variables. In general, the effects of the weights are small—less than a 1% change in values is typical. On the whole, the traditional adjustment scheme produces more

Table 11.3. Weighted and unweighted estimates from AHEAD and NSHS

		Weighted Estimates	
	Unweighted	Two-stage	Traditional
AHEAD			
Mean number of limitations in daily activities	0.19	0.20	0.17
Percentage with limitation re bathing	5.7	6.0	5.3
Percentage with heart conditions	31.7	31.8	32.0
Percentage with other health conditions	29.8	29.9	30.7
Mean costs of supplementary insurance	$631	$648	$591
Mean income from social security	$643	$648	$658
NSHS			
Percentage with alcohol dependence	15.0	14.9	14.4
Percentage with antisocial personality	3.2	3.2	3.2
Percentage with major depression	18.0	17.8	17.5
Percentage with drug dependence	7.8	7.9	7.5
Percentage with dysthymia	7.1	7.2	6.7
Percentage with generalized anxiety disorder	5.1	5.2	5.1
Percentage with panic disorder	3.4	3.3	3.5
Percentage with simple phobia	10.8	10.6	10.9
Percentage with social phobia	13.1	12.9	13.0
Percentage below poverty level	12.6	12.1	12.0

movement in the estimated values than do the two-stage models, although the differences are quite small, within traditional levels of uncertainty from sampling variability. In these two cases used to illustrate postsurvey adjustment for nonresponse, neither the traditional nor the two-stage adjustments have noticeable effects on simple estimates of means and percentages. (For that reason, the examples do not permit us to illustrate how one might evaluate large effects of adjustment, but merely illustrate the steps in the adjustment process.)

It is common that the use of postsurvey adjustment weights increases the variance of survey statistics. A common estimate of the inflation of variance due to weighting is merely $1 + CV^2$, a simple function of the coefficient of variation of the weight variable. This can be interpreted as a factor for the inflation of variance of the weights, assuming no covariance of weight values and values of the variate itself.

The overall values of these inflation factors are 1.11 for AHEAD and 1.23 for the NSHS. These are somewhat lower than those of the traditional weighting class adjustment (1.77 and 1.54, respectively). One can measure the separate effects of the two adjustment steps on the variance in the two-stage approach. In these two surveys, there is greater inflation of variance by the cooperation adjustment (stage 2) than the contact adjustment. We would expect that from the high contact rates relative to the cooperation rate.

These analyses offer an example of how the conceptual work of this book might be implemented in the adjustment step for nonresponse. One limitation of the analyses, we should note, is that simple random sample estimators of variance are used for the coefficients above. We expect that these underestimate the true standard errors that would be expected under the complex sample design used for these surveys.

11.9.6   Conclusions and Next Steps

This section has presented an approach to postsurvey adjustment with propensity estimation that differs from the past literature in two ways: a) the specification of the propensity model was motivated by theoretical perspectives on survey participation and the data collection process was altered to provide predictors for propensity and proxy indicators of key dependent variables, and b) the modeling procedure had two stages, first modeling contactability, then likelihood of participation, given contact.

The empirical data used to illustrate this technique showed relatively minor changes due to the nonresponse adjustment, whether traditional weighting class adjustments or the new techniques were used. We can speculate on the survey circumstances that would lead to the two-stage models being more effective adjustment tools than single models. These would be situations in which those not providing an interview once contacted have higher likelihoods of being contacted initially or vice versa. That is, the correlates of noncontact and nonparticipation, given contact, are different from the correlates of overall likelihood of participation. For example, we have seen a tendency for negative correlates of contact to be positive correlates of cooperation (e.g., young households), and we suspect that, in general, there are many situations when the circumstance might apply. Thus the two-stage approach deserves consideration at the postsurvey adjustment step.

The potential power of the two-stage technique will depend on the ability of the researcher to enrich the data collection effort with additional observations, beyond those in the questionnaire. These measures must be available both on respondents and nonrespondents to be useful for modeling purposes. The measures in this example are preliminary versions of what might be produced.

11.10 SUMMARY

We end this book by reminding ourselves of the beginning. We mounted this research because the existing literature was overwhelmingly atheoretical, based largely on bivariate analyses of socio-demographic correlates from single surveys. The lack of a comprehensive theory hampered the development of multivariate hypotheses. The bivariate nature of the literature prevented understanding of how multiple influences combine to produce overall survey participation. The literature also ignored insights about the role of the householder–interviewer interaction in influencing survey participation. At the same time, interviewers were telling us that they needed to adapt methods to different householders, to acknowledge diverse concerns among them.

One implication of the perspective taken in this book is that survey design features may differentially affect different populations. In other words, the decision to participate may be affected by characteristics of the survey design and topic as well as by characteristics of the populations sampled. This implies that there will exist different nonresponse errors for the same nonresponse rate in different circumstances (or alternatively, the nonresponse rate is not a good indicator of nonresponse error). We shouldn't expect there to be a simple relationship between nonresponse error and nonresponse rate. Only when the mechanisms influencing the decision to participate are correlated with the survey variables will this happen.

Ideally we need to identify systematic linkages between these three sets of variables—some attribute of the survey request (e.g., interviewer, incentive, topic of questionnaire, agency of data collection), the householder's decision to participate, and the survey variable of interest. When values on the survey variables themselves cause survey participation, nonignorable nonresponse arises, and adjustment hopes are greatly diminished. If, however, attributes of the survey interviewer's protocol produce the participation decision, measuring those attributes can produce useful adjustment models.

In interviewer-assisted surveys, we suspect the relative frequencies of nonresponse arising because the householders have particular values on the survey variables is low. After all, there are many survey protocols in which the householder has only a very vague notion about the survey questions prior to making the participatory decision. However, we expect that this might be true in mailed questionnaire surveys where the decision to participate is fully informed by the content of the questions.

This research points us in the direction of developing recruitment strategies for interviewers that create adaptive protocols, changing to fit the needs of each householder. Much of the focus turns away from tight scripting of the first few words of the interviewer, to the second conversational turn of the interviewer. It moves from urging the interviewer to convey all positive aspects of the survey in one breath, to making salient those attributes of the request that are most responsive to the householder's informational needs.

Because of the need to tailor one's approach, documentation of lessons learned about the householder are important for reduction of nonresponse during data collection. This documentation for field purposes also forms the basis for a data record on key characteristics of the householder and interviewer–householder interaction, which can be used in postsurvey adjustment for nonresponse.

There is much research remaining to exploit the theoretical perspectives taken in this book. We are most interested in measuring the extent of applicability to telephone surveys, where we see the greatest threat being the inability of interviewers to maintain interaction in order to customize their approach to respondents, and the leanness of sampling frames for telephone numbers (limiting variables useful for postsurvey adjustment). Many of the ideas in Chapter 10, Section 10.3 appear worthy of study.

We believe that some of the lessons of this research may have application in establishment surveys, as well as household surveys. There are certainly large differences between the two populations. They include the existence of corporate policies specifically on release of proprietary information to external inquiries, of specific roles designated to interact with the external environment, of authority hierarchies where decisions to act are removed from the actors performing the work, of dispersion of information into specialty units (e.g., personnel, accounting, sales), etc. However, at some point in the interaction with the sample unit, interviewer–employee interactions often determine the decision to cooperate. In those interactions, the lessons of maintaining interaction and tailoring to individual concerns appear relevant. On the favorable side, the protocol of business communication (even on the telephone) appears to facilitate frank expression of concerns, facilitating tailoring behavior by the interviewer.

None of the empirical data in this book came from commercial surveys. We suspect that the principles underlying the work apply, but some common design decisions of commercial surveys affect their application. For example, some commercial surveys use restricted contact rules, generating higher noncontact rates. Some surveys use a single-contact rule, which eliminates the ability to tailor behavior to events in the prior contacts. We suspect that commercial survey requests might be especially susceptible to cognitive script errors by the householder. That is, a call from an organization that sounds like a company may be especially burdened with householders' suspecting a sales call, a request for money, etc. It is interesting in this regard that surveys in that sector have recently become interested in the use of an "I'm not selling anything" utterance early in the interaction.

We speculate that the concepts presented in the book apply in different cultural settings (e.g., outside the United States). However, the importance of different factors should vary by the survey climate. For example, during a open public debate in a society about privacy of personal records on health or employment, we suspect that householders would have concerns about confidentiality of the data provided to the interviewer. Interviewers would need quick and intelligible answers to those concerns in order to gain participation of many. In cultures in which universities are labeled as places of radical leftist activity, householders might use that attribute to help judge the attraction of their participating in an academic survey.

The common theme to all this commentary is that multiple influences acting on the householder's decision to participate are highly susceptible to situational factors. Attempting to address all of them in a brilliantly and tightly scripted introduction by an interviewer is near impossible in heterogeneous populations. Instead, design features can be chosen to make some more salient than others. Interviewers need to elicit the concerns of individual sample householders and address them to gain their participation. In free societies, this will always be an imperfect action. Thus, we judge that elimination of nonresponse through survey design is generally impossible. Hence the interviewer needs to be employed as both an agent to collect survey data and also to collect data useful for postsurvey adjustment. The articulation of that role for each survey is a new design task for researchers.

References

Abelson, R.P. (1981). "Psychological Status of the Script Concept." *American Psychologist,* 36, 7, 715–729.

Abramson, P.R. (1983). *Political Attitudes in America.* San Francisco: W.H. Freeman.

Abramson, P.R., and Aldrich, J.H. (1982). "The Decline of Electoral Participation in America." *American Political Science Review,* 76, 502–521.

Aiello, J.R., and Baum, A. (eds.) (1979). *Residential Crowding and Design.* New York: Plenum Press.

Allen, I.L., and Colfax, J.D. (1968). "Respondents' Attitudes Toward Legitimate Surveys in Four Cities." *Journal of Marketing Research,* 5, 431–433.

American Statistical Association (1974). "Report on the ASA Conference on Surveys of Human Populations." *American Statistician,* 28, 1, 30–34.

Argyle, M. (1992). *The Social Psychology of Everyday Life.* New York: Routledge.

Armstrong, J.S. (1975). "Monetary Incentives in Mail Surveys." *Public Opinion Quarterly,* 39, 111–116.

Atchley, R.C. (1969). "Respondents vs. Refusers in an Interview Study of Retired Women: An Analysis of Selected Characteristics." *Journal of Gerontology,* 24, 42–47.

Axelrod, M., and Cannell, C.F. (1959). "A Research Note on an Attempt to Predict Interviewer Effectiveness." *Public Opinion Quarterly,* 23, 4, 571–576.

Baim, J. (1991). "Response Rates: A Multinational Perspective." *Marketing and Research Today,* 19, 2, 114–119.

Bank, M.J.A., and Landsmeer, M.M. (1990). "The Human Factor in Data Collection." *Proceedings of the EMAC/ESOMAR Symposium on New Ways in Marketing and Marketing Research.* Amsterdam: ESOMAR.

Barker, R.F. (1987). "A Demographic Profile of Marketing Research Interviewers." *Journal of the Market Research Society,* 29, 3, 279–292.

Barnes, R., and Birch, F. (1975). "The Census as an Aid in Estimating the Characteristics of Non-response in the GHS." *New Methodology Series,* No. NM 1, London: Office of Population Censuses and Surveys.

Baum, A., and Valins, S. (1977). *Architecture and Social Behavior: Psychological Studies of Social Density.* Hillsdale, NJ: Erlbaum.

Baum, A., and Valins, S. (1979). "Architectural Mediation of Residential Density and Control: Crowding and the Regulation of Social Contact." In L. Berkowitz (ed.), *Advances in Experimental Social Psychology,* Vol. 12. New York: Academic Press, pp. 131–175.

Baumgartner, R., and Rathbun, P. (1997). "Prepaid Monetary Incentives and Mail Survey Response Rates." Paper presented at the annual conference of the American Association for Public Opinion Research, Norfolk, VA.

Baxter, R. (1964). "An Inquiry into the Misuse of the Survey Technique by Sales Solicitors." *Public Opinion Quarterly,* 28, 124–134.

Belyea, M., and Zingraff, M.T. (1988). "Fear of Crime and Residential Location." *Rural Sociology,* 53, 4, 475–486.

Bennett, S.E., and Bennett, L.M. (1986). "Political Participation." In S. Long (ed.), *Annual Review of Political Science,* Vol. 1, Norwood, NJ: Ablex, pp. 157–204.

Benson, S., Booman, W.P., and Clark, K.E. (1951). "A Study of Interview Refusals." *Journal of Applied Psychology,* 35, 116–119.

Benus, J., and Ackerman, J.C. (1971). "The Problem of Nonresponse in Sample Surveys." In J. B. Lansing *et al.* (eds.), *Working Papers on Survey Research in Poverty Areas.* Ann Arbor: Survey Research Center, University of Michigan.

Bergman, L.R., Hanve, R., and Rapp, J. (1978). "Why Do Some People Refuse to Participate in Interview Surveys?" *Statitisk Tidskrift,* 5, 341–356.

Berk, M.L., Mathiowetz, N.A., Ward, E.P., and White, A.A. (1987). "The Effect of Prepaid and Promised Incentives: Results of a Controlled Experiment." *Journal of Official Statistics,* 3, 4, 449–457.

Berlin, M., Mohadjer, L., Waksberg, J., Kolstad, A., Kirsch, I., Rock, D., and Yamamoto, K. (1992). "An Experiment in Monetary Incentives." *Proceedings of the Section on Survey Research Methods,* American Statistical Association, pp. 393–398.

Birnbaum, Z.W., and Sirken, M.G. (1950). "On the Total Error Due to NonInterview and to Random Sampling." *International Journal of Opinion and Attitude Research,* 4, 2, 171–191.

Blau, P.M. (1964). *Exchange and Power in Social Life.* New York: Wiley.

Bogen, K. (1996). "The Effect of Questionnaire Length on Response Rates—A Review of the Literature." Paper presented at the annual conference of the American Association for Public Opinion Research, Salt Lake City, Utah.

Booker, H.S., and David, S.T. (1952). "Differences in Results Obtained by Experienced and Inexperienced Interviewers." *Journal of the Royal Statistical Society,* Series A, 114, 232–257.

Botman, S.L., and Thornberry, O.T. (1992). "Survey Design Features Correlates of Nonresponse." *Proceedings of the Section on Survey Research Methods,* American Statistical Association, pp. 309–314.

Bower, G.H., Black, J.B., and Turner, T.J. (1979). "Scripts in Memory for Text," *Cognitive Psychology,* 11, 177–220.

Bradburn, N.M. (1978). "Respondent Burden." Paper presented at the annual meetings of the American Statistical Association, San Diego, CA.

Bradburn, N.M. (1992). "A Response to the Nonresponse Problem." *Public Opinion Quarterly,* 56, 3, 391–397.

Braver, S.L., and Cialdini, R.B. (1994). "Consistency of Nonresponse to Unrelated Survey Requests." Paper presented at the Fifth International Workshop on Household Survey Nonresponse, Ottawa, Canada.

Brehm, J. (1993). *The Phantom Respondents; Opinion Surveys and Political Representation.* Ann Arbor: University of Michigan Press.

Brehm, J. (1994). "Stubbing Our Toes for a Foot in the Door? Prior Contact, Incentives and Survey Response." *International Journal of Public Opinion Research,* 6, 1, 45–63.

Brehm, J.W., and Cole, A. (1966). "Effect of a Favor Which Reduces Freedom." *Journal of Personality and Social Psychology,* 3, 420–426.

Brenner, M. (1982). "Response-effects of "Role-restricted" Characteristics of the Interviewer." In Dijkstra, W., and van der Zouwen, J. (eds.), *Response Behaviour in the Survey-Interview.* London: Academic Press, pp. 131–165.

Brown, M. (1994). "What Price Response?." *Journal of the Market Research Society,* 36, 3, 227–244.

Brown, P.R., and Bishop, G.F. (1982). "Who Refuses and Resists in Telephone Surveys? Some New Evidence." Paper presented at the annual conference of the Midwest Association for Public Opinion Research.

Brunner, G.A., and Carroll, S.J. (1969). "The Effect of Prior Notification on the Refusal Rate in Fixed Address Surveys." *Journal of Marketing Research,* 9, 42–44.

Bryk, A.S., and Raudenbush, S.W. (1992). *Hierarchical Linear Models: Applications and Data Analysis Methods.* Newbury Park, CA: Sage.

Butz, W.P. (1985). "Data Confidentiality and Public Perceptions: The Case of the European Censuses." *Proceedings of the Section on Survey Research Methods,* American Statistical Association, pp. 90–97.

Caccioppo, J.T., and Petty, R.E. (1985). "Central and Peripheral Routes to Persuasion: the Role of Message Repetition." In Alwitt, L.F., and Mitchell, A.A. (eds.), *Psychological Processes and Advertising Effects.* Hillsdale, NJ: Erlbaum, pp. 91–111.

Campanelli, P., Sturgis, P., and Purdon, S. (1997). "Can You Hear Me Knocking: An Investigation into the Impact of Interviewers on Survey Response Rates." London: The Survey Methods Centre at SCPR.

Cannell, C.F., Groves, R.M., Magilavy, L., Mathiowetz, N.A., and Miller, P.V. (1987). *An Experimental Comparison of Telephone and Personal Health Surveys.* National Center for Health Statistics: Technical Series 2, No. 106.

Cartwright, A. (1959). "The Families and Individuals Who Did Not Cooperate on a Sample Survey." *The Milbank Memorial Fund Quarterly,* 37, 347–368.

Chaiken, S. (1980). "Heuristic Versus Systematic Information Processing and the Use of Source Versus Message Cues in Persuasion." *Journal of Personality and Social Psychology,* 37, 1387–1397.

Chaiken, S. (1987). "The Heuristic Model of Persuasion." In Zanna, M.P. *et al.* (eds.), *Social Influence: The Ontario Symposium,* Volume 5. Hillsdale, NJ: Erlbaum.

Champion, D.J., and Sear, A.M. (1969). "Questionnaire Response Rate: A Methodological Analysis." *Social Forces,* 47, 335–339.

Christianson, A. (1991). "Nonresponse Research Within the Swedish TV Audience Surveys 1969–1985." Paper presented at the Second International Workshop on Household Survey Nonresponse, Washington, D.C.

Cialdini, R.B. (1984). *Influence: The New Psychology of Modern Persuasion.* New York: Quill.

Cialdini, R.B. (1990). "Deriving Psychological Concepts Relevant to Survey Participation from the Literatures on Compliance, Helping, and Persuasion." Paper presented at the First International Workshop on Household Survey Nonresponse, Stockholm.

Cialdini, R.B., Vincent, J.E., Lewis, S.K., Catalan, J., Wheeler, D., and Darby, B.L. (1975). "Reciprocal Concessions Procedure for Inducing Compliance: The Door-in-the-face Technique." *Journal of Personality and Social Psychology,* 31, 206–215.

Cialdini, R.B., Braver, S.L., and Wolf, W.S. (1991). "A New Paradigm for Experiments on the Causes of Survey Nonresponse." Paper presented at the Second International Workshop on Household Survey Nonresponse, Washington, D.C.

Cialdini, R.B., Braver, S.L., and Wolf, W.S. (1992). "Who Says No to Legitimate Survey Requests? Evidence from a New Method for Studying the Causes of Survey Nonresponse." Paper presented at the Third International Workshop on Household Survey Nonresponse, The Hague.

Collins, M., Sykes, W., Wilson, P., and Blackshaw, N. (1988). "Nonresponse: The UK Experience." Chapter 13 in Groves, R.M., Biemer, P.P., Lyberg, L.E., Massey, J.T., Nicholls, W.L., and Waksberg, J. (eds.), *Telephone Survey Methodology.* New York: Wiley.

Comstock, G.W., and Helsing, K.J. (1973). "Characteristics of Respondents and Nonrespondents to a Questionnaire for Estimating Community Mood." *American Journal of Epidemiology,* 97, 4, 233–239.

Conway, M.M. (1991). *Political Participation in the United States,* 2nd edition. Washington, DC: Congressional Quarterly.

Council for Marketing and Opinion Research (1996). *Respondent Cooperation and Industry Image Survey.* Port Jefferson, NY: CMOR.

Couper, M.P. (1997). "Survey Introductions and Data Quality." *Public Opinion Quarterly,* 61, 2, 317–338.

Couper, M.P., and Burt, G. (1994). "Interviewer Attitudes Toward Computer-Assisted Personal Interviewing (CAPI)." *Social Science Computer Review,* 12, 1, 38–54.

Couper, M.P., and Groves, R.M. (1992a). "The Role of the Interviewer in Survey Participation." *Survey Methodology,* 18, 2, 263–271.

Couper, M.P., and Groves, R.M. (1992b). "Survey-Census Match Procedures." Unpublished memorandum. Washington, DC: Bureau of the Census.

Couper, M.P., and Groves, R.M. (1995). "Introductions in Telephone Surveys and Nonresponse." Paper presented at the Workshop on Interaction in the Standardized Survey Interview, Amsterdam.

Couper, M.P., and Groves, R.M. (1996). "Household-level Determinants of Survey Nonresponse." In M. Braverman, and J.K. Slater (eds.), *Advances in Survey Research,* San Francisco: Jossey-Bass.

Couper, M.P., Groves, R.M., and Raghunathan, T.E. (1996). "Nonresponse in the Second Wave of a Longitudinal Survey." Paper presented at the Seventh International Workshop on Household Survey Nonresponse, Rome.

Couper, M.P., Mathiowetz, N.A., and Singer, E. (1995). "Related Households, Mail Handling, and Returns to the 1990 Census." *International Journal of Public Opinion Research,* 7, 2, 172–177.

Couper, M.P., Singer, E., and Kulka, R.A. (1997). "Participation in the Decennial Census: Politics, Privacy, Pressures." *American Politics Quarterly,* 26, 59–80.

Couper, M.P., and Tremblay, A. (1991). "A Comparative Analysis of Interviewer Behavior Across Organizations." Paper presented at the Second International Workshop on Household Survey Nonresponse, Washington, DC.

Covington, J., and Taylor, R.B. (1991). "Fear of Crime in Urban Residential Neighborhoods: Implications of Between- and Within-neighborhood Sources for Current Models." *Sociological Quarterly,* 32, 2, 231–249.

Cummings, E., and Henry, W.E. (1961). *Growing Old: The Process of Disengagement.* New York: Basic Books.

Dalenius, T. (1988). *Controlling Invasion of Privacy in Surveys.* Stockholm: Statistics Sweden.

Davis, J.A., and Smith, T.W. (1990). *General Social Surveys, 1972–1990: Cumulative Codebook.* Chicago: National Opinion Research Center.

de Heer, W.F. (1992). "International Survey on Nonresponse: Report on the First Data-collection Round." Paper presented at the Third International Workshop on Household Survey Nonresponse, The Hague.

de Heer, W.F., and Israëls, A.Z. (1992). "Response Trends in Europe." Paper presented at the Joint Statistical Meetings of the American Statistical Association, Boston.

DeMaio, T.J. (1980). "Refusals: Who, Where and Why?." *Public Opinion Quarterly,* 44, 223–233.

Deming, W.E. (1953). "On a Probability Mechanism to Attain an Economic Balance Between the Resultant Error of Response and the Bias of Nonresponse." *Journal of the American Statistical Association,* 48, 264, 743–772.

Dillman, D. (1978). *Mail and Telephone Surveys.* New York: Wiley.

Dillman, D. (1991). "The Design and Administration of Mail Surveys." *Annual Review of Sociology,* 17, 225–249.

Dillman, D.A., Gallegos, J.G., and Frey, J.H. (1976). "Reducing Refusal Rates for Telephone Interviews." *Public Opinion Quarterly,* 40, 1, 66–78.

Dohrenwend, B., and Dohrenwend, B.P. (1968). "Sources of Refusals in Surveys." *Public Opinion Quarterly,* 32, 1, 74–83.

Dufour, J., Simard, M., and Mayda, F. (1995). "The First Year of Computer-assisted Interviewing for the Canadian Labour Force Survey: An Update." Paper presented at the Sixth International Workshop on Household Survey Nonresponse, Helsinki.

Dunkelberg, W.C., and Day, G.S. (1973). "Nonresponse Bias and Callbacks in Sample Surveys." *Journal of Marketing Research,* 10, May, 160–168.

Durbin, J., and Stuart, A. (1951). "Differences in Response Rates of Experienced and Inexperienced Interviewers." *Journal of the Royal Statistical Society,* Series A, 114, 163–206.

Eagly, A.H., and Chaiken, S. (1984). "Cognitive Theories of Persuasion." In Berkowitz, L. (ed.), *Advances in Experimental Social Psychology,* Volume 17. San Diego: Academic Press, pp. 267–359.

Ekholm, A., and Laaksonen, S. (1990). "Reweighting by Nonresponse Modeling in the Finnish Household Survey." Helsinki: Department of Statistics (Research Report No. 68).

Ekholm, A., and Laaksonen, S. (1991). "Weighting via Response Modeling in the Finnish Household Budget Survey." *Journal of Official Statistics,* 7, 3, 325–337.

Eltinge, J.L., and Rope, D.J. (1994). "Use of Decennial Census Match Data to Estimate

Noninterview and Attition Models for the U.S. Consumer Expenditure Survey." Paper presented at the Fifth International Workshop on Household Survey Nonresponse, Ottawa, Canada.

Everett, S.E., and Everett, S.C. (1989). "Effects of Interviewer Affiliation and Sex Upon Telephone Survey Refusal Rates." Paper presented at the annual conference of the Midwest Association for Public Opinion Research, Chicago.

Fay, R.L., Bates, N., and Moore, J. (1991). "Lower Mail Response in the 1990 Census: A Preliminary Interpretation." *Proceedings of the Bureau of Census Annual Research Conference.* Washington, DC: U.S. Bureau of the Census, pp. 3–32.

Ferber, R., and Sudman, S. (1974). "A Comparison of Alternative Procedures for Collecting Consumer Expenditure Data for Frequently Purchased Products." *Journal of Marketing Research,* 11, 2, 128–135.

Fischer, C. (1973). "On Urban Alienation and Anomie: Powerlessness and Social Isolation." *American Sociological Review,* 38, 311–326.

Fischer, C. (1982). *To Dwell Among Friends: Personal Networks in Town and City.* Chicago: University of Chicago Press.

Fitzgerald, R., and Fuller, L. (1982). "I Hear You Knocking but You Can't Come In: The Effects of Reluctant Respondents and Refusers on Sample Survey Estimates." *Sociological Methods and Research,* 11, 1, 3–32.

Forsman, G. (1993). "Sampling Individuals Within Households in Telephone Surveys." Paper presented at the annual conference of the American Association for Public Opinion Research, St. Charles, IL.

Foster, K., and Bushnell, D. (1994). "Non-response Bias on Government Surveys in Great Britain." Paper presented at the Fifth International Workshop on Household Survey Nonresponse, Ottawa, Canada.

Fowler, F.J., and Mangione, T.W. (1990). *Standardized Survey Interviewing; Minimizing Interviewer-Related Error.* Newbury Park: Sage.

Franck, K.A. (1980). "Friends and Strangers: The Social Experience of Living in Urban and Non-urban Settings." *Journal of Social Issues,* 36, 3, 52–71.

Frankel, J., and Sharp, L.M. (1981). "Measurement of Respondent Burden." *Statistical Reporter,* 81, 4, 105–111.

Freedman, L.J., and Fraser, S.C. (1966). "Compliance Without Pressure: The Foot-in-the-door Technique," *Journal of Personality and Social Psychology,* 4, 195–202.

Glenn, N.D. (1969). "Aging, Disengagement, and Opinionation." *Public Opinion Quarterly,* 33, 17–33.

Glorioux, I. (1993). "Social Interaction and the Social Meanings of Action: A Time Budget Approach." *Social Indicators Research,* 30, 149–173.

Goetz, E.G., Ryler, T.R., and Cook, F.L. (1984). "Promised Incentives in Media Research: A Look at Data Quality, Sample Representativeness, and Response Rate." *Journal of Marketing Research,* 21, 148–154.

Goldstein, H. (1987). *Multilevel Models in Educational and Social Research.* New York: Oxford University Press.

Gonzalez, M.E., Kasprzyk, D., and Scheuren, F. (1995). "Exploring Nonresponse in U.S. Federal Government Surveys." *Seminar on New Directions in Statistical Methodology.* Washington, DC: Office of Management and Budget (Statistical Policy Working Paper No. 23). pp. 603–624.

Gonzenbach, W.J., and Jablonski, P. (1993). "The Non-Solitation Statement: A Methodological Consideration for Survey Introductions." Paper presented at the annual conference of the American Association for Public Opinion Research, St. Charles, IL.

Goolsby, J.R., Lagace, R.R., and Boorom, M.L. (1992). "Psychological Adaptiveness and Sales Performance." *Journal of Personal Selling and Sales Management,* 12, 2, 51–66.

Gouldner, A.W. (1960). "The Norm of Reciprocity: A Preliminary Statement." *American Sociological Review,* 25, 161–178.

Gower, A.R. (1979). "Non-response in the Canadian Labour Force Survey." *Survey Methodology,* 5, 1, 29–58.

Goyder, J. (1986). "Surveys on Surveys: Limitations and Potentialities." *Public Opinion Quarterly,* 50, 1, 27–41.

Goyder, J. (1987). *The Silent Minority; Nonrespondents on Sample Surveys.* Boulder, CO: Westview Press.

Goyder, J., and Leiper, J.M. (1985). "The Decline in Survey Response: A Social Values Interpretation." *Sociology,* 19, 1, 55–71.

Goyder, J., Lock, J., and McNair, T. (1992). "Urbanization Effects on Survey Nonresponse: A Test Within and Across Cities." *Quality and Quantity,* 26, 39–48.

Groves, R.M. (1989). *Survey Errors and Survey Costs.* New York: Wiley.

Groves, R.M. (1990). "Theories and Methods of Telephone Surveys." *Annual Review of Sociology,* 16, 221–240.

Groves, R.M., Cialdini, R.B., and Couper, M.P. (1992). "Understanding the Decision to Participate in a Survey." *Public Opinion Quarterly,* 56, 4, 475–495.

Groves, R.M., and Couper, M.P. (1993a). "Unit Nonresponse in Demographic Surveys." *Proceedings of the Bureau of the Census 1993 Annual Research Conference.* Washington: Bureau of the Census, pp. 593–619.

Groves, R.M., and Couper, M.P. (1993b). "Multivariate Analysis of Nonresponse in Personal Visit Surveys." *Proceedings of the Section on Survey Research Methods, American Statistical Association.* Alexandria: ASA, pp. 514–519.

Groves, R.M., and Fultz, N.H. (1985). "Gender Effects among Telephone Interviewers in a Survey of Economic Attitudes." *Sociological Methods and Research,* 14, 1, 31–52.

Groves, R.M., and Kahn, R.L. (1979). *Surveys by Telephone: A National Comparison with Personal Interviews.* New York: Academic Press.

Guadagnoli, E., and Cunningham, S. (1989). "The Effects of Nonresponse and Late Response on a Survey of Physician Attitudes." *Evaluation and the Health Professions,* 13, 3, 318–328.

Hagenaars, J.A., and Heinen, T.G. (1982). "Effects of Role-independent Interviewer Characteristics on Responses." In Dijkstra, W., and van der Zouwen, J. (eds.), *Response Behaviour in the Survey-Interview.* London: Academic Press, pp. 91–130.

Hansen, M.H., and Hurwitz, W.M. (1958). "The Problem of Nonresponse in Sample Surveys." *Journal of the American Statistical Association,* December, 517–529.

Hartley, H.O. (1974). "Multiple Frame Methodology and Selected Applications." *Sankhya, Series C,* 36, 3, 99–118.

Hawkins, D.F. (1975). "Estimation of Nonresponse Bias." *Sociological Methods and Research,* 3, 461–488.

Heckman, J.J. (1979). "Sample Selection Bias as a Specification Error." *Econometrica,* 47, 1, 153–161.

Heiskanen, M., and Laaksonen, S. (1995). "Non-response at the SLC and the Depression Trap." Paper presented at the Sixth International Workshop on Household Survey Nonresponse, Helsinki.

Heeringa, S. (1992). "National Survey of Americans' Mental Health, Study Sample Design." Ann Arbor, MI: Survey Research Center, University of Michigan (unpublished report).

Herzog, A.R., and Rodgers, W.L. (1988). "Age and Response Rates to Interview Sample Surveys." *Journal of Gerontology: Social Sciences,* 43, S200–S205.

Hidiroglou, M.A., Drew, J.D., and Gray, G.B. (1993). "A Framework for Measuring and Reducing Nonresponse in Surveys." *Survey Methodology,* 19, 1, 81–94.

Hill, D.H. (1978). "Home Production and the Residential Electric Load Curve." *Resources and Energy,* 1, 339–358.

Hippler, H.-J., Schwarz, N., and Sudman, S. (eds.) (1987). *Social Information Processing and Survey Methodology.* New York: Springer-Verlag.

Hoinville, G., Jowell, R., and Associates (1977). *Survey Research Practice.* London: Heinemann.

Homans, G.C. (1961). *Social Behavior.* New York: Harcourt, Brace, and World.

Hosmer, D.W., and Lemeshow, S. (1989). *Applied Logistic Regression.* New York: Wiley Interscience.

House, J.S., and Wolf, S. (1978). "Effects of Urban Residence on Interpersonal Trust and Helping Behavior." *Journal of Personality and Social Psychology,* 36, 9, 1029–1043.

Hox, J.J. (1995). *Applied Multilevel Analysis,* 2nd Edition. Amsterdam: TT-Publikaties.

Hox, J.J., de Leeuw, E.D., and Vorst, H. (1995). "Survey Participation as Reasoned Action: A Behavioral Paradigm for Survey Nonresponse?" *Bulletin de Méthodologie Sociologique,* 47, 52–67.

Hyman, H. (1954). *Interviewing in Social Research.* Chicago: University of Chicago Press.

Inderfurth, G.P. (1972). *Investigation of Census Bureau Interviewer Characteristics, Performance, and Attitudes: A Summary.* Working Paper No. 3, Washington, DC: U.S. Bureau of the Census.

Iutcovich, J., and Cox, H. (1990). "Fear of Crime among the Elderly—Is it Justified?" *Journal of Applied Psychology,* 7, 63–76.

Jackson, D.N. (1976). *Jackson Personality Inventory Manual.* Goshen, NY: Research Psychologists Press.

James, J.M., and Bolstein, R. (1990). "The Effect of Monetary Incentives and Follow-up Mailings on the Response Rate and Response Quality in Mail Surveys." *Public Opinion Quarterly,* 54, 346–361.

Jay, G.M., Liang, J., Liu, X., and Sugisawa, H. (1993). "Patterns of Nonresponse in a National Survey of Elderly Japanese." *Journal of Gerontology: Social Sciences,* 48, 3, S143–S152.

Jennings, M.K., and Markus, G.B. (1988). "Political Involvement in the Later Years: A Longitudinal Survey." *American Journal of Political Science,* 32, 302–316.

Johnson, R., and Price, Y. (1988). "The Relationship of the Jackson Personality Traits of U.S. Census Bureau Current Survey Interviewers to Work Performance and Turnover." Washington, DC: U.S. Bureau of the Census (unpublished internal memorandum).

Juster, F.T., and Stafford, F.P. (eds.) (1985). *Time, Goods, and Well-being.* Ann Arbor, MI: Institute for Social Research, University of Michigan.

Kahn, R.L., and Cannell, C.F. (1957). *The Dynamics of Interviewing.* New York: Wiley.

Kalton, G. (1983). *Compensating for Missing Survey Data.* Ann Arbor: Institute for Social Research, University of Michigan.

Kalton, G., and Citro, C.F. (1993). "Panel Surveys: Adding the Fourth Dimension." *Survey Methodology,* 19: 205–215.

Kalton, G., Lepkowski, J.M., Montanari, G.E., and Maligalig, D. (1990). "Characteristics of Second Wave Nonrespondents in a Panel Survey." *Proceedings of the Survey Research Methods Section,* American Statistical Association, pp. 462–467.

Kanuk, L., and Berenson, C. (1975). "Mail Surveys and Response Rates: A Literature Review." *Journal of Marketing Research,* 12, 440–453.

Kemsley, W.F.F. (1975). "Family Expenditure Survey. A Study of Differential Response Based on a Comparison of the 1971 Sample with the Census." *Statistical News,* 31, 16–21.

Kemsley, W.F.F. (1976). "National Food Survey. A Study of Differential Response Based on a Comparison of the 1971 Sample with the Census." *Statistical News,* 35, 18–22.

Kennickell, A.B. (1997). "Analysis of Nonresponse Effects in the 1995 Survey of Consumer Finances." Paper presented at the annual meetings of the American Statistical Association, Anaheim, CA.

Kessler, R.C., McGonagle, K.A., Zhao, S., Nelson, C.B., Hughes, M., Eshleman, S., Wittchen, H.-U., and Kendler, K.S. (1994). "Lifetime and 12-Month Prevalence of DSM-III-R Psychiatric Disorders in the United States." *Archives of General Psychiatry,* 51, 8–19.

Knack, S. (1990). "Why We Don't Vote—or Say 'Thank You'." *Wall Street Journal.* December 31.

Knack, S. (1992). "Civic Norms, Social Sanctions, and Voter Turnout." *Rationality and Society,* 4, 2, 133–156.

Koch, A. (1991). "Zum Zusammenhang von Interviewermerkmalen und Ausschöpfungsquoten." *ZUMA Nachrichten No. 28.* Mannheim, Germany: Zentrum für Umfragen, Methoden und Analysen.

Kojetin, B., Tucker, C., and Cashman, E. (1994). "Response to a Government Survey as Political Participation: The Relation of Economic and Political Conditions to Refusal Rates in the Current Population Survey." Paper presented at the annual conference of the American Association for Public Opinion Research, Danvers, MA.

Korte, C., and Kerr, N. (1975). "Responses to Altruistic Opportunities in Urban and Nonurban Settings." *Journal of Social Psychology,* 95, 183–184.

Korte, C., Ypma, I., and Toppen, A. (1975). "Helpfulness in Dutch Society as a Function of Urbanization and Environmental Input Level." *Journal of Personality and Social Psychology,* 32, 996–1003.

Krause, N. (1993). "Neighborhood Deterioration and Social Isolation in Later Life." *International Journal of Aging and Human Development,* 36, 1, 9–38.

Kriastiansson, K.E. (1980). "A Non-response Study in the Swedish Labour Force Surveys." In Methodological Studies from the Research Institute for Statistics on Living Conditions. Stockholm, Sweden, National Central Bureau of Statistics (report no. 10E).

Kulka, R.A., Holt, N.A., Carter, W., and Dowd, K.L. (1991). "Self Reports of Time Pressures, Concerns for Privacy, and Participation in the 1990 Mail Census." *Proceedings of the Bureau of the Census Annual Research Conference.* Washington, DC: U.S. Bureau of the Census, pp. 33–54.

Lassk, F.G., Kennedy, K.N., Powell, C.M., and Lagace, R.R. (1992). "Psychological Adaptiveness and Sales Managers' Job Performance." *Journal of Social Behavior and Personality,* 7, 4, 611–620.

Lavin, S.A. (1989). "Results of the Nonsampling Error Measurement Survey of Field Interviewers." Washington, DC: U.S. Bureau of the Census (unpublished report).

Lavrakas, P.J. (1982). "Fear of Crime and Behavioral Restrictions in Urban and Suburban Neighborhoods." *Population and Environment,* 5, 4, 242–264.

Lavrakas, P.J., and Merkle, D.M. (1991). "A Reversal of Roles: When Respondents Question Interviewers." Paper presented at the Midwest Association for Public Opinion Research conference, Chicago.

Lehtonen, R. (1995). "Unit Nonresponse and Interviewer Effect in Two Different Interviewing Schemes for the 1995 FHS Survey." Paper presented at the Sixth International Workshop on Household Survey Nonresponse, Helsinki.

Lengacher, J.E., Sullivan, C.M., Couper, M.P., and Groves, R.M. (1995). "Once Reluctant, Always Reluctant? Effects of Differential Incentives on Later Survey Participation in a Longitudinal Study." *Proceedings of the Section on Survey Research Methods,* American Statistical Association, pp. 1029–1034.

Lennox, R.D., and Wolfe, R.N. (1984). "Revision of the Self-monitoring Scale." *Journal of Personality and Social Psychology,* 46, 6, 1349–1364.

Lepkowski, J.M. (1988). "The Treatment of Wave Nonresponse in Panel Surveys." In Kasprzyk, D., Duncan, G., and Singh, M.P. (eds.), *Panel Survey Design and Analysis.* New York: Wiley.

Lepkowski, J.M., and Groves, R.M. (1986). "A Mean Squared Error Model for Dual Frame, Mixed Mode Survey Design." *Journal of the American Statistical Association,* 81, 396, 930–937.

Lessler, J.T., and Kalsbeek, W.D. (1992). *Nonsampling Error in Surveys.* New York: Wiley.

Levy, M., and Sharma, A. (1994). "Adaptive Selling: The Role of Gender, Age, Sales Experience, and Education." *Journal of Business Research,* 31, 39–47.

Lievesley, D. (1988). "Unit Non-response in Interview Surveys." Unpublished working paper, London: Social and Community Planning Research.

Lin, I-F., and Schaeffer, N.C. (1995). "Using Survey Participants to Estimate the Impact of Nonparticipation." *Public Opinion Quarterly,* 59, 2, 236–258.

Lindström, H.L. (1983). *Non-Response Errors in Sample Surveys.* Urval, No. 16. Orebro: Statistics Sweden.

Lindström, H.L. (1986). *A General View of Nonresponse Bias in Some Sample Surveys of the Swedish Population.* Report No. 23. Orebro: Statistics Sweden.

Lindström, H.L., and Dean, P. (1986). *Nonresponse Rates in 1970–1985 in Surveys of Individuals and Households.* Report No. 243. Orebro: Statistics Sweden.

Liska, A.E., and Baccaglini, W. (1990). "Feeling Safe by Comparison." *Social Problems,* 37, 3, 360–374.

Little, R., and Rubin, D.B. (1987). *Statistical Analysis with Missing Data.* New York: Wiley.

Luevano, P. (1994). "Response Rates in the National Election Studies, 1948–1992." NES Technical Report #44. Ann Arbor: Center for Political Studies.

Luiten, A., and de Heer, W. (1993). "Controlling the Fieldwork: Influence on Response-Results." Paper presented at the Fourth International Workshop on Household Survey Nonresponse, Bath, England.

Luiten, A., and de Heer, W. (1994). "International Questionnaire and Itemlist 'Fieldwork Strategy'." Paper presented at the Fifth International Workshop on Household Survey Nonresponse, Ottawa.

Luppes, M. (1993). "A Content Analysis of Advance Letters from Budget Surveys of Seven Countries." Paper presented at the Fourth International Workshop on Household Survey Nonresponse, Bath, England.

Luppes, M. (1994). "Interpretation and Evaluation of Advance Letters." Paper presented at the Fifth International Workshop on Household Survey Nonresponse, Ottawa.

Lutyńska, K. (1987). "Questionnaire Studies in Poland in the 1980s: Analysis of Refusals to Give an Interview." *Polish Sociological Bulletin,* 3, 43–53.

Lyberg, L., and Dean, P. (1992). "Methods for Reducing Nonresponse Rates: A Review." Paper presented at the annual conference of the American Association for Public Opinion Research, St. Petersburg, FL.

McFarlane Smith, J. (1972). *Interviewing in Marketing and Social Research.* London: Routledge and Kegan Paul.

Madow, W.G., Nisselson, H., and Olkin, I. (1983). *Incomplete Data in Sample Surveys,* Vol. I: Report and Case Studies. New York: Academic Press.

Market Research Society (1976). "Response Rates in Sample Surveys. Report of a Working Party of the Market Research Society's Research and Development Committee." *Journal of the Market Research Society,* 18, 3, 113–142.

Market Research Society (1981). "Report of the Second Working Party on Respondent Cooperation: 1977–1980." *Journal of the Market Research Society,* 23, 3–25.

Marton, A. (1995). "Nonresponse in the Hungarian Household Surveys." Paper presented at the Sixth International Workshop on Household Survey Nonresponse, Helsinki.

Mathiowetz, N.A., DeMaio, T.J., and Martin, E. (1991). "Political Alienation, Voter Registration and the 1990 Census." Paper presented at the annual conference of the American Association for Public Opinion Research, Phoenix, AZ.

Maynard, D.W., Schaeffer, N.C., and Cradock, R.M. (1993). "Declinations of the Request to Participate in the Survey Interview." Working Paper No. 93–23, Madison: University of Wisconsin, Center for Demography and Ecology.

Maynard, D.W., Schaeffer, N.C., and Cradock, R.M. (1995). "A Preliminary Analysis of 'Gatekeeping' as a Feature of Declinations to Participate in the Survey Interview." Unpublished paper, Madison: University of Wisconsin.

McCarthy, D.P., and Saegert, S. (1979). "Residential Density, Social Overload, and Social Withdrawal." In J.R. Aiello and A. Baum (eds.). *Residential Crowding and Design.* New York: Plenum Press, pp. 55–75.

McCrossan, L. (1991). *A Handbook for Interviewers,* 3rd Edition. London: Office of Population Censuses and Surveys, Social Survey Division (HMSO No. MI34).

McCrossan, L. (1992). "Training and Management of Interviewers on British Government Surveys." Paper presented at the Third International Workshop on Household Survey Nonresponse, The Hague.

McCrossan, L. (1993). "Respondent-Interviewer Interactions in Survey Introductions." Paper presented at the Fourth International Workshop on Household Survey Nonresponse, Bath, England.

McGuire, W.J. (1985). "Attitudes and Attitude Change." In Lindzey, G. and Aronson, E. (eds.), *Handbook of Social Psychology,* 3rd Edition, Volume 2. New York: Random House, pp. 233–346.

Meier, E. (1991). "Response Rate Trends in Britain." *Marketing and Research Today,* 19, 2, 120–123.

Mercer, J.R., and Butler, E.W. (1967). "Disengagement of the Aged Population and Response Differentials in Survey Research." *Social Forces,* 46, 1, 89–96.

Miethe, T.D., and Lee, G.R. (1984). "Fear of Crime Among Older People: A Reassessment of the Predictive Power of Crime-Related Factors." *Sociological Quarterly,* 25, 397–415.

Miethe, T.D., Hughes, M., and McDowall, D. (1991). "Social Change and Crime Rates: An Evaluation of Alternative Theoretical Approaches." *Social Forces,* 70, 1, 165–185.

Milgram, S. (1970). "The Experience of Living in Cities." *Science,* 67, 1461–1468.

Miller, H.W., Kennedy, J., and Bryant, E.E. (1972). "A Study of the Effect of Remuneration Upon Response in a Health and Nutrition Examination Survey." *Proceedings of the Section on Social Statistics,* American Statistical Association, pp. 370–375.

Miller-Steiger, D., and Groves, R.M. (1997). "Interviewer Training Techniques: Current Practice in Survey Organizations." Ann Arbor, MI: Survey Research Center, University of Michigan (report to the National Agricultural Statistics Service).

Moore, D.L., Hausknecht, D., and Thamodaran, K. (1986). "Time Compression, Response Opportunity, and Persuasion." *Journal of Consumer Research,* 13, 85–99.

Morton-Williams, J. (1991). *Obtaining Co-operation in Surveys—The Development of a Social Skills Approach to Interviewer Training in Introducing Surveys.* Working Paper No. 3. London: Joint Centre for Survey Methods.

Morton-Williams, J. (1993). *Interviewer Approaches.* Aldershot: Dartmouth.

Morton-Williams, J., and Young, P. (1987). "Obtaining the Survey Interview—An Analysis of Tape Recorded Doorstep Introductions." *Journal of the Market Research Society,* 29, 1, 35–54.

Mowen, J.C., and Cialdini, R.B. (1980). "On Implementing the Door-in-the-face Compliance Technique in a Business Context." *Journal of Marketing Research,* 17, 253–258.

National Academy of Sciences (1979). *Panel on Privacy and Confidentiality as Factors in Survey Response.* Washington, DC: National Academy Press.

Nederhof, A.J. (1983). "The Effects of Material Incentives in Mail Surveys: Two Studies." *Public Opinion Quarterly,* 47, 1, 103–111.

Norris, P. (1987). "The Labour Force Survey: A Study of Differential Response According to Demographic and Socio-economic Characteristics." *Statistical News,* 79, 20–23.

O'Neil, M.J. (1979). "Estimating the Nonresponse Bias Due to Refusals in Telephone Surveys." *Public Opinion Quarterly,* 43, 218–232.

O'Neil, M.J., Groves, R.M., and Cannell, C.F. (1979). "Telephone Interview Introductions and Refusal Rates: Experiments in Increasing Respondent Cooperation." *Proceedings of the Section on Survey Research Methods,* American Statistical Association, pp. 252–255.

Oakes, G. (1990). *The Soul of the Salesman: The Moral Ethos of Personal Sales.* Atlantic Highlands, NJ: Humanities Press International.

Oksenberg. L., Coleman, L., and Cannell, C.F. (1986). "Interviewers' Voices and Refusal Rates in Telephone Surveys." *Public Opinion Quarterly,* 50, 1, 97–111.

Oldendick, R.W., Bishop, G.W., Sorenson, S., and Tuchfarber, A.J. (1988). "A Comparison of the Kish and Last Birthday Methods of Respondent Selection in Telephone Surveys." *Journal of Official Statistics,* 4, 307–318.

Parker, K.D., and Ray, M.C. (1990). "Fear of Crime: An Assessment of Related Factors." *Sociological Spectrum,* 10, 29–40.

Paul, E.C., and Lawes, M. (1982). "Characteristics of Respondent and Non-respondent Households in the Canadian Labour Force Survey." *Survey Methodology,* 8, 48–85.

Peneff, J. (1988). "The Observers Observed: French Survey Researchers at Work." *Social Problems,* 35, 5, 520–535.

Pennell, S.G. (1990). "Evaluation of Advance Letter Experiment in 1988 New York Reproductive Health Survey." Report to the Centers for Disease Control. Ann Arbor, MI: Survey Research Center.

Petroni, R.J., and Allen, T.M. (1992). "Mover Nonresponse Adjustment Research for the SIPP." Paper presented at the Third International Workshop on Household Survey Nonresponse, The Hague.

Petty, R.E., and Cacioppo, J.T. (1984). "The Effects of Involvement on Response to Argument Quantity and Quality." *Journal of Personality and Social Psychology,* 46, 69–81.

Petty, R.E., and Cacioppo, J.T. (1986). *Communication and Persuasion: Central and Peripheral Routes to Attitude Change.* New York: Springer-Verlag.

Petty, R.E., Wells, G.L., and Brock, T.C. (1976). "Distraction Can Enhance or Reduce Yielding to Propaganda: Thought Disruption Versus Effort Justification." *Journal of Personality and Social Psychology,* 34, 874–884.

Pinkleton, B., Reagan, J., Aaronson, D., and Ramo, E. (1994). "Does 'I'm Not Selling Anything' Increase Response Rates in Telephone Surveys?" Paper presented at the Annual Conference of the American Association for Public Opinion Research, Danvers, MA.

Presser, S., Blair, J., and Triplett, T. (1992). "Survey Sponsorship, Response Rates, and Response Effects." *Social Science Quarterly,* 73, 3, 699–702.

Purdon, S., Campanelli, P., and Sturgis, P. (1996). "Studying Interviewers' Calling Strategies as a Way to Improve Response Rates." Paper presented at the Fourth International ISA Conference on Social Science Methodology, Colchester, UK.

Rao, P.S.R.S. (1983). "Nonresponse and Double Sampling: Randomization Approach." In Madow, W.G., Olkin, I., and Rubin, D.B. (eds.), *Incomplete Data in Sample Surveys,* Volume 2. New York: Academic Press, pp. 97–105.

Rauta, I. (1985). "A Comparison of the Census Characteristics of Respondents and Nonrespondents to the 1981 General Household Survey (GHS)." *Statistical News,* 71, 12–15.

Redpath, B. (1986). "Family Expenditure Survey: A Second Study of Differential Response, Comparing Census Characteristics of FES Respondents and Non-respondents." *Statistical News,* 72, 13–16.

Redpath, B., and Elliot, D. (1988). "National Food Survey: A Second Study of Differential Response, Comparing Census Characteristics of NFS Respondents and Non-respondents; Also a Comparison of NFS and FES Response Bias." *Statistical News,* 80, 6–10.

Regan, D.T. (1971). "Effects of a Favor and Liking on Compliance." *Journal of Experimental Social Psychology,* 7, 627–639.

Remington, T.D. (1992). "Telemarketing and Declining Survey Response Rates." *Journal of Advertising Research,* 32, RC6–RC7.

Robinson, J.P., and Godbey, G. (1997). *Time for Life: The Surprising Way Americans Use Their Time*. University Park, PA: Penn State Press.

Rosenstone, S.J., and Hansen, J.M. (1993). *Mobilization, Participation, and Democracy in America*. New York: Macmillan.

Rubin, D.B. (1987). *Multiple Imputation for Nonresponse in Surveys*. New York: Wiley.

Rucker, R.E. (1990). "Urban Crime: Fear of Victimization and Perceptions of Risk." *Free Inquiry in Creative Sociology*, 18, 2, 151–160.

Salmon, C.T., and Nichols, J.S. (1983). "The Next-Birthday Method of Respondent Selection." *Public Opinion Quarterly*, 47, 270–276.

Sampson, R.J. (1988). "Local Friendship Ties and Community Attachment in Mass Society: A Multilevel Systemic Approach." *American Sociological Review*, 53, 766–779.

Sampson, R.J., and Groves, W.B. (1989). "Community Structure and Crime: Testing Social-disorganization Theory." *American Journal of Sociology*, 94, 4, 774–802.

Schaeffer, N.C. (1990). "Conversation with a Purpose—or Conversation? Interaction in the Standardized Interview." Chapter 19 in Biemer, P.P., Groves, R.M., Lyberg. L.E., Mathiowetz, N.A. and Sudman, S. (eds.), *Measurement Errors in Surveys*. New York: Wiley.

Schiffenbauer, A. (1979). "Designing for High-density Living." In J.R. Aiello and A. Baum (eds.), *Residential Crowding and Design.* New York: Plenum Press, pp. 229–240.

Schwartz, A. (1964). "Interviewing and the Public." *Public Opinion Quarterly*, 28, 135–142.

Schwartz, S.H., and Clausen, G.T. (1970). "Responsibility, Norms, and Helping in an Emergency." *Journal of Personality and Social Psychology*, 15, 4, 283–293.

Schyberger, B.W. (1967). "A study of Interviewer Behavior." *Journal of Marketing Research*, 4, 32–35.

Scott, A.J., and Wild, C.J. (1986). "Fitting Logistic Models under Case-control or Choice Based Sampling." *Journal of the Royal Statistical Society, Series B*, 48, 170–182.

Scott, A.J., and Wild, C.J. (1989). "Selection Based on the Response Variable in Logistic Regression." In Skinner, C.J., Holt, D., and Smith, T.M.F. (eds). *Analysis of Complex Surveys.* New York: Wiley, pp. 191–205.

Scott, J.C. (1978). "Trends in Response Rates in SRC Surveys." Unpublished report, Ann Arbor: Survey Research Section.

Shah, B.V., Barnwell, B.G., and Bieler, G.S. (1996). *SUDAAN User's Manual, Release 7.0.* Research Triangle Park, NC: Research Triangle Institute.

Sheatsley, P.B. (1951). "An Analysis of Interviewer Characteristics and Their Relationship to Performance. Part III." *International Journal of Opinion and Attitude Research*, 5, 191–220.

Shettle, C.F., Guenther, P.M., Kasprzyk, D., and Gonzalez, M.E. (1994). "Investigating Nonresponse in Federal Surveys." *Proceedings of the Survey Research Methods Section, American Statistical Association*, pp. 972–976.

Singer, E., Frankel, M.R., and Glassman, M.B. (1983). "The Effect of Interviewer Characteristics and Expectation on Response." *Public Opinion Quarterly*, 47, 68–83.

Singer, E., Mathiowetz, N.A., and Couper, M.P. (1993). "The Impact of Privacy and Confidentiality Concerns on Survey Participation: The Case of the 1990 Census." *Public Opinion Quarterly*, 57, 465–482.

Singer, E., and Kohnke-Aguirre, L. (1979). "Interviewer Expectation Effects: A Replication and Extension." *Public Opinion Quarterly*, 43, 2, 245–260.

Singer, E., Gebler, N., Raghunathan, T., Van Hoewyk, J., and McGonagle, K. (1996). "The Effects of Incentives on Response Rates in Personal, Telephone, and Mixed-Mode Surveys: Results from a Meta Analysis." Paper presented at the annual conference of the American Association of Public Opinion Research, Salt Lake City, Utah.

Singh, B. (1983). "Nonresponse and Double Sampling: Bayesian Approach." In Madow, W.G., Olkin, I., and Rubin, D.B. (eds.), *Incomplete Data in Sample Surveys,* Volume 2. New York: Academic Press, pp. 107–119.

Skogan, W.G., and Maxfield, M.G. (1981). *Coping With Crime: Individual and Neighborhood Reactions.* Beverly Hills: Sage.

Smith, D.A., and Jarjoura, G.R. (1989). "Household Characteristics, Neighborhood Composition and Victimization Risk." *Social Forces,* 68, 2, 621–640.

Smith, T.W. (1983). "The Hidden 25 Percent: An Analysis of Nonresponse on the 1980 General Social Survey." *Public Opinion Quarterly,* 47, 386–404.

Smith, T.W. (1984). "Estimating Nonresponse Bias with Temporary Refusals." *Sociological Perspectives,* 27, 4, 473–489.

Smith, T.W. (1995). "Trends in Non-response Rates." *International Journal of Public Opinion Research,* 7, 2, 157–171.

Snyder, M. (1974). "Self-monitoring of Expressive Behavior." *Journal of Personality and Social Psychology,* 30, 4, 526–537.

Snyder, M. (1980). "The Many Me's of the Self-monitor." *Psychology Today,* March, 33–40.

Soldo, B.J., Hurd, M.D., Rodgers, W.L., and Wallace, R.B. (1997). "Asset and Health Dynamics Among the Oldest Old: An Overview of the AHEAD Study." *Journals of Gerontology: Psychological and Social Sciences,* 52B, 1–20.

Southwell, P.L. (1985). "Alienation and Nonvoting in the United States: A Refined Operationalization." *Western Political Quarterly,* 38, 663–674.

Spiro, R.L., and Weitz, B.A. (1990). "Adaptive Selling: Conceptualization, Measurement, and Nomological Validity." *Journal of Marketing Research,* 27, 61–69.

Steblay, N.M. (1987). "Helping Behavior in Rural and Urban Environments: A Meta-analysis." *Psychological Bulletin,* 102, 3, 346–356.

Steeh, C.G. (1981). "Trends in Nonresponse Rates, 1952–1979." *Public Opinion Quarterly,* 45, 40–57.

Stokols, D. (1976). "The Experience of Crowding in Primary and Secondary Environments." *Environment and Behavior,* 8, 1, 49–86.

Suchman, L., and Jordan, B. (1990). "Interactional Troubles in Face-to-face Survey Interviews." *Journal of the American Statistical Association,* 85, 232–241.

Sudman, S., and Bradburn, N.M. (1974). *Response Effects in Surveys.* Chicago: Aldine.

Sudman, S., Bradburn, N.M., and Schwarz, N. (1996). *Thinking about Answers.* San Francisco: Jossey-Bass.

Sudman, S., Bradburn, N.M., Blair, E., and Stocking, C. (1977). "Modest Expectations: The Effect of Interviewer Prior Expectations on Response." *Sociological Methods and Research,* 6, 177–182.

Sugiyama, M. (1991). "Responses and Non-responses." Paper presented at the Seminar on Quality of Information in Sample Surveys, Paris.

Sykes, W., and Collins, M. (1988). "Effects of Mode of Interview: Experiments in the UK."

In Groves, R.M., Biemer, P.P., Lyberg, L.E., Massey, J.T., Nicholls, W.L., and Waksberg, J. (eds.), *Telephone Survey Methodology.* New York: Wiley.

Szymanski, D.M. (1988). "Determinants of Selling Effectiveness: The Importance of Declarative Knowledge to the Personal Selling Concept." *Journal of Marketing,* 52, 65–77.

Tanner, J.F. (1994). "Adaptive Selling at Trade Shows." *Journal of Personal Selling and Sales Management,* 14, 2, 15–23.

Thibaut, J.W., and Kelley, H.H. (1959). *The Social Psychology of Groups.* New York: Wiley.

Tittle, C.R., and Stafford, M.C. (1992). "Urban Theory, Urbanism, and Suburban Residence." *Social Forces,* 70, 3, 725–744.

Traugott, M.W., and Goldstein, K. (1993). "Evaluating Dual Frame Samples and Advance Letters as a Means of Increasing Response Rates." *Proceedings of the Survey Research Methods Section,* American Statistical Association, pp. 1284–1286.

Traugott, M.W., Groves, R.M., and Lepkowski, J.M. (1987). "Using Dual Frame Designs to Reduce Nonresponse in Telephone Surveys." *Public Opinion Quarterly,* 51, 1, 48–57.

Trice, A.D., and Layman, W.H. (1984). "Improving Guest Surveys." *The Cornell Hotel and Restaurant Administration Quarterly,* 25, 10–13.

Tuckel, P., and Feinberg, B.M. (1991). "The Answering Machine Poses Many Questions for Telephone Survey Researchers." *Public Opinion Quarterly,* 55, 2, 200–217.

Tuckel, P., and O'Neill, H. (1995). "A Profile of Answering Machine Owners and Screeners." *Proceedings of the Section on Survey Research Methods,* American Statistical Association, pp. 1157–1162.

Tuckel, P., and Shukers, T. (1997). "The Effect of Different Introductions and Answering Machine Messages on Response Rates." Paper presented at the annual conference of the American Association for Public Opinion Research, Norfolk, VA.

Tucker, C., and Kojetin, B. (1994). "Measuring the Effects of the CPS Transition on Nonresponse." Paper presented at the Fifth International Workshop on Household Survey Nonresponse, Ottawa.

Tucker, C., Haugh, G., Johnson-Herring, S., Krieger, S., and Vigliano, K. (1991). "The Impact of Political and Economic Conditions on Refusal Rates to the Current Population Survey and the Consumer Expenditure Survey." Paper presented at the Second International Workshop on Household Survey Nonresponse, Washington, D.C.

U.S. Bureau of the Census (1988). *County and City Data Book, 1988.* Washington, DC: Bureau of the Census.

U.S. Bureau of the Census (1990). *1990 Decennial Census Geographical Handbook.* Washington, DC: Bureau of the Census (Report D–519).

U.S. Bureau of the Census (1993). *Statistical Abstracts of the United States: 1993.* 113th Edition. Washington, DC: Bureau of the Census.

U.S. Bureau of the Census (1995). *Statistical Abstracts of the United States: 1995.* 115th Edition. Washington, DC: Bureau of the Census.

Walker Research, Inc. (1992). *Industry Image Survey,* 10th Edition. Indianapolis, IN: Walker Research, Inc.

Weatherford, M.S. (1991). "Mapping the Ties That Bind: Legitimacy, Representation, and Alienation." *Western Political Quarterly,* 44, 251–276.

Weaver, C.N., Holmes, S.L., and Glenn, N.D. (1975). "Some Characteristics of Inaccessible Respondents in a Telephone Survey." *Journal of Applied Psychology,* 60, 260–262.

Weeks, M.F., Jones, B.L., Folsom, R.E., and Benrud, C.H. (1980). "Optimal Times to Contact Sample Households." *Public Opinion Quarterly,* 44, 1, 101–114.

Weeks, M.F., Kulka, R.A., and Pierson, S.A. (1987). "Optimal Call Scheduling for a Telephone Survey." *Public Opinion Quarterly,* 51, 540–549.

Weiner, J., and Brehm, J.W. (1966). "Buying Behavior as a Function of Verbal and Monetary Inducements." Reported in Brehm, J.W. (1966). *A Theory of Psychological Reactance.* New York: Academic Press.

Weitz, B.A. (1981). "Effectiveness in Sales Interactions: A Contingency Framework." *Journal of Marketing,* 45, 85–103.

Weitz, B.A., Sujan, H., and Sujan, M. (1986). "Knowledge, Motivation, and Adaptive Behavior: A Framework for Improving Selling Effectiveness." *Journal of Marketing,* 50, 174–191.

White, A., and Freeth, S. (1996). "Improving Advance Letters." Paper presented at the Seventh International Workshop on Household Survey Nonresponse, Rome.

Wilcox, J.B. (1977). "The Interaction of Refusal and Not-at-home Sources of Nonresponse Bias." *Journal of Marketing Research,* 14, 592–597.

Willimack, D.K., Schuman, H., Pennell, B.E., and Lepkowski, J.M. (1995). "Effects of a Prepaid Nonmonetary Incentive on Response Rates and Response Quality in a Face-to-face Survey." *Public Opinion Quarterly,* 59, 78–92.

Wilson, T.C. (1985). "Settlement Type and Interpersonal Estrangement: A Test of the Theories of Wirth and Gans." *Social Forces,* 64, 1, 139–150.

Winship, C., and Mare, R.D. (1992). "Models for Sample Selection Bias." *Annual Review of Sociology,* Vol. 18, pp 327–350.

Wiseman, F., and McDonald, P. (1979). "Noncontact and Refusal Rates in Consumer Telephone Surveys." *Journal of Marketing Research,* 16, 478–484.

Wright, T., and Tsao, H.J. (1983). "A Frame on Frames: An Annotated Bibliography." In T. Wright (ed.), *Statistical Methods and the Improvement of Data Quality.* New York: Academic Press, pp. 25–72.

Yu, J., and Cooper, H. (1983). "A Quantitative Review of Research Design Effects on Response Rates to Questionnaires." *Journal of Marketing Research,* 20, 36–44.

Index

WILEY SERIES IN PROBABILITY AND STATISTICS

ESTABLISHED BY WALTER A. SHEWHART AND SAMUEL S. WILKS

Editors

*Vic Barnett, Ralph A. Bradley, Noel A. C. Cressie, Nicholas I. Fisher,
Iain M. Johnstone, J. B. Kadane, David G. Kendall, David W. Scott,
Bernard W. Silverman, Adrian F. M. Smith, Jozef L. Teugels;
J. Stuart Hunter, Emeritus*

Probability and Statistics Section

*ANDERSON · The Statistical Analysis of Time Series
ARNOLD, BALAKRISHNAN, and NAGARAJA · A First Course in Order Statistics
BACCELLI, COHEN, OLSDER, and QUADRAT · Synchronization and Linearity:
    An Algebra for Discrete Event Systems
BASILEVSKY · Statistical Factor Analysis and Related Methods: Theory and
    Applications
BERNARDO and SMITH · Bayesian Statistical Concepts and Theory
BILLINGSLEY · Convergence of Probability Measures
BOROVKOV · Asymptotic Methods in Queuing Theory
BRANDT, FRANKEN, and LISEK · Stationary Stochastic Models
CAINES · Linear Stochastic Systems
CAIROLI and DALANG · Sequential Stochastic Optimization
CONSTANTINE · Combinatorial Theory and Statistical Design
COVER and THOMAS · Elements of Information Theory
CSÖRGŐ and HORVÁTH · Weighted Approximations in Probability Statistics
CSÖRGŐ and HORVÁTH · Limit Theorems in Change Point Analysis
DETTE and STUDDEN · The Theory of Canonical Moments with Applications in
    Statistics, Probability, and Analysis
*DOOB · Stochastic Processes
DRYDEN and MARDIA · Statistical Analysis of Shape
DUPUIS and ELLIS · A Weak Convergence Approach to the Theory of Large Deviations
ETHIER and KURTZ · Markov Processes: Characterization and Convergence
FELLER · An Introduction to Probability Theory and Its Applications, Volume 1,
    Third Edition, Revised; Volume II, *Second Edition*
FULLER · Introduction to Statistical Time Series, *Second Edition*
FULLER · Measurement Error Models
GELFAND and SMITH · Bayesian Computation
GHOSH, MUKHOPADHYAY, and SEN · Sequential Estimation
GIFI · Nonlinear Multivariate Analysis
GUTTORP · Statistical Inference for Branching Processes
HALL · Introduction to the Theory of Coverage Processes
HAMPEL · Robust Statistics: The Approach Based on Influence Functions
HANNAN and DEISTLER · The Statistical Theory of Linear Systems
HUBER · Robust Statistics
IMAN and CONOVER · A Modern Approach to Statistics
JUREK and MASON · Operator-Limit Distributions in Probability Theory
KASS and VOS · Geometrical Foundations of Asymptotic Inference
KAUFMAN and ROUSSEEUW · Finding Groups in Data: An Introduction to Cluster
    Analysis

*Now available in a lower priced paperback edition in the Wiley Classics Library.

Applied Probability and Statistics Section

*Now available in a lower priced paperback edition in the Wiley Classics Library.

*Now available in a lower priced paperback edition in the Wiley Classics Library.

*Now available in a lower priced paperback edition in the Wiley Classics Library.

Texts and References Section

*Now available in a lower priced paperback edition in the Wiley Classics Library.

WILEY SERIES IN PROBABILITY AND STATISTICS

ESTABLISHED BY WALTER A. SHEWHART AND SAMUEL S. WILKS

Editors
Robert M. Groves, Graham Kalton, J. N. K. Rao, Norbert Schwarz, Christopher Skinner

Survey Methodology Section

*Now available in a lower priced paperback edition in the Wiley Classics Library.